State-of-the-Art Metabolomics and Lipidomics in Life Sciences: Methods and Applications

State-of-the-Art Metabolomics and Lipidomics in Life Sciences: Methods and Applications

Editors

Cora Weigert
Xinyu Liu

Basel • Beijing • Wuhan • Barcelona • Belgrade • Novi Sad • Cluj • Manchester

Editors
Cora Weigert
University Hospital Tübingen
Tübingen
Germany

Xinyu Liu
Dalian Institute of Chemical Physics,
Chinese Academy of Science
Dalian
China

Editorial Office
MDPI
St. Alban-Anlage 66
4052 Basel, Switzerland

This is a reprint of articles from the Special Issue published online in the open access journal *Metabolites* (ISSN 2218-1989) (available at: https://www.mdpi.com/journal/metabolites/special_issues/KFY472EF75).

For citation purposes, cite each article independently as indicated on the article page online and as indicated below:

Lastname, A.A.; Lastname, B.B. Article Title. *Journal Name* **Year**, *Volume Number*, Page Range.

ISBN 978-3-7258-0103-9 (Hbk)
ISBN 978-3-7258-0104-6 (PDF)
doi.org/10.3390/books978-3-7258-0104-6

© 2024 by the authors. Articles in this book are Open Access and distributed under the Creative Commons Attribution (CC BY) license. The book as a whole is distributed by MDPI under the terms and conditions of the Creative Commons Attribution-NonCommercial-NoDerivs (CC BY-NC-ND) license.

Contents

About the Editors vii

Preface ix

Xinyu Liu and Cora Weigert
State-of-the-Art Metabolomics and Lipidomics in Life Sciences: Methods and Applications
Reprinted from: Metabolites 2024, 14, 8, doi:10.3390/metabo14010008 1

Jinzhi Xu, Lina Zhou, Xiaojing Du, Zhuoran Qi, Sinuo Chen, Jian Zhang, et al.
Transcriptome and Lipidomic Analysis Suggests Lipid Metabolism Reprogramming and
Upregulating *SPHK1* Promotes Stemness in Pancreatic Ductal Adenocarcinoma Stem-like Cells
Reprinted from: Metabolites 2023, 13, 1132, doi:10.3390/metabo13111132 7

Yifei Zhan, Huaiyan Wang, Zeying Wu and Zhongda Zeng
Study on the Common Molecular Mechanism of Metabolic Acidosis and Myocardial Damage
Complicated by Neonatal Pneumonia
Reprinted from: Metabolites 2023, 13, 1118, doi:10.3390/metabo13111118 29

Tianfu Wei, Jifeng Liu, Shurong Ma, Mimi Wang, Qihang Yuan, Anliang Huang, et al.
A Nucleotide Metabolism-Related Gene Signature for Risk Stratification and Prognosis
Prediction in Hepatocellular Carcinoma Based on an Integrated Transcriptomics and
Metabolomics Approach
Reprinted from: Metabolites 2023, 13, 1116, doi:10.3390/metabo13111116 44

**Sijia Zheng, Lina Zhou, Miriam Hoene, Andreas Peter, Andreas L. Birkenfeld,
Cora Weigert, et al.**
A New Biomarker Profiling Strategy for Gut Microbiome Research: Valid Association
of Metabolites to Metabolism of Microbiota Detected by Non-Targeted Metabolomics in
Human Urine
Reprinted from: Metabolites 2023, 13, 1061, doi:10.3390/metabo13101061 62

**Yixuan Guo, Shuangshuang Wei, Mengdi Yin, Dandan Cao, Yiling Li, Chengping Wen and
Jia Zhou**
Gas Chromatography–Mass Spectrometry Reveals Stage-Specific Metabolic Signatures of
Ankylosing Spondylitis
Reprinted from: Metabolites 2023, 13, 1058, doi:10.3390/metabo13101058 74

Xiaojing Jia, Chunyan Hu, Xueyan Wu, Hongyan Qi, Lin Lin, Min Xu, et al.
Evaluating the Effects of Omega-3 Polyunsaturated Fatty Acids on Inflammatory Bowel Disease
via Circulating Metabolites: A Mediation Mendelian Randomization Study
Reprinted from: Metabolites 2023, 13, 1041, doi:10.3390/metabo13101041 87

Liming Gu, Wenli Wang, Yifeng Gu, Jianping Cao and Chang Wang
Metabolomic Signatures Associated with Radiation-Induced Lung Injury by Correlating Lung
Tissue to Plasma in a Rat Model
Reprinted from: Metabolites 2023, 13, 1020, doi:10.3390/metabo13091020 103

Ming Yang, Jichun Jiang, Lei Hua, Dandan Jiang, Yadong Wang, Depeng Li, et al.
Rapid Detection of Volatile Organic Metabolites in Urine by High-Pressure Photoionization
Mass Spectrometry for Breast Cancer Screening: A Pilot Study
Reprinted from: Metabolites 2023, 13, 870, doi:10.3390/metabo13070870 120

Shan Zhang, Xujiang Shan, Linchi Niu, Le Chen, Jinjin Wang, Qinghua Zhou, et al.
The Integration of Metabolomics, Electronic Tongue, and Chromatic Difference Reveals the Correlations between the Critical Compounds and Flavor Characteristics of Two Grades of High-Quality Dianhong Congou Black Tea
Reprinted from: *Metabolites* **2023**, *13*, 864, doi:10.3390/metabo13070864 133

Runze Ouyang, Sijia Zheng, Xiaolin Wang, Qi Li, Juan Ding, Xiao Ma, et al.
Crosstalk between Breast Milk N-Acetylneuraminic Acid and Infant Growth in a Gut Microbiota-Dependent Manner
Reprinted from: *Metabolites* **2023**, *13*, 846, doi:10.3390/metabo13070846 150

Jun Zeng, Jingwen Hao, Zhiqiang Yang, Chunyu Ma, Longhua Gao, Yue Chen, et al.
Anti-Allergic Effect of Dietary Polyphenols Curcumin and Epigallocatechin Gallate via Anti-Degranulation in IgE/Antigen-Stimulated Mast Cell Model: A Lipidomics Perspective
Reprinted from: *Metabolites* **2023**, *13*, 628, doi:10.3390/metabo13050628 168

Xiaoshan Sun, Zhen Jia, Yuqing Zhang, Xinjie Zhao, Chunxia Zhao, Xin Lu and Guowang Xu
A Strategy for Uncovering the Serum Metabolome by Direct-Infusion High-Resolution Mass Spectrometry
Reprinted from: *Metabolites* **2023**, *13*, 460, doi:10.3390/metabo13030460 186

Yilan Ding, Shuangyuan Wang and Jieli Lu
Unlocking the Potential: Amino Acids' Role in Predicting and Exploring Therapeutic Avenues for Type 2 Diabetes Mellitus
Reprinted from: *Metabolites* **2023**, *13*, 1017, doi:10.3390/metabo13091017 199

Shuling He, Lvyun Sun, Jiali Chen and Yang Ouyang
Recent Advances and Perspectives in Relation to the Metabolomics-Based Study of Diabetic Retinopathy
Reprinted from: *Metabolites* **2023**, *13*, 1007, doi:10.3390/metabo13091007 220

About the Editors

Cora Weigert

Prof. Dr. Cora Weigert is from the Institute for Clinical Chemistry and Pathobiochemistry, the Department for Diagnostic Laboratory Medicine, University Hospital Tübingen. She is an expert in diabetes-related research on exercise metabolism, skeletal muscle, insulin resistance and mitochondrial respiration.

Xinyu Liu

Associate Prof. Dr. Xinyu Liu comes from the CAS Key Laboratory of Separation Science for Analytical Chemistry, Metabolomics Research Center, Dalian Institute of Chemical Physics, Chinese Academy of Sciences. Her main research fields are in chromatography–mass spectrometry-based novel metabolomics and exposomics methods development and their applications in major diseases, as well as the crosstalk between the host metabolism and gut microbiota.

Preface

This Special Issue celebrates Prof. Guowang Xu and Prof. Rainer Lehmann on their long-standing, fruitful, Sino–German scientific cooperation and friendship. Their interdisciplinary research on the field of high-resolution analytical chemistry and life sciences has existed for decades.

Here, two reviews and twelve original research articles were accepted for publication in this Special Issue. These articles address the use of metabolomics or lipidomics to study different aspects of life science research, from microbes, plants and animals to humans, and new biomarkers, protocols, strategies or bioinformatic tools are reported.

We sincerely hope that this Special Issue will provide a valuable resource for other researchers in these fields. We would like to sincerely thank all the authors for their contributions to this Special Issue.

Cora Weigert and Xinyu Liu
Editors

 metabolites

Editorial

State-of-the-Art Metabolomics and Lipidomics in Life Sciences: Methods and Applications

Xinyu Liu [1,2,*] and Cora Weigert [3,4,5,*]

1. CAS Key Laboratory of Separation Science for Analytical Chemistry, Dalian Institute of Chemical Physics, Chinese Academy of Sciences, Dalian 116023, China
2. Liaoning Province Key Laboratory of Metabolomics, Dalian 116023, China
3. Department for Diagnostic Laboratory Medicine, Institute for Clinical Chemistry and Pathobiochemistry, University Hospital Tübingen, 72076 Tüebingen, Germany
4. Institute for Diabetes Research and Metabolic Diseases, Helmholtz Zentrum München, University of Tübingen, 72076 Tübingen, Germany
5. German Center for Diabetes Research (DZD), 90451 Nuremberg, Germany
* Correspondence: liuxy2012@dicp.ac.cn (X.L.); cora.weigert@med.uni-tuebingen.de (C.W.); Tel.: +86-411-84379532 (X.L.); +49-7071-2985670 (C.W.)

Citation: Liu, X.; Weigert, C. State-of-the-Art Metabolomics and Lipidomics in Life Sciences: Methods and Applications. *Metabolites* **2024**, *14*, 8. https://doi.org/10.3390/metabo14010008

Received: 5 December 2023
Accepted: 12 December 2023
Published: 21 December 2023

Copyright: © 2023 by the authors. Licensee MDPI, Basel, Switzerland. This article is an open access article distributed under the terms and conditions of the Creative Commons Attribution (CC BY) license (https:// creativecommons.org/licenses/by/ 4.0/).

This Special Issue was initiated to celebrate and congratulate Prof. Guowang Xu and Prof. Rainer Lehmann on their long-standing, fruitful Sino-German scientific cooperation and their close friendship. Their interdisciplinary interaction in the field of high-resolution analytical chemistry and life sciences has existed for decades. The first joint publications addressing biomedical topics and the development of new capillary electrophoresis approaches date back to 1998 [1,2]. Two reviews and twelve original research articles were accepted for publication in this Special Issue. The articles address different aspects of life science research, from microbes, plants and animals to humans applying metabolomics or lipidomics, and new biomarkers, protocols, strategies or bioinformatic tools are reported on.

The Sino-German scientific cooperation between the Dalian Institute of Chemical Physics of the Chinese Academy of Sciences and the University of Tuebingen was originally initiated in the 1970s by Prof. Peichang Lu, the pioneer of chromatography in China, and Prof. Ernst Bayer, a pioneer of gas chromatography in Europe. Since then, a lively, regular scientific exchange between Dalian and Tuebingen developed. In the mid-nineties, Prof. Guowang Xu and Prof. Rainer Lehmann started their joint research activities during Prof. Xu's two-year research stay in Tuebingen as a Max-Planck-Institute fellow.

First, as mentioned above, their joint bioanalytical interest was the development of new applications of capillary electrophoresis for the analysis of human body fluids for diagnostic purposes. Later on, high-resolution mass spectrometric profiling in a biomedical context increasingly became the focus of their joint research interest. In particular, the comprehensive investigation of metabolite and lipid profiles by metabolomics and lipidomics analyses has remained a core theme of their cooperation. In addition to various body fluids, other sample materials like biopsies from various tissues, human primary cell culture samples, etc., were analyzed to investigate pathomechanisms and to identify diagnostic biomarkers of metabolic diseases (prediabetes and diabetes) and various cancer diseases. Additionally, multi-omics approaches were applied in systems medicine studies. An important aspect of their work has been addressing sample quality, considering the error-prone preanalytical phase from "bed to bench", including the identification of a sample quality biomarker allowing analytical (bio)chemists to assess the quality of blood samples without knowledge about the preceding blood collection and blood handling process. Other cornerstones of their joint research activities include the development of new methods and analytical strategies to either facilitate or increase the identification of metabolites or the coverage of metabolomics and lipidomics approaches, as well as the optimization of sample preparation procedures.

In total, more than 50 joint publications in high-ranking international journals like *Diabetes Care*, *Clinical Chemistry*, *Analytical Chemistry*, etc., have been published by this Sino-German collaboration. The research performed through this collaboration has been regularly supported by joint funds from the national science foundation China (NSFC), the German research foundation (DFG), the Humboldt foundation, as well as from the Chinese Academy of Sciences (CAS). The funding has enabled the exchange of many young scientists between Dalian and Tuebingen, which has built a solid base for continuous and successful cooperation between the CAS Key Laboratory of Separation Science for Analytical Chemistry from the DICP in Dalian and the Institute for Clinical Chemistry and Pathobiochemistry at the University Hospital in Tuebingen.

The 14 publications in this Special Issue cover several thematic aspects of the Sino-German scientific cooperation and are briefly highlighted in the following paragraphs in the order of appearance in the Special Issue.

Jinzhi Xu et al. (2023) (contribution 1) identified lipid metabolism reprogramming due to stemness acquisition in pancreatic ductal adenocarcinoma (PDAC) cancer stem cells (CSCs) using lipidomes combined with the transcriptome analysis of PDAC tumor-repopulating cells (TRCs, a novel CSCs model). The results were supported by the analysis of data obtained from PDAC patients from The Cancer Genome Atlas (TCGA) database. The research also highlighted SPHK1 (sphingosine kinases 1) as the important enzyme involved in the up-regulation of sphingolipid metabolism. Hence, SPHK1 may have a relevant role in PDAC CSCs and may be an appropriate therapeutic target candidate.

Yifei Zhan et al. (2023) (contribution 2) discovered a common molecular mechanism of metabolic acidosis and myocardial damage in neonatal pneumonia. Applying UPLC-HRMS-based untargeted metabolomics, a total of 23 and 21 differential metabolites were found in the comparison of serum samples of pneumonia and samples of pneumonia with the two complications, respectively. The 14 identical molecules found in both disease states were found to be related to sphingolipid, porphyrin, and glycerophospholipid metabolisms, which offers valuable information for the rapid and accurate identification, classification, staging and, most probably, disease diagnosis and therapy of complications of neonatal pneumonia.

Tianfu Wei et al. (2023) (contribution 3) integrated multidisciplinary data to provide an in-depth view of the relationships between genes and metabolites. With this strategy, a prediction model of hepatocellular carcinoma based on nucleotide metabolism was developed. A pattern showing the nucleotide metabolism of patients with hepatocellular carcinoma was elucidated, which created new perspectives for the clinical treatment of hepatocellular carcinoma.

Sijia Zheng et al. (2023) (contribution 4) established valid associations of metabolites in human urine with the metabolism of gut microbiota, which serves as a new biomarker profiling strategy in gut microbiome research. Considering that bowel evacuation was a simple and efficient approach to reveal gut microbiota-related metabolites, a non-targeted modifying group-assisted metabolomics approach was used to investigate urine samples collected in two independent experiments at various time points, before and after laxative use, to drastically reduce the gut microbiome. Additionally, fasting over the same time period was performed as a control experiment. Finally, the levels of 331 urinary metabolite ions were significantly affected by the depletion of the fecal microbiome, including 100 with specific modifying groups, 32 of which were structurally elucidated. The applied strategy has the potential to generate a microbiome-associated metabolite map of urine, and presumably other body fluids as well.

Yixuan Guo et al. (2023) (contribution 5) aims to identify metabolic differences in ankylosing spondylitis (AS) patients at different stages of the disease using an untargeted metabolomics approach based on the gas chromatography–mass spectrometry of serum. The findings of the study indicate that patients in acute stage and remission stage have specific metabolic characteristics. In particular, 2–hydroxybutanoate and hexadecanoate had good efficacy with regard to the stage division of AS. This research may contribute to the understanding of the pathogenesis of ankylosing spondylitis and to the staging treatment of this chronic disease.

Xiaojing Jia et al. (2023) (contribution 6) explored the impact of omega-3 polyunsaturated fatty acids (PUFAs) on inflammatory bowel disease (IBD) using Mendelian randomization. Increased genetically predicted eicosapentaenoic acid (EPA) concentrations are associated with decreased IBD risk, mediated through lower linoleic acid and histidine metabolites. To date, limited evidence has supported the effects of total omega-3, α-linolenic acid and docosahexaenoic acid on IBD risk. Robust colocalization in the *fatty acid desaturase 2 (FADS2)* region suggests *FADS2* gene mediation. Overall, the study highlights EPA as the key active component of omega-3 PUFAs in reducing IBD risk and suggests *FADS2* gene involvement, offering insights for targeted intervention strategies.

Liming Gu et al. (2023) (contribution 7) identified a panel of potential plasma metabolic markers for radiation-induced lung injury (RILI) via correlation analysis between the lung tissue and plasma metabolic features and evaluated the radiation injury levels within 5 days following whole-thorax irradiation (WTI) in a rat model. Moreover, the data imply disorders of the urea cycle, intestinal microbiota metabolism and mitochondrial dysfunction. This research unveils metabolic traits associated with WTI, providing new perspectives on potential therapeutic measures.

Ming Yang et al. (2023) (contribution 8) developed a rapid and highly sensitive detection method for volatile organic metabolites (VOMs) in urine based on the integration of high-pressure photoionization mass spectrometry (HPPI-TOFMS) and dynamic purge-injection technique. Nine differential metabolites in the urine samples between breast cancer patients and healthy controls were successfully identified through statistical analysis of the HPPI MS data. The results demonstrate the good sensitivity and specificity of the method and provide a promising avenue for the development of a new non-invasive diagnostic tool for breast cancer.

The work of Shan Zhang et al. (2023) (contribution 9) compared the differences in the sensory features and chemical profiles of the two grades of premium Dianhong congou black tea (DCT) produced in southwest China and identified the correlations of critical non-volatile compounds and flavor characteristics in these DCTs. This study also highlighted the promising perspective of the integration of metabolomics, electronic tongues, chromatic differences and human sensory evaluation in the analysis of food flavor.

Runze Ouyang et al. (2023) (contribution 10) illustrated the gut microbiota-dependent crosstalk between breast milk N-acetylneuraminic acid (Neu5Ac) and infant growth. The research demonstrates the negative association between breast milk Neu5Ac and infant obesity risk. The data show Neu5Ac-related alterations to infant gut microbiota and bile acid metabolism, and similar associations were found in mice colonized with infant-derived microbiota. Finally, this study identified the mediator between breast milk Neu5Ac and the risk of infant obesity.

Jun Zeng et al. (2023) (contribution 11) analyzed the antiallergic activities of two representative dietary polyphenols, curcumin and epigallocatechin gallate (EGCG), and elucidated their effects on the cellular lipidome in the progression of degranulation. This work contributes to further understanding of the molecular mechanism of antigen stimulation and curcumin/EGCG involvement in antianaphylaxis and helps to guide future attempts to use dietary polyphenols in this context.

Xiaoshan Sun et al. (2023) (contribution 12) introduces a novel serum metabolome characterization method employing direct-infusion high-resolution mass spectrometry, thereby addressing limitations in metabolite assignment. Different from conventional database search, this strategy utilizes a reaction network along with mass accuracy and isotopic pattern filters to achieve unequivocal formula assignments. The developed approach proved database-independent and rapid, assigning unique monoisotopic features in the serum. Its merits lie in the comprehensive and reliable formula assignment, exhibiting strong potential for large-scale metabolomics studies.

In a review by Yilan Ding et al. (2023) (contribution 13), the authors conducted an in-depth literature search on specific amino acids as potential biomarkers for the onset and progression of diabetes. They described underlying mechanisms, signaling pathways, and metabolic implications, providing valuable insights into preventive and therapeutic interventions. Additional clinical research and therapeutic strategies targeting distinct amino acids were discussed, aiming to prevent or slow down the progression to type 2 diabetes in the prediabetic stage.

Shuling He et al. (2023) (contribution 14) performed a systematic, comprehensive literature analysis of metabolomics approaches studying diabetic retinopathy. An overview providing insights into relevant metabolites and metabolic pathways was given. A gap in knowledge in the existing literature was detected with respect to data from large-cohort or multicenter studies, as well as data from platforms applying various analytical metabolomics approaches to study diabetic retinopathy. In addition, future metabolomics research directions with regard to diabetic retinopathy are discussed.

From these contributions and also based on the most recent developments in metabolomics research, we can draw the following conclusions:

1. Analytical challenges to investigate metabolomes and lipidomes in life sciences still exist, such as limitations in metabolite and lipid coverage, a still high number of unknowns in non-targeted approaches and further improvements in analytical sensitivity. However, based on the ongoing continuous improvements and new developments in chromatographic and mass spectrometric techniques, an increasing amount of information on metabolomes can be gathered.
2. Metabolomics and lipidomics analyses can be applied to evaluate the important metabolic effects and functions of molecules and nourishments like dietary polyphenols, polyunsaturated fatty acids, etc., to evaluate the interfering effects of drugs and lifestyle on health.
3. In biomedical and disease-related research fields, metabolomics analysis enables us to study metabolic reprogramming, define potential prospective or prognostic diagnostic biomarkers or subclassify patients in precision medicine by metabolite pattern to allow individualized treatment. However, to draw reliable conclusions based on valid results which reflect the situation in the population, samples of large-scale multi-center studies should be investigated. In this context, a big analytical challenge which needs to be solved is the application of very robust, sensitive and highly reproducible metabolomics methodology, suitable for the analysis of thousands of samples with different pre-analytical quality levels. The comparability of data between laboratories is a fundamental requirement, and quantitative metabolomics will become increasingly important.
4. Metabolomics analysis contributes substantially to deepening the understanding of metabolic mechanisms on the cellular level via investigations like isotope labeling experiments to study dynamics of metabolism or via single-cell metabolomics, which has recently become a hot topic in the breakdown of cell heterogenicity in tissues to elucidate cell type-specific metabolic functions.
5. Traditionally, to study tissue metabolism, homogenates are analyzed, leading to the spatial loss of information. Now, to eliminate this limitation, mass spectrometric imaging or laser microdissection instruments are in use, which are increasingly bringing spatial metabolomics into focus. It is foreseeable that in the near future, with improvements in sensitivity, resolution and speed, spatial and single-cell metabolomics will create new perspectives and play a significant role in the fields of life sciences and biomedical research.
6. Finally, it should be emphasized that a person´s health state is influenced by various factors, and the environment will decide a person´s health. Therefore, a recent research direction, i.e., the combinational use of metabolomics and exposomics, is a

fascinating new research field and will be one of the key strategies used to shed light on the assessment of risks and causes of disease. Additionally, and highly relevant in this context, multi-omics applications are already well established in the study of complex diseases, especially chronic diseases, and contribute to our understanding of them. However, currently, a significant challenge is the evaluation and interpretation of these data, as well as the integration of these multi-omics data to support clinical decision-making processes. For this purpose, important contributions from the application of artificial intelligence approaches can be expected from interdisciplinary collaborations.

7. In summary, metabolomics and lipidomics are now well established and frequently applied technologies in life sciences and health promotion, but many bottlenecks still need to be broken by intense, interdisciplinary interaction between scientists from different fields, including analytical (bio)chemists, biologists, bioinformaticians and clinicians.

Funding: This work was supported by the DFG/NSFC Sino-German mobility program (M-0257) to G.X. and R.L., the foundation from the Youth Innovation Promotion Association CAS (2021186) to X.L., and the Innovation program (DICP ZZBS201804) of Science and Research from the DICP to G.X.

Acknowledgments: As Guest Editors of the Special Issue entitled "State-of-the-Art Metabolomics and Lipidomics in Life Sciences: Methods and Applications", we would like to express our deep appreciation to Rainer Lehmann and Guowang Xu, who gave good suggestions for this Special Issue, and all authors whose valuable work was published under this issue and thus contributed to the success of the edition. Each contribution was included in the simple introduction of this corresponding paper.

Conflicts of Interest: The authors declare no conflict of interest.

List of Contributions

1. Xu, J.; Zhou, L.; Du, X.; Qi, Z.; Chen, S.; Zhang, J.; Cao, X.; Xia, J. Transcriptome and Lipidomic Analysis Suggests Lipid Metabolism Reprogramming and Upregulating SPHK1 Promotes Stemness in Pancreatic Ductal Adenocarcinoma Stem-like Cells. *Metabolites* **2023**, *13*, 1132. https://doi.org/10.3390/metabo13111132.
2. Zhan, Y.; Wang, H.; Wu, Z.; Zeng, Z. Study on the Common Molecular Mechanism of Metabolic Acidosis and Myocardial Damage Complicated by Neonatal Pneumonia. *Metabolites* **2023**, *13*, 1118. https://doi.org/10.3390/metabo13111118.
3. Wei, T.; Liu, J.; Ma, S.; Wang, M.; Yuan, Q.; Huang, A.; Wu, Z.; Shang, D.; Yin, P. A Nucleotide Metabolism-Related Gene Signature for Risk Stratification and Prognosis Prediction in Hepatocellular Carcinoma Based on an Integrated Transcriptomics and Metabolomics Approach. *Metabolites* **2023**, *13*, 1116. https://doi.org/10.3390/metabo13111116.
4. Zheng, S.; Zhou, L.; Hoene, M.; Peter, A.; Birkenfeld, A.; Weigert, C.; Liu, X.; Zhao, X.; Xu, G.; Lehmann, R. A New Biomarker Profiling Strategy for Gut Microbiome Research: Valid Association of *Metabolites* to Metabolism of Microbiota Detected by Non-Targeted Metabolomics in Human Urine. *Metabolites* **2023**, *13*, 1061. https://doi.org/10.3390/metabo13101061.
5. Guo, Y.; Wei, S.; Yin, M.; Cao, D.; Li, Y.; Wen, C.; Zhou, J. Gas Chromatography-Mass Spectrometry Reveals Stage-Specific Metabolic Signatures of Ankylosing Spondylitis. *Metabolites* **2023**, *13*, 1058. https://doi.org/10.3390/metabo13101058.
6. Jia, X.; Hu, C.; Wu, X.; Qi, H.; Lin, L.; Xu, M.; Xu, Y.; Wang, T.; Zhao, Z.; Chen, Y.; Li, M.; Zheng, R.; Lin, H.; Wang, S.; Wang, W.; Bi, Y.; Zheng, J.; Lu, J. Evaluating the Effects of Omega-3 Polyunsaturated Fatty Acids on Inflammatory Bowel Disease via Circulating Metabolites: A Mediation Mendelian Randomization Study. *Metabolites* **2023**, *13*, 1041. https://doi.org/10.3390/metabo13101041.
7. Gu, L.; Wang, W.; Gu, Y.; Cao, J.; Wang, C. Metabolomic Signatures Associated with Radiation-Induced Lung Injury by Correlating Lung Tissue to Plasma in a Rat Model. *Metabolites* **2023**, *13*, 1020. https://doi.org/10.3390/metabo13091020.

8. Yang, M.; Jiang, J.; Hua, L.; Jiang, D.; Wang, Y.; Li, D.; Wang, R.; Zhang, X.; Li, H. Rapid Detection of Volatile Organic Metabolites in Urine by High-Pressure Photoionization Mass Spectrometry for Breast Cancer Screening: A Pilot Study. *Metabolites* **2023**, *13*, 870. https://doi.org/10.3390/metabo13070870.
9. Zhang, S.; Shan, X.; Niu, L.; Chen, L.; Wang, J.; Zhou, Q.; Yuan, H.; Li, J.; Wu, T. The Integration of Metabolomics, Electronic Tongue, and Chromatic Difference Reveals the Correlations between the Critical Compounds and Flavor Characteristics of Two Grades of High-Quality Dianhong Congou Black Tea. *Metabolites* **2023**, *13*, 864. https://doi.org/10.3390/metabo13070864.
10. Ouyang, R.; Zheng, S.; Wang, X.; Li, Q.; Ding, J.; Ma, X.; Zhuo, Z.; Li, Z.; Xin, Q.; Lu, X.; Zhou, L.; Ren, Z.; Mei, S.; Liu, X.; Xu, G. Crosstalk between Breast Milk N-Acetylneuraminic Acid and Infant Growth in a Gut Microbiota-Dependent Manner. *Metabolites* **2023**, *13*, 846. https://doi.org/10.3390/metabo13070846.
11. Zeng, J.; Hao, J.; Yang, Z.; Ma, C.; Gao, L.; Chen, Y.; Li, G.; Li, J. Anti-Allergic Effect of Dietary Polyphenols Curcumin and Epigallocatechin Gallate via Anti-Degranulation in IgE/Antigen-Stimulated Mast Cell Model: A Lipidomics Perspective. *Metabolites* **2023**, *13*, 628. https://doi.org/10.3390/metabo13050628.
12. Sun, X.; Jia, Z.; Zhang, Y.; Zhao, X.; Zhao, C.; Lu, X.; Xu, G. A Strategy for Uncovering the Serum Metabolome by Direct-Infusion High-Resolution Mass Spectrometry. *Metabolites* **2023**, *13*, 460. https://doi.org/10.3390/metabo13030460.
13. Ding, Y.; Wang, S.; Lu, J. Unlocking the Potential: Amino AcidsRole in Predicting and Exploring Therapeutic Avenues for Type 2 Diabetes Mellitus. *Metabolites* **2023**, *13*, 1017. https://doi.org/10.3390/metabo13091017.
14. He, S.; Sun, L.; Chen, J.; Ouyang, Y. Recent Advances and Perspectives in Relation to the Metabolomics-Based Study of Diabetic Retinopathy. *Metabolites* **2023**, *13*, 1007. https://doi.org/10.3390/metabo13091007.

References

1. Xu, G.; Lehmann, R.; Schleicher, E.; Häring, H.U.; Liebich, H. Advantages in the analysis of UDP-sugars by capillary electrophoresis-comparison of the conventional HPLC method with two new capillary electrophoretic micro-procedures. *Biomed. Chromatogr.* **1998**, *12*, 113–115. [CrossRef]
2. Liebich, H.M.; Xu, G.; Di-Stefano, C.; Lehmann, R. Capillary electrophoresis of urinary normal and modified nucleosides of cancer patients. *J. Chromatogr. A* **1998**, *793*, 341–347. [CrossRef]

Disclaimer/Publisher's Note: The statements, opinions and data contained in all publications are solely those of the individual author(s) and contributor(s) and not of MDPI and/or the editor(s). MDPI and/or the editor(s) disclaim responsibility for any injury to people or property resulting from any ideas, methods, instructions or products referred to in the content.

Article

Transcriptome and Lipidomic Analysis Suggests Lipid Metabolism Reprogramming and Upregulating *SPHK1* Promotes Stemness in Pancreatic Ductal Adenocarcinoma Stem-like Cells

Jinzhi Xu [1,†], Lina Zhou [2,†], Xiaojing Du [3,†], Zhuoran Qi [1], Sinuo Chen [1], Jian Zhang [1], Xin Cao [1,4,*] and Jinglin Xia [1,5,*]

[1] National Medical Center and National Clinical Research Center for Interventional Medicine, Liver Cancer Institute, Zhongshan Hospital, Fudan University, 180 Fenglin Road, Shanghai 200032, China
[2] CAS Key Laboratory of Separation Science for Analytical Chemistry, Dalian Institute of Chemical Physics, Chinese Academy of Sciences, Dalian 116023, China
[3] Endoscopy Center, Shanghai East Hospital, Tongji University School of Medicine, Shanghai 200120, China
[4] Institute of Clinical Science, Zhongshan Hospital, Fudan University, Shanghai 200032, China
[5] Key Laboratory of Diagnosis and Treatment of Severe Hepato-Pancreatic Diseases of Zhejiang Province, The First Affiliated Hospital of Wenzhou Medical University, Wenzhou 325000, China
* Correspondence: caox@fudan.edu.cn (X.C.); xia.jinglin@zs-hospital.sh.cn (J.X.)
† These authors contributed equally to this work.

Abstract: Cancer stem cells (CSCs) are considered to play a key role in the development and progression of pancreatic ductal adenocarcinoma (PDAC). However, little is known about lipid metabolism reprogramming in PDAC CSCs. Here, we assigned stemness indices, which were used to describe and quantify CSCs, to every patient from the Cancer Genome Atlas (TCGA-PAAD) database and observed differences in lipid metabolism between patients with high and low stemness indices. Then, tumor-repopulating cells (TRCs) cultured in soft 3D (three-dimensional) fibrin gels were demonstrated to be an available PDAC cancer stem-like cell (CSLCs) model. Comprehensive transcriptome and lipidomic analysis results suggested that fatty acid metabolism, glycerophospholipid metabolism, and, especially, the sphingolipid metabolism pathway were mostly associated with CSLCs properties. *SPHK1* (sphingosine kinases 1), one of the genes involved in sphingolipid metabolism and encoding the key enzyme to catalyze sphingosine to generate S1P (sphingosine-1-phosphate), was identified to be the key gene in promoting the stemness of PDAC. In summary, we explored the characteristics of lipid metabolism both in patients with high stemness indices and in novel CSLCs models, and unraveled a molecular mechanism via which sphingolipid metabolism maintained tumor stemness. These findings may contribute to the development of a strategy for targeting lipid metabolism to inhibit CSCs in PDAC treatment.

Keywords: pancreatic ductal adenocarcinoma; cancer stem-like cells; lipid metabolism reprogramming; sphingolipid metabolism; *SPHK1*

1. Introduction

Pancreatic cancer is one of the most fatal cancers, ranking the seventh leading cause of cancer-related deaths worldwide [1], and has been predicted to become the second most common cause of cancer-related deaths by 2030 [2]. As the most common histological type, pancreatic ductal adenocarcinoma (PDAC) accounts for the majority of the incidence and mortality of pancreatic cancer cases [3]. Its strong heterogeneity endows PDAC with the feature of high lethality, which is thought to be closely related to a small group of cells that are characterized by self-renewal, unique plasticity and metabolism, and high proliferative capacity [4,5], known as cancer stem cells (CSCs) [6,7]. Therapy resistance of PDAC

CSCs [8] is also mainly responsible for the limited survival benefit of chemotherapeutic agents, targeted therapy and immunotherapy for PDAC patients [9,10]. Thus, an in-depth understanding of CSCs in PDAC is urgently needed and may provide a foundation to explore new therapeutic strategies for clinical practice.

Metabolic reprogramming is one of the major hallmarks of tumorigenesis [11]. PDAC cells rely on altered metabolism pathways, including enhanced aerobic glycolysis [12,13], deregulation of lipid metabolism [14,15], raised branched-chain amino acids and glutamine routes [16,17], and increased nucleotide metabolism [18,19], to support their unlimited proliferation and metastasis [20]. Recent studies suggest that due to the heterogeneity of tumors, unique metabolism characteristics play a distinctive role in maintaining the pluripotency and tumorigenic capacity in PDAC CSCs [21,22]. PDAC CSCs are supposed to facilitate the metabolic flip from glycolytic to oxidative [21,23]. In addition, glutamine dependence is not limited to PDAC cells, as CSCs also rely on glutamine metabolism to promote tumor growth [24,25]. It is well acknowledged that CSCs can reprogram their cellular metabolism to support their continuous proliferation and tumorigenesis [20,26,27], while the understanding of the lipid metabolism disorder in CSCs is limited and unilateral. Several lipid metabolites [28,29] or lipid-metabolism-related genes (LMRGs) [30–32] have separately been reported to play important roles in maintaining the stemness and enhancing tumor metastasis. It is worth noting that tumor metabolic remodeling is a dynamic process; thus, studying changes in metabolic pathways may be more appropriate than directly studying the role of specific metabolite differences. Observed changes due to stemness acquisition in CSCs often encompass large numbers of structurally related lipids, and recent developments in technologies, such as lipidomic and machine learning, enable researchers to explore the lipid metabolism pathways and more comprehensively underline altered lipid metabolism in tumorigenesis [33]. In an attempt to explain the lipid metabolism characteristics of PDAC CSCs, only one institution has carried out a proteomic analysis and subsequently a comprehensive proteomic and lipidomic report on pancreatic cancer stem-like cells (CSLCs). They have reported that fatty acid synthesis, especially biosynthesis of unsaturated FAs, and mevalonate pathways, with downregulation of LDHA (Lactate Dehydrogenase A) and upregulation of genes involved in FA elongation, are essential in PDAC CSLCs [34,35]. Despite multi-omics characterization of PDAC CSCs suggesting the importance of lipid metabolic alterations, explorations on further characterization are still deficient. On the one hand, studies of lipid metabolism in PDAC CSCs were performed only on one traditional CSCs model. On the other hand, it appears that CSCs are extremely reliant on the enzymes involved in the lipid metabolism, but there is currently no research on transcriptomics that combines the stemness phenotypes of the patient's tumor and the CSCs model.

In this study, we explored the difference in the lipid metabolism in patients with high and low stemness indices using single sample gene set enrichment analysis (ssGSEA) algorithms based on the data from the Cancer Genome Atlas (TCGA). Then, to overcome the limitation of the traditional CSCs model by sorting CSC of PDAC based on the identified "stem cell surface markers", we used 3D soft fibrin-gel as the culture medium to select malignant tumor cells with high tumorigenicity in PDAC by adjusting the mechanical stress, defined as tumor-repopulating cells (TRCs), which has been successfully applied in many tumors, such as liver cancer, melanoma, lung adenocarcinoma, etc. [36,37]. Lipidomic combined with transcriptome analysis has been carried out and the results suggested that fatty acid metabolism, sphingolipid metabolism and glycerophospholipid metabolism alterations were mostly observed in PDAC TRCs. Further investigations revealed that *SPHK1*, encoding the key enzyme SPHK1 (sphingosine kinases 1) to catalyze sphingosine to generate S1P (sphingosine-1-phosphate) in sphingolipid metabolism, contributed to promote the stemness of PDAC, which may be a promising therapeutic target in PDAC.

2. Materials and Methods

2.1. Patients' Data Collection and Analysis

The mRNA expression profiles and clinical features of PDAC patients were downloaded from the TCGA data portal (https://portal.gdc.cancer.gov/, accessed on 20 November 2022). The stemness indices were assigned to every PDAC patients from TCGA (tumor, n = 179), which were calculated by using ssGSEA algorithms [38] and one-class logistic regression machine learning (OCLR) algorithms [39]. The stemness gene set was obtained from Miranda's studies and applied to the ssGSEA algorithm to calculate ssGSEA-based stemness indices (Table S1) [38]. The mean (the standard error of the mean (SEM)) of the stemness indices using the ssGSEA algorithm was 2.069 (0.011), and patients with stemness indices of less than 2.058 (mean-SEM) and over 2.080 (mean + SEM) were classified into the low stemness group and the high stemness group, respectively. PDAC is formally staged using a tumor node metastasis (TNM) system based on the eighth edition of the American Joint Committee on Cancer Staging Manual [40]. Student's t test was used to assess the relationship of clinical information and stemness indices and the results were plotted using the "ggplot2" (http://cran.r-project.org/package=ggplot2, accessed on 25 November 2022) package. The Kaplan–Meier (K–M) curve was plotted using the "survival" package (https://cran.r-project.org/package=survival, accessed on 25 November 2022) to achieve the survival analysis of patients with high or low stemness indices. The "DESeq2" R package was employed to identify the differential expressed genes (DEGs) between patients with high or low stemness indices. Enrichment analysis of the DEGs was conducted as follows.

2.2. Enrichment Analysis

Kyoto Encyclopedia of Genes and Genomes (KEGG) and Gene Ontology (GO) analyses were performed by using the "clusterProfiler" [41] package of R and visualized by applying the "ggplot2" package. Enrichment gene sets, including c2.cp.kegg.v7.4.symbols and h.all.v7.5.1.symbols, were obtained from the Molecular Signatures Database (MSigDB) [42]. Gene set variation analysis (GSVA) was utilized to calculate the enrichment score of these oncogenic signatures [43]. The correlation between SPHK1 and pathway scores was analyzed via Spearman correlation. The "pheatmap" R package was used for clustering heatmaps with standardization processing "scale = row".

2.3. Cell Line and Cell Culture

PDAC cell lines (MiaPaCa-2, PANC−1) were obtained from the American Type Culture Collection (ATCC) and were preserved at the Liver Cancer Institute, Zhongshan Hospital, Fudan University (Shanghai, China). All cells passed conventional quality control tests, which was consistent with the findings reported by the ATCC. The culture conditions for the cell lines were complete medium, consisting of Dulbecco's modified Eagle's medium (DMEM; GNM12800-2, GENOM, Jiaxing, Zhejiang, China) supplemented with 10% fetal bovine serum (FBS; 10270-106, Gibco, Grand Island, NY, USA) and 1% penicillin-streptomycin (1719675, Gibco, Grand Island, NY, USA), in a humidified ThermoForma incubator (Thermo Fisher Scientific, Waltham, MA, USA) with 37 °C and 5% CO_2, as described in a previous study [44].

2.4. Culture of PDAC TRCs

A previous study showed that TRCs cultured in the 3D fibrin gels represented an available CSLCs [37]. Thus, we cultured PDAC-TRCs as previously described [36]. Specifically, MiaPaCa-2 or PANC−1 cells were trypsinized and resuspended in complete medium, and then mixed with an equal volume 2 mg/mL salmon fibrinogen (SEA-133, Sea Run Holdings Inc., Freeport, ME, USA) diluted with T7 buffer (50 mM Tris-HCl, 150 mM NaCl, pH 7.4). Next, 100 U/mL thrombin (SEA-135, Sea Run Holdings Inc., Freeport, ME, USA) diluted with T7 buffer was added at a 1:50 ratio to the cell suspension to form cell mixture. The complete medium was added to cell plates after incubation for 30 min in a humidified

ThermoForma incubator (Thermo Fisher Scientific, Waltham, MA, USA) with 37 °C and 5% CO_2, and the cells were sequentially cultured for 72 h. The resulting PDAC-TRCs were used for subsequent experiments.

2.5. Quantitative Reverse Transcription Polymerase Chain Reaction (qRT-PCR)

The mRNA of 2D−cultured cells or PDAC TRCs was extracted by using an RNAeasy™ kit (R0026; Beyotime Biotechnology, Shanghai, China) according to the manufacturer's recommended procedure, and then reversely transcribed into cDNA by using Hifair® V one-step RT-gDNA digestion SuperMix Kit (11141ES60, Yeasen, Shanghai, China) according to manufacturer's instructions. Then, the conditions of qRT−PCR were set as follows: initial denaturation at 95 °C for 5 min; and 40 cycles of 95 °C for 10 s and 60 °C for 30 s, which was performed using SYBR Green kit (11202ES08) on QuantStudio5 fluorescence quantitative PCR system (Applied Biosystems, Foster City, CA, USA). The sequences of all primers were displayed in Table 1. The $2^{\Delta\Delta CT}$ method was used to calculate the relative gene expression change with β-actin as the internal normalization. Each experiment was performed with three independent replicates, and the results were displayed as the mean ± SD.

Table 1. The sequences of all primers.

Name	Forward Primer	Reverse Primer
β-actin	CCACGAAACTACCTTCAACTCC	GTGATCTCCTTCTGCATCCTGT
Sox2	CCTACAGCATGTCCTACTCGCA	CTGGAGTGGGAGGAAGAGGTAAC
CD24	CTCCTACCCACGCAGATTTATTC	AGAGTGAGACCACGAAGAGAC
CD133	GTACAACGCCAAACCACGACT	CGCACACGCCACACAGTAA
ESA	CACCAGTCTTCTTACCAAACACG	AGTCCATTAGGCAGTATCTCCAAG
SPHK1	CAGCTCTTCCGGAGTCACGT	CGTCTCCAGACATGACCACCA

2.6. RNA Interference

To silence the expression of *SPHK1*, the cells were transfected with siRNA by applying riboFECT™ CP (C10511-05, RIBOBIO, Guangzhou, Guangdong, China). The targeted sequence of siSPHK1 was GAGGCUGAAAUCUCCUUCATT.

2.7. Western Blotting

Western blotting was performed as described in our previous study [45]. Rabbit monoclonal to SPHK1((ab302714)) was purchased from Abcam.

2.8. Transwell Assays

Transwell assays were used to assess the invasion and migration ability of PDAC TRCs. For migration assays, 10,000 cells were placed into the upper chamber with DMEM medium, while for invasion assays, 10,000 cells were plated into the upper chamber, which was precoated with Matrigel (356234, BD Biosciences, San Jose, CA, USA) diluted at 1:8 with DMEM medium. Then, 800 µL of DMEM medium containing 20% FBS was added to the lower chamber and the cells were cultured for 48 h. Then, the cells in the upper chamber were carefully removed. The cells passing through the membrane filter were stained with 0.1% crystal violet solution (V5265, Sigma, St. Louis, MO, USA) and recorded by using a microscope and counted using Image J software (National Institutes of Health, Bethesda, MD, USA). Each experiment was performed with three independent replicates, and the results were displayed as the mean ± SD.

2.9. Reagent and Intervention Process

S1P (HY-108496) was purchased from MCE. The preparation of stock solution and storage were conducted according to the manufacturer's recommended procedure. When PDAC TRCs transfected with si-SPHK1 were cultured in 3D gel for 5 days, exogenous S1P

(10 µM) was supplemented and the growth of TRCs was continuously observed. In the Transwell assays, 10 µM S1P was added to the lower chamber in the testing group.

2.10. Subcutaneous Tumors in Mice

Four-week-old male nude (*nu/nu*) mice were obtained from the Shanghai Institute of Material Medicine (Shanghai, China), Chinese Academy of Science. All mice were randomly allocated to 2D group or TRC group (*n* = 18 for each group). For subcutaneous tumors, single-cell suspensions of PANC−1 and PANC−1 TRCs were injected with gradient cell density (2×10^4, 2×10^5, 2×10^6, *n* = 6 for every group) on the right side of the armpit of the nude mice. The animal study protocols were performed in accordance with the Guide for the Care and Use of Laboratory Animals stipulated by the National Academy of Sciences and the National Institutes of Health (NIH publication 86-23, revised 1985) and approved by the Animal Care and Use Committee of Zhongshan Hospital, Fudan University, Shanghai, China (Approval No. 2020-135 and date of approval 2 November 2020).

2.11. RNA Seq

The total RNA of 2D−cultured cells or PDAC TRCs (three replicates for each cell type) was extracted by using TRIzol reagent (Invitrogen, Carlsbad, CA, USA) and the quantity and purity were monitored using NanoDrop ND-1000 (NanoDrop, Wilmington, DE, USA) as well as Bioanalyzer 2100 (Agilent, Santa Clara, CA, USA). OligodT-magnetic-beads (25-61005, Thermo Fisher, Waltham, CA, USA)-enriched mRNAs were fragmented. cDNAs were synthesized from the fragmented RNA using a Reverse Transcriptase (Invitrogen SuperScript™ II Reverse Transcriptase, Carlsbad, CA, USA), and then sequenced using Illumina Novaseq™ 6000 (LC Bio Technology Co., Ltd., Hangzhou, Zhejiang, China). The obtained RNA-Seq raw data were uploaded to the Sequence Read Archive (SRA) database of the National Center for Biotechnology Information (NCBI) (https://www.ncbi.nlm.nih.gov/, accessed on 20 October 2023) with the accession number PRJNA-1020096.

2.12. The Procedure for LC-MS-Based Lipidomic Analysis

The samples of 2D−cultured PANC−1 cells or PANC−1 TRCs (four biological replicates for each cell type) were collected and were added into 1 mL of pre-cooled methanol with an internal standard (1 µg/mL of tridecanoic acid and n-valine). After vortexing for 1 min, the mixtures were stored at −80 °C ThermoForma incubator (Thermo Fisher Scientific, Waltham, MA, USA). Then, the sample preparation, lipidomic data acquisition, data preprocessing, and peak annotation were performed as described in our previous study [46].

2.13. Bioinformatics Analysis of Lipidomic Data

The lipid profile levels obtained above were loaded into an open access tool BioPAN, on LIPID MAPS Lipidomic Gateway (https://lipidmaps.org/biopan/, accessed on 25 July 2022) [47]. BioPAN calculates statistical scores for all possible lipid pathways to predict which are active or suppressed in PANC−1-TRCs samples compared to the PANC−1 cells samples. In brief, BioPAN workflow utilizes Z-score, which takes into account both the mean and the standard deviation to assume normally distributed data of lipid subclasses and determines a reaction or pathway to be significantly modified at a *p*-value < 0.05 (equivalent to Z-score > 1.645). The calculation of the Z-score was detailed by Gaud et al. [48].

2.14. Statistical Analysis

All plots and statistical analyses were conducted using R 4.3.1 and GraphPad Prism 9.5.0. Student's *t*-tests (two-tailed) and one-way analysis of variance (ANOVA) were used to compare the means of two or more samples. The predictable value of SPHK1 expression was assessed using univariate and multivariate Cox analysis. As for the cellular experiments, each experiment was performed with at least three independent replicates, and the results are displayed as the mean ± SD. A *p*-value of less than 0.05 was considered statistically significant, unless otherwise indicated. *, $p < 0.05$; **, $p < 0.01$; ***, $p < 0.001$; ****, $p < 0.0001$; ns, not significant.

3. Results

3.1. Correlation between Stemness Indices via ssGSEA Algorithms and Clinicopathological Characteristics of PDAC Patients

At first, we calculated the stemness indices using ssGSEA algorithms for each patient in TCGA-PDAC patients ($n = 179$) using RNA-seq data (Table S1). Then, according to the stemness indices, we ranked the patients from low to high (Figure 1A) and investigated the relationship between the indices and clinicopathological features including age, sex, pathological grade of tumor, N stage, T stage and TNM stage (Figure 1B–G). The results showed that patients with a higher pathological grade (Figure 1D, G2 vs. G1, $p = 0.0018$, and G3 vs. G1, $p = 0.0016$) or patients diagnosed with a higher T stage (Figure 1F, T3/4 vs. T1/2, $p = 0.039$) had significantly higher stemness indices. Moreover, the patients were divided into high ($n = 82$) and low stemness groups ($n = 82$) according to the aforementioned method, and survival analysis was conducted to compare these two groups. The K-M curve results showed that patients in the high stemness group suffered shorter median OS (high stemness group vs. low stemness group, 17.0 vs. 34.8 months, $p = 0.0011$, Figure 1H) and median DFS (high stemness group vs. low stemness group, 13.1 vs. 20.4 months, $p = 0.0007$, Figure 1I). Cox multivariate analysis with significant factors obtained from the univariate analysis ($p < 0.05$) was carried out to further assess the relationship between tumor stemness and patients' OS (Table 2) and it was found that patients belonging to the low stemness group was an independent favorable prognosis factor for PDAC (HR = 0.594, 95% CI, 0.379–0.932, $p = 0.023$). However, the stemness indices using OCLR algorithms (Table S2) were not associated with patients' OS ($p = 0.15$) and tumor dedifferentiation, as reflected in the histopathological grade (Figure S1). Taken together, these data suggested that the stemness indices using the ssGSEA algorithms could effectively distinguish PDAC patients and were consistent with the degree of tumor dedifferentiation and prognosis. Thus, we assumed that these stemness indices could be used to better describe and quantify CSCs in patients' tumors.

Table 2. Univariate and multivariate Cox regression analysis determined the independent prognostic role of stemness.

Variable		n	Univariate Cox Analysis			Multivariate Cox Analysis		
			HR	95% CI	p	HR	95% CI	p
Age	Old (> 65)	86	1			NA		
	Young (≤65)	78	0.775	0.506–1.190	0.241			
Sex	Female	89	1			NA		
	Male	75	0.799	0.523–1.220	0.300			
TNM Stage	I	20	1			NA		
	II	134	2.11	0.965–4.620	0.062			
	NA	8						
Grade	G1	29	1			1		
	G2	86	1.980	0.987–3.960	0.055	1.501	0.746–3.018	0.255
	G3/4	47	2.590	1.250–5.340	0.010 *	1.807	0.876–3.726	0.109
	Gx	2						
Lymph node stage	N0	45	1			1		
	N1/2	114	2.100	1.230–3.580	0.007 *	1.875	1.052–3.343	0.033 *
	Nx	5						
Tumor stage	T1/2	28	1			1		
	T3/4	134	2.020	1.040–3.930	0.038 *	1.237	0.597–2.563	0.567
	Tx	2						
Stemness index	High	82	1			1		
	Low	82	0.486	0.313–0.756	0.001 *	0.594	0.379–0.932	0.023 *

HR, hazard ratio; CI, confidence interval; NA, not available; * means $p < 0.05$.

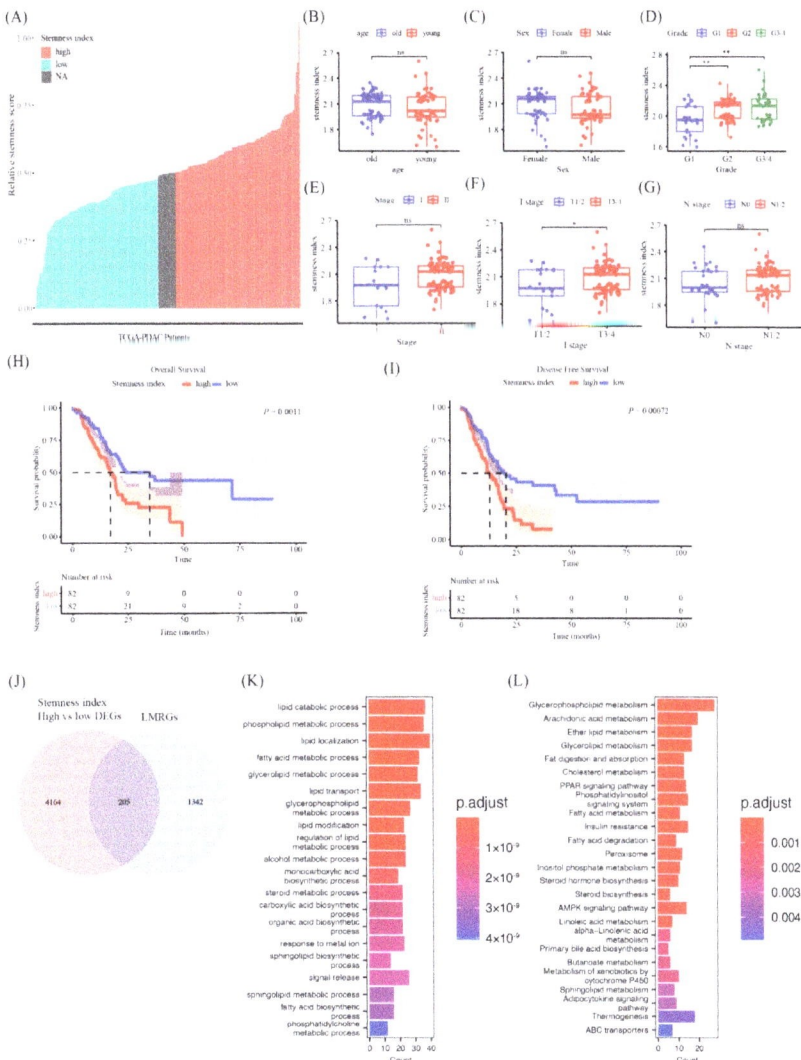

Figure 1. Correlation between stemness indices using ssGSEA and clinical features in PDAC patients. (**A**) An overview of the distribution of relative stemness indices in PDAC patients (n = 179) and the classification of stemness groups (high stemness > mean + SEM, n = 82; low stemness < mean—SEM, n = 82; NA~ mean ± SEM, n = 15). (**B–G**) Boxplots of stemness indices for PDAC patients stratified by clinical features including age, sex, pathological grade of tumor (2 Gx removed), N stage (5 Nx removed), T stage (2 Tx removed) and TNM stage (8 NA removed). OS K-M curve (**H**) and DFS K-M curve (**I**) showed the outcomes of PDAC patients in the high stemness group and low stemness group. (**J**) Venn diagram shows the overlapped genes between LMRGs and DEGs of the two stemness groups. (**K**) GO and (**L**) KEGG enrichment analysis of the overlapped genes. *, p < 0.05; and **, p < 0.01; Student's t-test. ssGSEA, single sample gene set enrichment analysis; PDAC, pancreatic ductal adenocarcinoma; DFS, disease-free survival; K-M curve, Kaplan–Meier curve; OS, overall survival; LMRGs, lipid-metabolism-related genes; DEGs, differential expressed genes; GO, gene ontology; KEGG, Kyoto Encyclopedia of Genes and Genomes.

3.2. Difference in Lipid Metabolism in Patients with High and Low Stemness Indices

More and more studies have shown that the dysregulation of lipid metabolism may be one of the most unique metabolic hallmarks of cancer, providing important targets for therapeutic interventions. To comprehensively elucidate the functional roles of deregulated lipid metabolic genes in PDAC patients, we selected 1543 LMRGs whose GO annotations included lipid-metabolism-related pathways. About 4.7% (205/4369) DEGs between tumors with high (n = 82) and low (n = 82) stemness indices are LMRGs (Figure 1J). To understand the characteristics of the lipid metabolism of PDAC, GO and KEGG enrichment analysis (Figure 1K,L and Tables S3 and S4) were performed and several specific genes and lipid metabolic pathways were identified, including fatty acid metabolism, glycerolipid metabolism, glycerophospholipid metabolism, and sphingolipid metabolism, etc. (Tables S3 and S4).

3.3. Characteristics of PDAC TRCs as an Available CSLCs Model

In order to better study the characteristics of the lipid metabolism of PDAC CSCs, we cultured human PDAC cell lines in 3D soft fiber gel to obtain PDAC TRCs, PANC−1 TRCs and MIA PaCa−2 TRCs, based on the method that our research team has previously confirmed to culture CLSCs in other tumor species [36,37]. PANC−1 TRCs and MIA PaCa−2 TRCs gradually formed clone spheres in 3D soft fiber gel, and the morphological changes from day 1 to day 5 are shown in Figure 2A. The qRT-PCR experiment results showed a significant increase in the expression of classic CSC surface markers CD133, CD24, ESA, and Sox2 in PANC−1 TRCs and MIA PaCa−2 TRCs compared to PANC−1 and MIA PaCa−2, respectively (Figure 2B). The transwell assays showed that PANC−1 TRCs and MIA PaCa−2 TRCs migrated to and invaded the lower chamber earlier than PANC−1 and MIA PaCa−2, and PANC−1 TRCs and MIA PaCa−2 TRCs exhibited more cell migration and invasion than their control groups within the same period of time (Figure 2C). These results proved that PANC−1 TRCs and MIA PaCa−2 TRCs captured a stronger tumorigenesis and metastasis ability than PANC−1 and MIA PaCa−2 in vitro.

In order to verify the malignant biology of TRCs in vivo, we constructed a subcutaneous tumor model in nude mice. PANC−1 cells and PANC−1 TRCs were inoculated with gradient cell density on the right side of the armpit near the back of the nude mice, and the tumorigenesis was observed daily. As shown in Table 3, the tumorigenesis rates of PANC−1 TRCs reached 83.3% at one month, while no tumor was observed in the 2×10^4 PANC−1 group (Figure 2D and Figure S2A). Moreover, it was observed that the tumor formation time was earlier, and the tumor volume in the PANC−1 TRC group was larger after the same observation time (30 days) (Figure 2D and Figure S2B), further confirming the notable self-renewal and tumorigenic properties of PDAC TRCs.

Table 3. Comparison of subcutaneous tumor development between PANC−1 and PANC−1 TRCs in nude mice.

Number of Cells	PANC−1 TRCs	PANC−1 Cells
2×10^6	100.0% (6/6)	66.7% (4/6)
2×10^5	100.0% (6/6)	66.7% (4/6)
2×10^4	83.3% (5/6)	0

3.4. Identification of Lipid Metabolism Pathways in PDAC TRCs via RNA-seq

Since PDAC TRCs were proved to present CSCs features, PANC−1 TRCs and MIA PaCa−2 TRCs were used to explore the lipid metabolism characteristics of PDAC CSCs in gene expression level. About 7125 DEGs between PANC−1 and PANC−1 TRCs, as well as 9999 DEGs between MIA PaCa−2 and MIA PaCa−2 TRCs, were detected via RNA-seq (Figure 3A,B). As shown in Figure 3C, genes in set 2 (n = 531) were the LMRGs among DEGs of PANC−1 and PANC−1 TRCs, and genes in set 3 (n = 666) were the LMRGs among DEGs of MIA PaCa−2 and MIA PaCa−2 TRCs. Genes in set 4 (n = 306) represented the overlapped genes between LMRGs set and the set 1 (common DEGs of TRCs and normal 2D−cultured cells, n = 3864). The top three altered KEGG pathways of PANC−1 TRCs LMRGs (Figure 3D and Table S5) were fatty acid metabolism, sphingolipid metabolism and fatty acid degradation pathways, and the related LMRGs' expression is shown in Figure 3F. When comparing MIA PaCa−2 TRCs with MIA PaCa−2, the top three altered KEGG pathways (Figure 3E and Table S6) were the glycerophospholipid metabolism, fatty acid metabolism and sphingolipid metabolism pathways, and the related LMRGs' expression is shown in Figure 3G. It was obvious that fatty acid metabolism, and sphingolipid metabolism were commonly detected as the most altered pathways. Overall, the lipid metabolism pathways' alteration in PDAC TRCs in different cell lines was similar, and the involved LMRGs may be different types of one genotype. In addition, the overlap of DEGs (set 4, n = 306, Figure 3C) with consistent trends in the two cell lines was analyzed for KEGG enrichment, and the results showed that glycerolipid metabolism, fat acid degradation, and sphingolipid metabolism pathways were the most significant changes in the lipid metabolism pathways (Figure S3 and Table S7). Despite numerous DEGs and lipid metabolic modifications in the common consistent trends set or between individual cell lines, each of our analysis identified sphingolipid metabolism as a key element regulating the phenotypes shift between TRCs and normal 2D−cultured cancer cells.

3.5. Alteration in Lipid Metabolism in PDAC TRCs via Lipidomic Analysis

Due to the similarity between the enrichment results of differential LMRGs in PANC−1 and PANC−1 TRC and the results of common differential LMRGs in the two cell lines, lipidomic analysis based on LC-MS was performed in PANC−1 TRC and PANC−1 to explore the differences in lipid metabolism products and further understand the lipid metabolism characteristics of PDAC CSCs. Principal component analysis (PCA) revealed a difference in lipidome in two groups (Figure S4). Thirteen types of lipids, including 435 lipid metabolites, were detected. In addition to varying trends in fatty acids with different chain lengths, it was also found that sphingosine (SPB), ceramide (Cer), phosphatidylethanolamine (PE), phosphatidylcholine (PC), phosphatidylglycerol (PG), and triglycerides (TG) significantly increased in TRCs, while dihydroceramide (dhCer), diglycerides (DG), and lysophosphatidylcholine (LPC) significantly decreased in TRCs. The lipidomic analysis was performed via BioPAN [48], which combined current knowledge of lipid metabolism and predicted genes to compare two biological conditions to identify activated or suppressed pathways using Z-score values (Table S8). The results of fatty acid metabolism were showed in Figure 4A. In general, palmitic acid (FA 16:0) and stearic acid (FA 18:0) were the most common FA in PANC−1 and PANC−1 TRC. Meanwhile, the longer chain FA and the extremely long chain fatty acids (FA 24:1) were found significantly increased in PANC−1 TRC. Consistent with this finding, the BioPAN network map of FA metabolism showed that the elongation of FA was the most significantly activated pathway in the FA metabolism pathway, including the monounsaturated fatty acids (FA (18:1) → FA (20:1) → FA (22:1) → FA (24:1), Z-score = 5.965), saturated fatty acids (FA (16:0) → FA (18:0) → FA (20:0) → FA (22:0) → FA (24:0) → FA (26:0) → FA (28:0), Z-score = 3.516), and polyunsaturated fatty acids [FA (20:4) → FA (22:4) → FA (24:4) → FA (24:5) → FA (24:6), Z-score = 2.737]. In sphingolipids metabolism (Figure 4B), active reaction chains (dhCer → Cer → SPB, Z-score = 5.171; SM → Cer → SPB, Z-score = 4.704) and suppressed reaction chains (SPB → Cer → SM, Z-score = 4.577) jointly lead to a significant accumulation of sphingosine in PANC−1 TRC. The reaction chain of PE generated by DG and PS in the glycophoric metabolism reaction is activated, while the reaction chain of PE as a substrate (PE → PC → LPC, Z-score = 4.224, PE → PC → PS, Z-score = 3.869) is suppressed, leading to an increase in PE in PANC−1 TRC (Figure 4C). Furthermore, we validated and analyzed the predicted genes in the BioPAN analysis with RNA-seq data. It was found that these elongations of very-long-chain FA genes *ELOVL2*, *ELOVL6*, and *ELOVL7* were significantly overexpressed in TRC groups. DEGS2 actively catalyzing dhCer → Cer → SPB reaction chains, and CERS5 suppressing the generation of Cer by SPB, and ASAH1 involved in both reactions were significantly overexpressed. These results were consistent with the genes predicted as active or suppressed in BioPAN. The results of metabolite differences (Z-score) in lipidomic, and consistency between changes in metabolic genes and metabolites, highlighted that the up-regulation of the sphingolipid metabolism pathway played the most special role in the lipid metabolic remodeling process of PDAC TRCs.

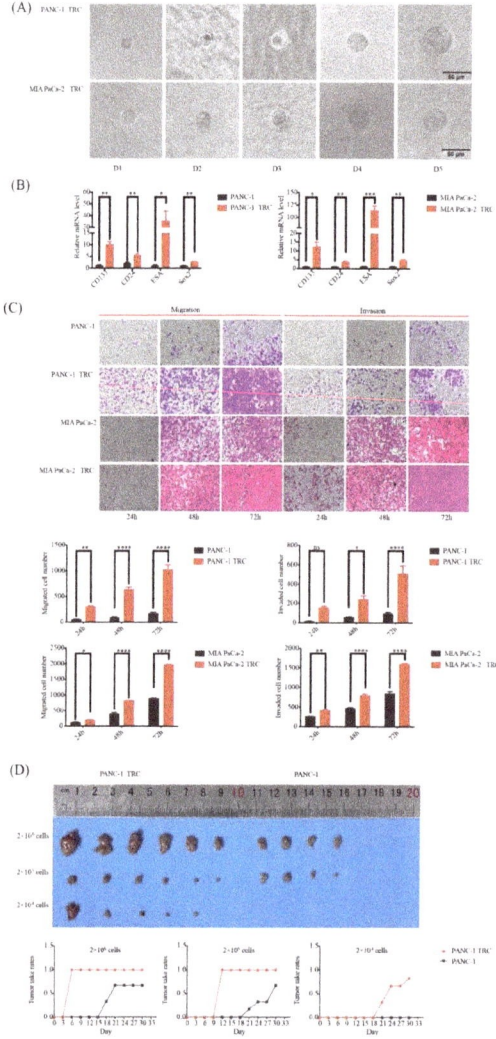

Figure 2. CLSCs' characteristics of PDAC TRCs. (**A**) Morphology of PDAC TRCs growing in 3D soft gel fibers at days 1–5. (**B**) q-RT PCR showed expression of CSCs surface markers in PDAC TRCs and normal PDAC cells (mean ± SD, n = 3, t-test). (**C**) Transwell experiment showed migration and invasion abilities of PDAC TRCs and normal PDAC cells (mean ± SD, n = 3, t-test). (**D**) Morphology of subcutaneous tumors and tumorigenesis ability in nude mice of PDAC TRCs and normal PDAC cells (n = 6 for every gradient cell density in each cell type). *, $p < 0.05$; **, $p < 0.01$; ***, $p < 0.001$; and ****, $p < 0.0001$; ns, not significant. CLSCs, cancer stem-like cells; TRCs, tumor-repopulating cells; ESA, erythropoiesis-stimulating agent; Sox2, sex-determining region Y-box 2; q-RT PCR, quantitative reverse transcription polymerase chain reaction.

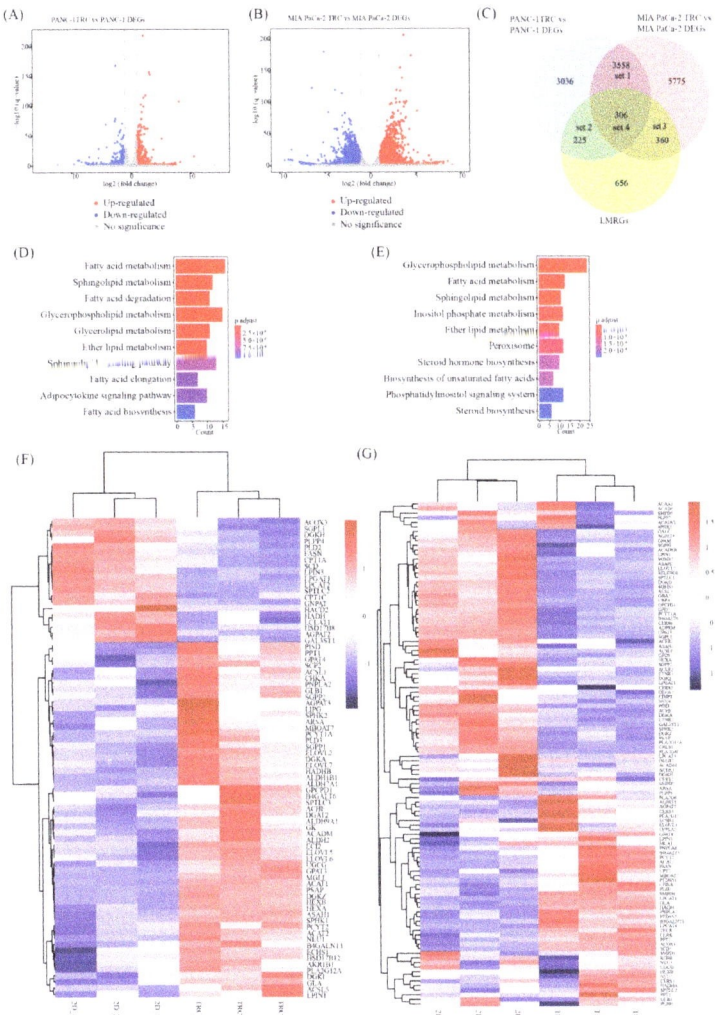

Figure 3. Identification of lipid metabolism reprogram pathways in PDAC TRCs via transcriptome analysis. (**A**) Volcano plots of DEGs of PANC−1 and PANC−1 TRCs (up-regulated genes in red and down-regulated in blue, n = 7125). (**B**) Volcano plots of DEGs of MIA PaCa−2 and MIA PaCa−2 TRCs (up-regulated genes in red and down-regulated in blue, n = 9999). (**C**) Venn diagram shows the overlapped genes between LMRGs and the DEGs of TRCs and the source normal 2D cells. Genes in set 2 (n = 531) were the LMRGs among DEGs of PANC−1 and PANC−1 TRCs and genes in set 3 (n = 666) were the LMRGs among DEGs of MIA PaCa−2 and MIA PaCa−2 TRCs. Genes in set 4 (n = 306) represented the overlapped genes between LMRGs set and the set 1 (common DEGs of PDAC TRCs and normal 2D−cultured cells, n = 3864) (**D**) Top 10 entries in KEGG enrichment pathway of genes in set2. (**E**) Top 10 entries in KEGG enrichment pathway of genes in set 3. (**F**) Heatmap of genes enriched in top 3 entries in KEGG enrichment pathway of genes in set 2 (relative high expression in red and relative low expression in blue). (**G**) Heatmap of genes enriched in top 3 entries in KEGG enrichment pathway of genes in set 3 (relative high expression in red and relative low expression in blue). PDAC, pancreatic ductal adenocarcinoma; TRCs, tumor-repopulating cells; DEGs, differential expressed genes; LMRGs, lipid-metabolism-related genes.

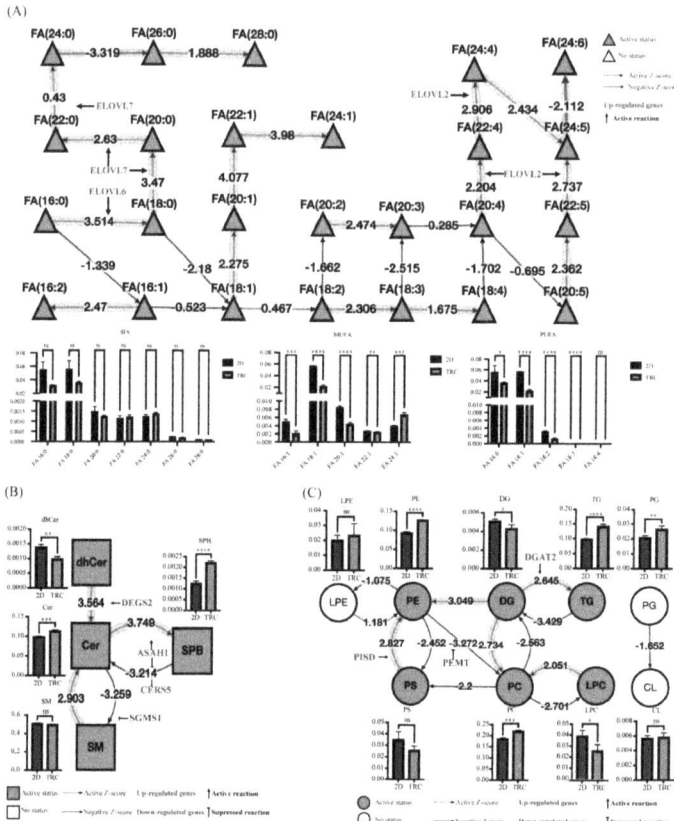

Figure 4. Lipid network generated using BioPAN software combined with alternation metabolites and related genes in PANC−1 TRCs compared to the normal 2D−cultured PANC−1 cells. (**A**) BioPAN fatty acids networks. FA graphs exported from BioPAN tool for PANC−1 TRCs compared to PANC−1. Green nodes correspond to active FAs and green shaded arrows to active pathways. Reactions with a positive Z score have green arrows, while negative Z scores are colored purple. Pathways options: PANC−1 TRCs condition of interest, PANC−1 control condition, lipid type, active status, subclass level, reaction subset of lipid data, p value 0.05, and no paired data. LMRGs of DEGs in red (up−regulated) of PANC−1 TRCs compared to PANC−1 cells using RNA−seq were consistent with the genes predicted as active (arrow) in BioPAN. (**B**,**C**) BioPAN lipid networks. Lipid network graphs exported from BioPAN for PANC−1 TRCs compared to PANC−1. Green nodes [glycerophospholipid metabolism in circle (**B**) and sphingolipid metabolism in square (**C**)] correspond to active lipids and green shaded arrows to active pathways. Reactions with a positive Z score have green arrows while negative Z scores are colored purple. Pathways options: PANC−1 TRCs condition of interest, PANC−1 control condition, lipid type, active status, subclass level, reaction subset of lipid data, p value 0.05, and no paired data. LMRGs of DEGs in red (up−regulated) or in blue (down−regulated) of PANC−1 TRCs compared to PANC−1 cells using RNA−seq were consistent with the genes predicted as active (arrow) or suppressed (Long line truncated by short dash) in BioPAN. *, $p < 0.05$; **, $p < 0.01$; ***, $p < 0.001$; ****, $p < 0.0001$; ns, not significant; t-test. TRCs, tumor-repopulating cells; FA, fatty acid; MUFA, monounsaturated fatty acid; PUFA, polyunsaturated fatty acid; SFA, saturated fatty acid; Cer, ceramide; dhCer, dihydroceramide; SPB, sphingosine; SM, sphingomyelin; LPC, lysophosphatidylcholine; PE, phosphatidylethanolamine; PS, phosphoserine; LPE, lysophos-phatidylethanolamine; PC, phosphatidylcholine; CL, cardiolipin; TG, triglycerides; DG, diglycerides.

3.6. Identification of SPHK1 as a Key Lipid-Metabolism-Related Stemness Gene in PDAC

Taking into account the comprehensive integrated transcriptomic and lipidomic analysis, the up-regulation of the sphingolipid metabolism pathway was found to be the most significant lipid metabolic remodeling process of PDAC TRCs. This finding was also supported by patient data analysis. The LMRGs in DEGs of PDAC patients in the high stemness group and in the low stemness group were enriched in the sphingolipid metabolism biological process via GO analysis and the sphingolipid metabolism pathway via KEGG analysis (Figure 1J,K). To identify the key genes of PDAC CSCs' sphingolipid metabolism, the 25 common genes of sphingolipid metabolic process among GO enrichment results (Figures 5A and S3A) were further screened using the thresholds (|logFoldChange| > 1 and FDR < 0.05). A total of 14 LMRGs were significantly differently expressed between PANC−1 TRC and PANC−1 (Figure 5B), and 8 of them were significantly different in the high stemness group and in the low stemness group, among which only the expression differences in *SPHK1, SPTLC3, HEXB, GAL3ST1*, and *ASAH1* were consistent in the CSLCs model and patients' grouping by stemness indices (Figures 5C and S5). Moreover, survival analysis (Figure 5D) showed that patients with high expression of SPHK1 suffered a shorter median OS (p = 0.029) and shorter median disease-free survival (DSS, p = 0.0069) than those with low expression, and that patients with *SPTLC3* high expressed had poor median OS (p = 0.0086) and median DFS (p = 0.0015). No significant survival difference was found to be in association with the expression of *HEXB, GAL3ST1*, and *ASAH1* (Figure S5B). The expression level of SPHK1 (p = 0.007) was significantly positively correlated with the tumor proliferation signature instead of *SPTLC3* (p = 0.616) using ssGSEA analysis (Figure 5E). In addition, a positive correlation between the expression of SPHK1 and malignant biological signaling pathways of CSCs including TGF-beta, P53 pathways, and EMT markers [49] was observed as well (Figure 5F). Accordingly, *SPHK1* was considered as a key LMRG involved in stemness and prognosis in PDAC.

3.7. SPHK1 Promotes the Malignant Behaviors of PDAC-TRC by Promoting Stemness

Finally, the biologic function of SPHK1 was evaluated in PDAC-TRC. The expression of SPHK1 was silenced using siRNA in PANC−1 TRC as well as MIA PaCa−2 TRC, being validated via qRT-PCR and Western blotting (Figure 6A,B). Silencing SPHK1 significantly inhibited the clonogenicity of both PANC−1 TRC as well as MIA PaCa−2 TRC (Figure 6C), and significantly decreased the migration and invasion ability of PANC−1 TRC as well as MIA PaCa−2 TRC (Figure 6D). By the fifth day of cultivation of PDAC TRCs transfected with siSPHK1, exogenous supplementation of S1P was performed, which recovered the clonogenic ability of TRCs. Exogenous supplementation of S1P to normal 2D−cultured PANC−1 and MIA PaCa−2 cells also resulted in enhanced migration and invasion (Figure S6). These results suggest that SPHK1 played a crucial role in promoting malignant behaviors in PDAC TRC. In addition, we evaluated the effect of SPHK1 on promoting stemness in PDAC TRC. Silencing SPHK1 significantly decreased the expression of multiple CSCs biomarkers, such as CD133, CD24, Nanog and Sox2 (Figure 6E), which were up-regulated in PDAC TRC compared to 2D−cultured PDAC cells. Taken together, SPHK1 may drive the malignant behaviors of PDAC-TRC by promoting stemness.

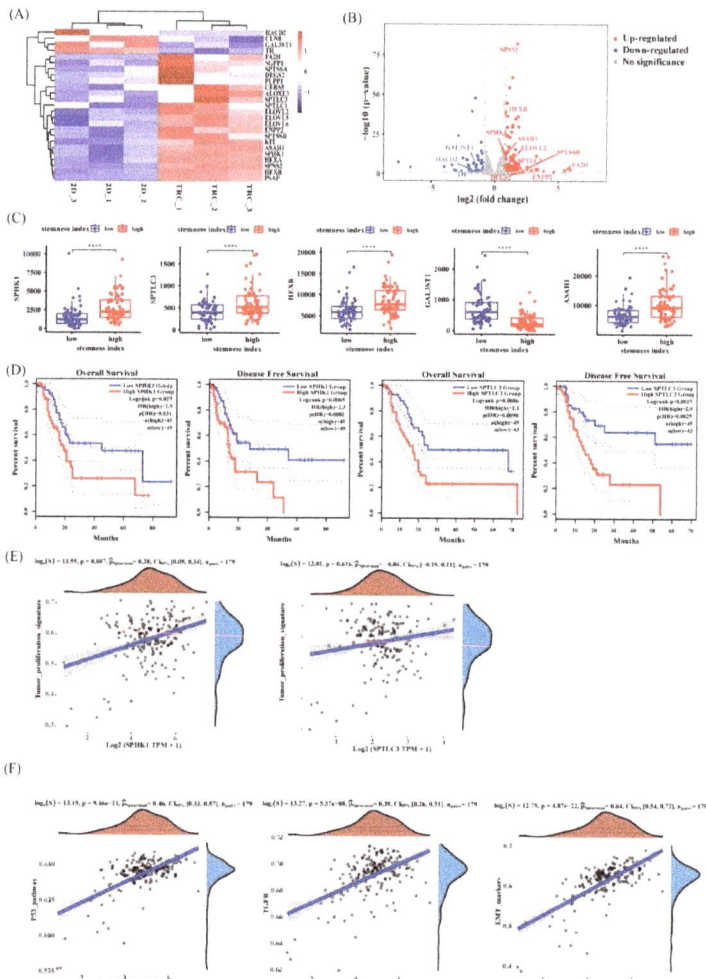

Figure 5. Identification of the key stemness LMRGs in PDAC. (**A**) Heatmap of gene expression (relative high expression in red and relative low expression in blue) which were enriched in sphingolipid metabolic process by GO enrichment analysis. (**B**) Volcano plot of the lipid-relative DEGs between PANC−1 TRC and PANC−1. Plots in red (up-regulated) or in blue (down-regulated) with gene names representing significant DEGs (|logFoldChange| > 1 and FDR < 0.05) enriched in sphingolipid metabolic process. (**C**) The correlation between the five genes (*SPHK1*, *SPTLC3*, *HEXB*, *GAL3ST1*, and *ASAH1*) and stemness indices by ssGSEA (****, $p < 0.0001$; t-test). (**D**) OS and DFS curves of PDAC patients from TCGA clustered by the expression of SPHK1 and SPTLC3 with quartile as group cutoff. (**E**) The correlation between the two genes (*SPHK1* and *SPTLC3*) and tumor proliferation signature using ssGSEA analysis. (**F**) The correlation between SPHK1 and TGF-beta, P53 pathways, and EMT markers using ssGSEA analysis ($n = 179$, Spearman correlation analysis). SPHK1, sphingo-sine kinases 1; PDAC, pancreatic ductal adenocarcinoma; GO, Gene Ontology; DEGs, differential expressed genes; SPTLC3, serine palmitoyltransferase 3; HEXB, beta-hexosaminidase; GAL3ST1, galac-tose-3-O-sulfotransferase 1; ASAH1, N-acylsphingosine amidohydrolase 1; ssGSEA, single sample gene set enrichment analysis; DFS, disease-free survival; OS, overall survival; TGF-beta, transforming growth factor-beta; EMT, epithelial mesenchymal transition.

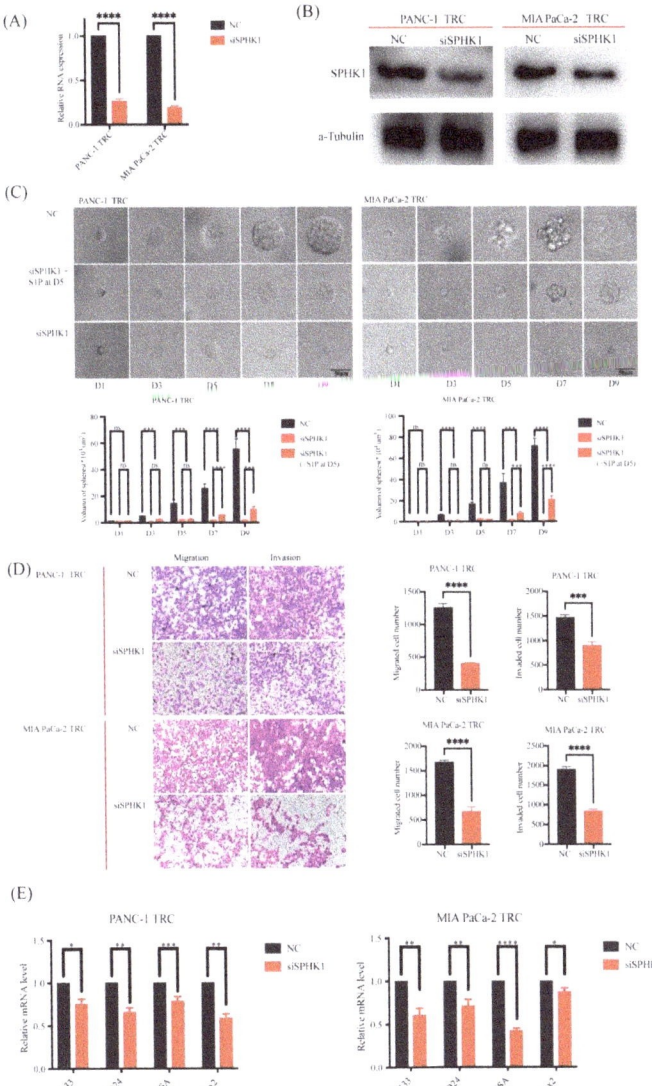

Figure 6. Effect of silencing SPHK1 by transfected siRNA to PDAC TRCs compared to the negative control (NC). (**A**) The mRNA level of SPHK1 in PDAC TRCs transfected with siSPHK1 and NC via qRT-PCR (mean ± SD, n = 3, t-test). (**B**) The expression of SPHK1 in PDAC TRCs transfected with siSPHK1 and NC via Western blotting. (**C**) Effect of silencing SPHK1 compared to NC on the colony growth of PDAC TRCs (mean ± SD, n = 3, t-test). (**D**) Effect of silencing SPHK1 compared to NC on the migration and invasion ability of PDAC TRCs via transwell assay (mean ± SD, n = 3, t-test). (**E**) Effect of silencing SPHK1 compared to NC on the expression of CSCs markers (ESA, CD133, Sox2 and CD24) detected via qRT-PCR (mean ± SD, n = 3, t-test). *, $p < 0.05$; **, $p < 0.01$; ***, $p < 0.001$; and ****, $p < 0.0001$; ns, not significant. SPHK1, sphingosine kinases 1; PDAC, pancreatic ductal adenocarcinoma; TRCs, tumor-repopulating cells; qRT-PCR, quantitative reverse transcription polymerase chain reaction; S1P, Sphingosine 1-phosphate; CSCs, cancer stem cells; ESA, erythropoiesis-stimulating agent; Sox2, sex-determining region Y-box 2.

4. Discussion

CSCs are thought to contribute to tumor heterogeneity, which is an essential and distinct feature of PDAC. The stemness indices calculated using the ssGSEA algorithm rather than OCLR algorithm were applied in our study to describe and quantify CSCs. Transcriptome and lipidomic analysis on PDAC TRCs, proved to be an available CSCs model, found that the up-regulation of the sphingolipid metabolism pathway played the most special role in the lipid metabolic remodeling process. Finally, we identified *SPHK1* as the key stemness gene involved in sphingolipid metabolism. This understanding of CSCs and lipid metabolism reprogramming paved the way for developing novel therapeutic strategies of PDAC, and SPHK1 might be an appropriate target candidate.

CSC represents a small group of cells with infinite proliferative capacity, which is considered the main cause of metastasis and therapeutic resistance [38]. Therefore, the targeted eradication of CSCs will be an important progress in PDAC treatment. However, it how to best define CSCs and the extent to which different tumor types can develop to tumor mass are still controversial. Despite these controversies, increasing evidence suggests that stem-cell-associated features, often referred to as "stemness", are biologically important in cancer [50], and are strongly related to poor outcomes in a wide variety of cancers [51,52]. An innovative OCLR on transcriptome was used to obtain the stemness indices (mRNAsi), which has been proven to stratify recognized undifferentiated BRCA, AML, and gliomas [39]. However, in our study, the stemness indices using OCLR failed to find an association with tumors' undifferentiated state and patients' outcomes. The possible reason may be related to Alex Miranda's findings in reproducing the OCLR algorithm, as the OCLR algorithm precludes an unbiased assessment of the relationship between stemness and tumor immunity [38]. Therefore, we adopted the ssGSEA algorithm mentioned in Alex Miranda' report to calculate the stemness indices. The results showed that higher stemness indices were correlated with more advanced clinical stages, a higher degree of oncogenic dedifferentiation, and worse outcomes, and that the classification of patients into high and low stemness group accordingly could be an independent prognostic predictor. Since the stemness indices using the ssGSEA algorithm can stratify recognized undifferentiated cancers, they were used to provide an approach to explore comprehensive lipid metabolism pathways on undifferentiated cancers in patients. Although it is currently unclear whether the stemness indices obtained from a large number of tumor samples represent a rare true CSCs population, our findings may advance the development for quantitating PDAC stemness, and provide a basis for the therapeutic targeting of the stemness phenotype itself.

In this study, we have mapped the specific lipid metabolism features of PDAC TRCs by combining the changes in lipid metabolites via lipidomic analysis and the expression of genes encoding metabolic enzymes via transcriptomic analysis. To understand the relationship of the stemness phenotype itself and the lipid metabolism, 3D soft fibrin gel [36] was used to culture PDAC TRCs by adjusting matrix stiffness, a significant physical property of ECM, which exerts a vital role in PDAC stemness regulation [53]. The results of malignant behaviors and overexpressed CSCs' makers verified PDAC TRCs as an available CSLCs model. Of note, we not only showed the enhancement of fatty acid prolongation in PDAC CSCs consistent with previous studies [34,35], but also found the unique changes in sphingolipid metabolism in PDAC CSCs for the first time. Sphingolipids are not only important structural components of biological membranes, but also bioactive molecules that play a predominant role in signal transduction, cell growth, differentiation, and programmed cell death and thus affect tumor suppression or survival [54]. De novo sphingolipid begins with the condensation of serine and palmitoyl-CoA by serine palmitoyltransferase (SPT) to form dhCer, and endogenous ceramides synthesized after dihydroceramide desaturation by dihydroceramide desaturase (DES). Ceramide is also generated via sphingomyelin hydrolysis by sphingomyelinases (SMases) and via glucosylceramide breakdown. In addition, the salvage pathway for ceramide generation utilizes the recycling of sphingosine by CERS1–6 [55,56]. Ceramide is a core molecule in sphingolipids' metabolism. Although cellular stress can induce the accumulation of ceramide and mediate cancer cell death [57], the active metabolism of ceramide has been confirmed in various tumors [58,59]. Ceramide can be converted to ceride-1-phosphate (C1P) [60] and SM [58], respectively, and is also utilized as a precursor for the generation of glycosphingolipids (GSL) including glucosylceramide and lactosylceramide [61]. Ceramide is hydroxylated by ceramidases (CDases) to yield sphingosine, which is phonologically late, by SPHK1 (also known as SK1) or SPHK2 (also known as SK2) to generate S1P [62]. In our study, we observed a slight increase in ceramide levels in TRCs, which may be related to the active metabolism of ceramide due to its role as a substrate for generation of sphingolipids with pro-survival functions. Different from the previous study's results that the increasing GSL [61] and C1P [60] in PDAC contribute to malignant metastasis and tumor progression, the significantly elevated sphingosine was observed in PDAC TRCs. The accumulation of sphingosine was the result of the significantly inhibited salvage pathway with the significantly activated sphingosine generation pathway. In addition, our study found a significant decrease in dhCer

It is hypothesized that CSCs, due to their unlimited proliferation, are in a long-term high demand for energy and cell division substances, resulting in significant changes in enzyme quantity instead of enzyme activity through which bio-reactions in the normal cells may be precisely regulated. Our results demonstrated that the significantly changing lipid reaction chain is coordinated with the trend of changes in key genes involved. For example, we observed the enhanced DEGS2's consumption amount of dhCer as reported in colorectal cancer [63], the significantly inhibited salvage pathway by overexpressed CERS5, and the activated sphingosine generation pathway by overexpressed ASAH1. Moreover, we found that *SPHK1*, the gene encoding the key enzyme catalyzing S1P from sphingosine, was observed to be overexpressed not only in PDAC TRCs but also in PDAC patients with high stemness indices, and overexpressed SPHK1 predicted worse prognosis of PDAC patients. The findings were consistent with the quantification of SPHK1 in PDAC specimens via immunohistochemistry, indicating high SPHK1 expression is independently associated with lymphatic invasion and unfavorable prognosis in PDAC patients [64]. Overexpression of SPHK1 facilitates the retention of endothelial progenitor cells at the progenitor stage [65] and promotes the proliferation of neural progenitor/stem cells [66]. And the involvement of the SPHK1 in CSC functioning has been recently investigated in several malignancies, including glioblastoma [67], melanoma [68], hepatocellular carcinoma [69], and breast adenocarcinoma [70]. Therefore, we hypothesized that SPHK1 plays an important role in maintaining the stemness of PDAC. The ssGSEA analysis demonstrated that the expression level of SPHK1 was significantly positively correlated with TGF-beta, P53, EMT, and tumor proliferation signals, in accordance with the results that SPHK1 are involved in CSCs markers expression, and the sphericity, migration, and invasion abilities of PDAC TRCs. Another study also demonstrated that SPHK1 upregulation may play a potential role in early neoplastic transformation of inflammatory lesions in long-standing chronic pancreatitis patients [71]. Mebendazolee was proved to be used as a potential therapeutic agent for treating PDAC, because it selectively inhibited SPHK1 more than SPHK2 and regulated the levels of sphingolipids [62]. In addition, the inhibitor of SPHK1 was reported to be effective in the combination treatment of PDAC [72], and can enhance the therapeutic effect of gemcitabine [73].

Nevertheless, limitations exist in this study. First, although this study has comprehensively considered the transcriptome of patients' tumors and the transcriptome and lipidomic characteristics at the cell level, our findings still need to be further verified in preclinical models such as PDX or PDO considering the unique tumor microenvironment of PDAC. Secondly, it is necessary to further explore the stemness phenotype of PDAC by combining single cell sequencing or metabolomics, which can more accurately reflect the role of lipid metabolism in PDAC. Finally, the treatment of CSCs remains at the theoretical level; therefore, targeted treatment of SPHK1 or sphingolipid metabolism should be more considered in combination therapy for exploration.

5. Conclusions

In this study, we explored the lipid metabolism reprogramming pathway in PDAC with high or low stemness indices. The sphingolipid metabolism pathway was associated with tumor stemness and SPHK1 was found to play an important role in promoting stemness and malignant behaviors in PDAC-TRC. Furthermore, SPHK1 was strongly correlated with patients' prognosis and a malignant-tumor-behavior-related signature in PDAC patients. These findings provide a novel strategy for targeting tumor lipid metabolism to inhibit CSCs in PDAC.

Supplementary Materials: The following supporting information can be downloaded at: https://www.mdpi.com/article/10.3390/metabo13111132/s1, Figure S1: Correlation between stemness indices by OCLR and clinical features in PDAC patients; Figure S2: Tumor bearing nude mice display and tumor volumes statistics; Figure S3: Supplementary enrichment analysis of PDAC TRCs' RNA-seq; Figure S4: Principal component analysis (PCA) of lipid profile; Figure S5: Key stemness genes in PDAC; Figure S6: Effect of exogenous supplementation of S1P on the migration and invasion ability of PDAC; Figure S7: Identification of lipid metabolic pathways in tumor tissue compared to normal tissue; Figure S8: Classification of stemness by 33% percentile and 66% percentile stemness indices and subsequent features analysis in PDAC patients; Figure S9: The correlation between the key genes enriched in sphingolipid metabolism biological process and stemness indices by top/bottom 1/3; Table S1: Stemness indices by ssGSEA algorithms; Table S2: Stemness index by OCLR; Table S3: GO enrichment results of differential expressed LMRGs between patients with high and low stemness indices stratified by mean indices ($n = 82$); Table S4: KEGG enrichment results of differential expressed LMRGs between patients with high and low stemness indices stratified by mean indices ($n = 82$); Table S5: KEGG enrichment results of differential expressed LMRGs between PANC−1 TRC and PANC−1; Table S6: KEGG enrichment results of differential expressed LMRGs between MIA PaCa−2 TRC and MIA PaCa−2; Table S7: KEGG enrichment results of differential expressed LMRGs of set 4; Table S8: The lipidomic analysis by BioPAN. Table S9: KEGG analysis of differential expressed LMRGs between tumor and normal tissue; Table S10: KEGG enrichment results of differential expressed LMRGs between patients with high and low stemness indices by the top 1/3 and the bottom 1/3 of indices ($n = 58$); Table S11: Univariate and multivariate Cox regression analysis of patients with stemness indices in the bottom 1/3 and top 1/3. The differential analysis comparing tumor tissues and normal tissue of PDAC patients (Figure S7 and Table S9) were described in the supplementary methods and results parts, as well as the analysis based on the classification of stemness by 33% percentile and 66% percentile stemness indices in PDAC patients (Figures S8 and S9 and Tables S10 and S11).

Author Contributions: J.X. (Jinzhi Xu), L.Z. and X.D. contributed to the bioinformatics analysis and cellular experiments. Z.Q., S.C. and J.Z. were responsible for the data download and figure layout. J.X. (Jinzhi Xu) wrote the manuscript. X.C. and J.X. (Jinglin Xia) provided the project idea, technical guidance, and guided the manuscript's revision. All authors have read and agreed to the published version of the manuscript.

Funding: This study was funded by the Natural Science Foundation of Shanghai (20ZR1410400), National Natural Science Foundation of China (81972233, 82173662, 81772590 and 81572395).

Institutional Review Board Statement: The animal study protocols were performed in accordance with the Guide for the Care and Use of Laboratory Animals stipulated by the National Academy of Sciences and the National Institutes of Health (NIH publication 86–23, revised 1985) and approved by the Animal Care and Use Committee of Zhongshan Hospital, Fudan University, Shanghai, China (Approval No. 2020-135 and date of approval 2 November 2020).

Informed Consent Statement: Not applicable.

Data Availability Statement: Human RNA-seq data were downloaded from TCGA (https://portal.gdc.cancer.gov/, accessed on 20 November 2022). Cancer metabolic gene sets and enrichment gene sets (c2.cp.kegg.v7.4.symbols and h.all.v7.5.1.symbols) were obtained from the MSigDB website. The obtained RNA-Seq raw data were uploaded to the Sequence Read Archive (SRA) database of the National Center for Biotechnology Information (NCBI) (https://www.ncbi.nlm.nih.gov/, accessed on 20 October 2023) with the accession number PRJNA-1020096.

Conflicts of Interest: The authors declare no conflict of interest.

References

1. Siegel, R.L.; Miller, K.D.; Fuchs, H.E.; Jemal, A. Cancer statistics, 2022. *CA Cancer J. Clin.* **2022**, *72*, 7–33. [CrossRef]
2. Ying, H.; Dey, P.; Yao, W.; Kimmelman, A.C.; Draetta, G.F.; Maitra, A.; DePinho, R.A. Genetics and biology of pancreatic ductal adenocarcinoma. *Genes Dev.* **2016**, *30*, 355–385. [CrossRef] [PubMed]
3. Ali, H.; Pamarthy, R.; Vallabhaneni, M.; Sarfraz, S.; Ali, H.; Rafique, H. Pancreatic cancer incidence trends in the United States from 2000-2017: Analysis of Surveillance, Epidemiology and End Results (SEER) database. *F1000Research* **2021**, *10*, 529. [CrossRef] [PubMed]
4. Walcher, L.; Kistenmacher, A.K.; Suo, H.; Kitte, R.; Dluczek, S.; Strauss, A.; Blaudszun, A.R.; Yevsa, T.; Fricke, S.; Kossatz-Boehlert, U. Cancer Stem Cells-Origins and Biomarkers: Perspectives for Targeted Personalized Therapies. *Front. Immunol.* **2020**, *11*, 1280. [CrossRef]
5. Yadav, A.K.; Desai, N.S. Cancer Stem Cells: Acquisition, Characteristics, Therapeutic Implications, Targeting Strategies and Future Prospects. *Stem Cell Rev. Rep.* **2019**, *15*, 331–355. [CrossRef] [PubMed]
6. Li, C.; Heidt, D.G.; Dalerba, P.; Burant, C.F.; Zhang, L.; Adsay, V.; Wicha, M.; Clarke, M.F.; Simeone, D.M. Identification of pancreatic cancer stem cells. *Cancer Res.* **2007**, *67*, 1030–1037. [CrossRef] [PubMed]
7. Li, C.; Lee, C.J.; Simeone, D.M. Identification of human pancreatic cancer stem cells. *Methods Mol. Biol.* **2009**, *568*, 161–173. [CrossRef]
8. Bisht, S.; Nigam, M.; Kunjwal, S.S.; Sergey, P.; Mishra, A.P.; Sharifi-Rad, J. Cancer Stem Cells: From an Insight into the Basics to Recent Advances and Therapeutic Targeting. *Stem Cells Int.* **2022**, *2022*, 9653244. [CrossRef]
9. Wood, L.D.; Canto, M.I.; Jaffee, E.M.; Simeone, D.M. Pancreatic Cancer: Pathogenesis, Screening, Diagnosis, and Treatment. *Gastroenterology* **2022**, *163*, 386–402.e1. [CrossRef]
10. Mukherji, R.; Debnath, D.; Hartley, M.L.; Noel, M.S. The Role of Immunotherapy in Pancreatic Cancer. *Curr. Oncol.* **2022**, *29*, 6864–6892. [CrossRef]
11. Hanahan, D. Hallmarks of Cancer: New Dimensions. *Cancer Discov.* **2022**, *12*, 31–46. [CrossRef]
12. Viale, A.; Pettazzoni, P.; Lyssiotis, C.A.; Ying, H.; Sanchez, N.; Marchesini, M.; Carugo, A.; Green, T.; Seth, S.; Giuliani, V.; et al. Oncogene ablation-resistant pancreatic cancer cells depend on mitochondrial function. *Nature* **2014**, *514*, 628–632. [CrossRef]
13. Ying, H.; Kimmelman, A.C.; Lyssiotis, C.A.; Hua, S.; Chu, G.C.; Fletcher-Sananikone, E.; Locasale, J.W.; Son, J.; Zhang, H.; Coloff, J.L.; et al. Oncogenic Kras maintains pancreatic tumors through regulation of anabolic glucose metabolism. *Cell* **2012**, *149*, 656–670. [CrossRef]
14. Philip, B.; Roland, C.L.; Daniluk, J.; Liu, Y.; Chatterjee, D.; Gomez, S.B.; Ji, B.; Huang, H.; Wang, H.; Fleming, J.B.; et al. A high-fat diet activates oncogenic Kras and COX2 to induce development of pancreatic ductal adenocarcinoma in mice. *Gastroenterology* **2013**, *145*, 1449–1458. [CrossRef] [PubMed]
15. Gabitova-Cornell, L.; Surumbayeva, A.; Peri, S.; Franco-Barraza, J.; Restifo, D.; Weitz, N.; Ogier, C.; Goldman, A.R.; Hartman, T.R.; Francescone, R.; et al. Cholesterol Pathway Inhibition Induces TGF-beta Signaling to Promote Basal Differentiation in Pancreatic Cancer. *Cancer Cell* **2020**, *38*, 567–583.e11. [CrossRef] [PubMed]
16. Xu, R.; Yang, J.; Ren, B.; Wang, H.; Yang, G.; Chen, Y.; You, L.; Zhao, Y. Reprogramming of Amino Acid Metabolism in Pancreatic Cancer: Recent Advances and Therapeutic Strategies. *Front. Oncol.* **2020**, *10*, 572722. [CrossRef] [PubMed]
17. Rossmeislova, L.; Gojda, J.; Smolkova, K. Pancreatic cancer: Branched-chain amino acids as putative key metabolic regulators? *Cancer Metastasis Rev.* **2021**, *40*, 1115–1139. [CrossRef]
18. Pavlova, N.N.; Thompson, C.B. The Emerging Hallmarks of Cancer Metabolism. *Cell Metab.* **2016**, *23*, 27–47. [CrossRef]
19. Hu, C.M.; Tien, S.C.; Hsieh, P.K.; Jeng, Y.M.; Chang, M.C.; Chang, Y.T.; Chen, Y.J.; Chen, Y.J.; Lee, E.Y.P.; Lee, W.H. High Glucose Triggers Nucleotide Imbalance through O-GlcNAcylation of Key Enzymes and Induces KRAS Mutation in Pancreatic Cells. *Cell Metab.* **2019**, *29*, 1334–1349.e10. [CrossRef]
20. Ogunleye, A.O.; Nimmakayala, R.K.; Batra, S.K.; Ponnusamy, M.P. Metabolic Rewiring and Stemness: A Critical Attribute of Pancreatic Cancer Progression. *Stem Cells* **2023**, *41*, 417–430. [CrossRef] [PubMed]
21. Valle, S.; Alcala, S.; Martin-Hijano, L.; Cabezas-Sainz, P.; Navarro, D.; Munoz, E.R.; Yuste, L.; Tiwary, K.; Walter, K.; Ruiz-Canas, L.; et al. Exploiting oxidative phosphorylation to promote the stem and immunoevasive properties of pancreatic cancer stem cells. *Nat. Commun.* **2020**, *11*, 5265. [CrossRef]
22. Nimmakayala, R.K.; Leon, F.; Rachagani, S.; Rauth, S.; Nallasamy, P.; Marimuthu, S.; Shailendra, G.K.; Chhonker, Y.S.; Chugh, S.; Chirravuri, R.; et al. Metabolic programming of distinct cancer stem cells promotes metastasis of pancreatic ductal adenocarcinoma. *Oncogene* **2021**, *40*, 215–231. [CrossRef] [PubMed]
23. Isayev, O.; Rausch, V.; Bauer, N.; Liu, L.; Fan, P.; Zhang, Y.; Gladkich, J.; Nwaeburu, C.C.; Mattern, J.; Mollenhauer, M.; et al. Inhibition of glucose turnover by 3-bromopyruvate counteracts pancreatic cancer stem cell features and sensitizes cells to gemcitabine. *Oncotarget* **2014**, *5*, 5177–5189. [CrossRef]
24. Wang, V.M.; Ferreira, R.M.M.; Almagro, J.; Evan, T.; Legrave, N.; Zaw Thin, M.; Frith, D.; Carvalho, J.; Barry, D.J.; Snijders, A.P.; et al. CD9 identifies pancreatic cancer stem cells and modulates glutamine metabolism to fuel tumour growth. *Nat. Cell Biol.* **2019**, *21*, 1425–1435. [CrossRef]

25. Li, D.; Fu, Z.; Chen, R.; Zhao, X.; Zhou, Y.; Zeng, B.; Yu, M.; Zhou, Q.; Lin, Q.; Gao, W.; et al. Inhibition of glutamine metabolism counteracts pancreatic cancer stem cell features and sensitizes cells to radiotherapy. *Oncotarget* **2015**, *6*, 31151–31163. [CrossRef] [PubMed]
26. Zhang, Y.; Beachy, P. Cellular and molecular mechanisms of Hedgehog signalling. *Nat. Rev. Mol. Cell Biol.* **2023**, *24*, 668–687. [CrossRef]
27. Martin-Perez, M.; Urdiroz-Urricelqui, U.; Bigas, C.; Benitah, S. The role of lipids in cancer progression and metastasis. *Cell Metab.* **2022**, *34*, 1675–1699. [CrossRef]
28. Lin, L.; Ding, Y.; Wang, Y.; Wang, Z.; Yin, X.; Yan, G.; Zhang, L.; Yang, P.; Shen, H. Functional lipidomics: Palmitic acid impairs hepatocellular carcinoma development by modulating membrane fluidity and glucose metabolism. *Hepatology* **2017**, *66*, 432–448. [CrossRef] [PubMed]
29. Ubellacker, J.M.; Tasdogan, A.; Ramesh, V.; Shen, B.; Mitchell, E.C.; Martin-Sandoval, M.S.; Gu, Z.; McCormick, M.L.; Durham, A.B.; Spitz, D.R.; et al. Lymph protects metastasizing melanoma cells from ferroptosis. *Nature* **2020**, *585*, 113–118. [CrossRef] [PubMed]
30. Pascual, G.; Avgustinova, A.; Mejetta, S.; Martin, M.; Castellanos, A.; Attolini, C.S.; Berenguer, A.; Prats, N.; Toll, A.; Hueto, J.A.; et al. Targeting metastasis-initiating cells through the fatty acid receptor CD36. *Nature* **2017**, *541*, 41–45. [CrossRef]
31. Wang, T.; Fahrmann, J.F.; Lee, H.; Li, Y.J.; Tripathi, S.C.; Yue, C.; Zhang, C.; Lifshitz, V.; Song, J.; Yuan, Y.; et al. JAK/STAT3-Regulated Fatty Acid beta-Oxidation Is Critical for Breast Cancer Stem Cell Self-Renewal and Chemoresistance. *Cell Metab.* **2018**, *27*, 136–150.e5. [CrossRef]
32. Sun, Q.; Yu, X.; Peng, C.; Liu, N.; Chen, W.; Xu, H.; Wei, H.; Fang, K.; Dong, Z.; Fu, C.; et al. Activation of SREBP-1c alters lipogenesis and promotes tumor growth and metastasis in gastric cancer. *Biomed. Pharmacother.* **2020**, *128*, 110274. [CrossRef]
33. Austin, B.K.; Firooz, A.; Valafar, H.; Blenda, A.V. An Updated Overview of Existing Cancer Databases and Identified Needs. *Biology* **2023**, *12*, 1152. [CrossRef] [PubMed]
34. Di Carlo, C.; Sousa, B.C.; Manfredi, M.; Brandi, J.; Dalla Pozza, E.; Marengo, E.; Palmieri, M.; Dando, I.; Wakelam, M.J.O.; Lopez-Clavijo, A.F.; et al. Integrated lipidomics and proteomics reveal cardiolipin alterations, upregulation of HADHA and long chain fatty acids in pancreatic cancer stem cells. *Sci. Rep.* **2021**, *11*, 13297. [CrossRef]
35. Brandi, J.; Dando, I.; Pozza, E.D.; Biondani, G.; Jenkins, R.; Elliott, V.; Park, K.; Fanelli, G.; Zolla, L.; Costello, E.; et al. Proteomic analysis of pancreatic cancer stem cells: Functional role of fatty acid synthesis and mevalonate pathways. *J. Proteom.* **2017**, *150*, 310–322. [CrossRef] [PubMed]
36. Liu, J.; Tan, Y.; Zhang, H.; Zhang, Y.; Xu, P.; Chen, J.; Poh, Y.C.; Tang, K.; Wang, N.; Huang, B. Soft fibrin gels promote selection and growth of tumorigenic cells. *Nat. Mater.* **2012**, *11*, 734–741. [CrossRef] [PubMed]
37. Qi, F.; Qin, W.; Zhang, Y.; Luo, Y.; Niu, B.; An, Q.; Yang, B.; Shi, K.; Yu, Z.; Chen, J.; et al. Sulfarotene, a synthetic retinoid, overcomes stemness and sorafenib resistance of hepatocellular carcinoma via suppressing SOS2-RAS pathway. *J. Exp. Clin. Cancer Res.* **2021**, *40*, 280. [CrossRef]
38. Miranda, A.; Hamilton, P.T.; Zhang, A.W.; Pattnaik, S.; Becht, E.; Mezheyeuski, A.; Bruun, J.; Micke, P.; de Reynies, A.; Nelson, B.H. Cancer stemness, intratumoral heterogeneity, and immune response across cancers. *Proc. Natl. Acad. Sci. USA* **2019**, *116*, 9020–9029. [CrossRef] [PubMed]
39. Malta, T.M.; Sokolov, A.; Gentles, A.J.; Burzykowski, T.; Poisson, L.; Weinstein, J.N.; Kaminska, B.; Huelsken, J.; Omberg, L.; Gevaert, O.; et al. Machine Learning Identifies Stemness Features Associated with Oncogenic Dedifferentiation. *Cell* **2018**, *173*, 338–354.e15. [CrossRef] [PubMed]
40. Margonis, G.A.; Pulvirenti, A.; Morales-Oyarvide, V.; Buettner, S.; Andreatos, N.; Kamphues, C.; Beyer, K.; Wang, J.; Kreis, M.E.; Cameron, J.L.; et al. Performance of the 7th and 8th Editions of the American Joint Committee on Cancer Staging System in Patients with Intraductal Papillary Mucinous Neoplasm-Associated PDAC: A Multi-institutional Analysis. *Ann. Surg.* **2023**, *277*, 681–688. [CrossRef] [PubMed]
41. Wu, T.; Hu, E.; Xu, S.; Chen, M.; Guo, P.; Dai, Z.; Feng, T.; Zhou, L.; Tang, W.; Zhan, L.; et al. clusterProfiler 4.0: A universal enrichment tool for interpreting omics data. *Innovation* **2021**, *2*, 100141. [CrossRef]
42. Liberzon, A.; Subramanian, A.; Pinchback, R.; Thorvaldsdottir, H.; Tamayo, P.; Mesirov, J.P. Molecular signatures database (MSigDB) 3.0. *Bioinformatics* **2011**, *27*, 1739–1740. [CrossRef]
43. Hanzelmann, S.; Castelo, R.; Guinney, J. GSVA: Gene set variation analysis for microarray and RNA-seq data. *BMC Bioinform.* **2013**, *14*, 7. [CrossRef]
44. Liang, C.; Shi, S.; Qin, Y.; Meng, Q.; Hua, J.; Hu, Q.; Ji, S.; Zhang, B.; Xu, J.; Yu, X.J. Localisation of PGK1 determines metabolic phenotype to balance metastasis and proliferation in patients with SMAD4-negative pancreatic cancer. *Gut* **2020**, *69*, 888–900. [CrossRef]
45. Du, X.; Qi, Z.; Xu, J.; Guo, M.; Zhang, X.; Yu, Z.; Cao, X.; Xia, J. Loss of GABARAPL1 confers ferroptosis resistance to cancer stem-like cells in hepatocellular carcinoma. *Mol. Oncol.* **2022**, *16*, 3703–3719. [CrossRef]
46. Zhou, L.; Wang, Z.; Hu, C.; Zhang, C.; Kovatcheva-Datchary, P.; Yu, D.; Liu, S.; Ren, F.; Wang, X.; Li, Y.; et al. Integrated Metabolomics and Lipidomics Analyses Reveal Metabolic Reprogramming in Human Glioma with IDH1 Mutation. *J. Proteome Res.* **2019**, *18*, 960–969. [CrossRef] [PubMed]

47. Nguyen, A.; Rudge, S.A.; Zhang, Q.; Wakelam, M.J. Using lipidomics analysis to determine signalling and metabolic changes in cells. *Curr. Opin. Biotechnol.* **2017**, *43*, 96–103. [CrossRef] [PubMed]
48. Gaud, C.; Sousa, B.C.; Nguyen, A.; Fedorova, M.; Ni, Z.; O'Donnell, V.B.; Wakelam, M.J.O.; Andrews, S.; Lopez-Clavijo, A.F. BioPAN: A web-based tool to explore mammalian lipidome metabolic pathways on LIPID MAPS. *F1000Research* **2021**, *10*, 4. [CrossRef] [PubMed]
49. Rodriguez-Aznar, E.; Wiesmuller, L.; Sainz, B., Jr.; Hermann, P.C. EMT and Stemness-Key Players in Pancreatic Cancer Stem Cells. *Cancers* **2019**, *11*, 1136. [CrossRef]
50. Kreso, A.; Dick, J.E. Evolution of the cancer stem cell model. *Cell Stem Cell* **2014**, *14*, 275–291. [CrossRef]
51. Ng, S.W.; Mitchell, A.; Kennedy, J.A.; Chen, W.C.; McLeod, J.; Ibrahimova, N.; Arruda, A.; Popescu, A.; Gupta, V.; Schimmer, A.D.; et al. A 17-gene stemness score for rapid determination of risk in acute leukaemia. *Nature* **2016**, *540*, 433–437. [CrossRef]
52. Smith, B.A.; Balanis, N.G.; Nanjundiah, A.; Sheu, K.M.; Tsai, B.L.; Zhang, Q.; Park, J.W.; Thompson, M.; Huang, J.; Witte, O.N.; et al. A Human Adult Stem Cell Signature Marks Aggressive Variants across Epithelial Cancers. *Cell Rep.* **2018**, *24*, 3353–3366.e5. [CrossRef]
53. Gong, T.; Wu, D.; Pan, H.; Sun, Z.; Yao, X.; Wang, D.; Huang, Y.; Li, X.; Guo, Y.; Lu, Y. Biomimetic Microenvironmental Stiffness Boosts Stemness of Pancreatic Ductal Adenocarcinoma via Augmented Autophagy. *ACS Biomater. Sci. Eng.* **2023**, *9*, 5347–5360. [CrossRef] [PubMed]
54. Janneh, A.H.; Atkinson, C.; Tomlinson, S.; Ogretmen, B. Sphingolipid metabolism and complement signaling in cancer progression. *Trends Cancer* **2023**, *9*, 782–787. [CrossRef]
55. Hannun, Y.A.; Obeid, L.M. Principles of bioactive lipid signalling: Lessons from sphingolipids. *Nat. Rev. Mol. Cell Biol.* **2008**, *9*, 139–150. [CrossRef]
56. Morad, S.A.; Cabot, M.C. Ceramide-orchestrated signalling in cancer cells. *Nat. Rev. Cancer* **2013**, *13*, 51–65. [CrossRef]
57. Obeid, L.M.; Linardic, C.M.; Karolak, L.A.; Hannun, Y.A. Programmed cell death induced by ceramide. *Science* **1993**, *259*, 1769–1771. [CrossRef]
58. Carpinteiro, A.; Becker, K.A.; Japtok, L.; Hessler, G.; Keitsch, S.; Pozgajova, M.; Schmid, K.W.; Adams, C.; Muller, S.; Kleuser, B.; et al. Regulation of hematogenous tumor metastasis by acid sphingomyelinase. *EMBO Mol. Med.* **2015**, *7*, 714–734. [CrossRef]
59. Schiffmann, S.; Sandner, J.; Birod, K.; Wobst, I.; Angioni, C.; Ruckhaberle, E.; Kaufmann, M.; Ackermann, H.; Lotsch, J.; Schmidt, H.; et al. Ceramide synthases and ceramide levels are increased in breast cancer tissue. *Carcinogenesis* **2009**, *30*, 745–752. [CrossRef] [PubMed]
60. Kuc, N.; Doermann, A.; Shirey, C.; Lee, D.D.; Lowe, C.W.; Awasthi, N.; Schwarz, R.E.; Stahelin, R.V.; Schwarz, M.A. Pancreatic ductal adenocarcinoma cell secreted extracellular vesicles containing ceramide-1-phosphate promote pancreatic cancer stem cell motility. *Biochem. Pharmacol.* **2018**, *156*, 458–466. [CrossRef] [PubMed]
61. Horejsi, K.; Jin, C.; Vankova, Z.; Jirasko, R.; Strouhal, O.; Melichar, B.; Teneberg, S.; Holcapek, M. Comprehensive characterization of complex glycosphingolipids in human pancreatic cancer tissues. *J. Biol. Chem.* **2023**, *299*, 102923. [CrossRef]
62. Limbu, K.R.; Chhetri, R.B.; Oh, Y.S.; Baek, D.J.; Park, E.Y. Mebendazole Impedes the Proliferation and Migration of Pancreatic Cancer Cells through SK1 Inhibition Dependent Pathway. *Molecules* **2022**, *27*, 8127. [CrossRef]
63. Guo, W.; Zhang, C.; Feng, P.; Li, M.; Wang, X.; Xia, Y.; Chen, D.; Li, J. M6A methylation of DEGS2, a key ceramide-synthesizing enzyme, is involved in colorectal cancer progression through ceramide synthesis. *Oncogene* **2021**, *40*, 5913–5924. [CrossRef] [PubMed]
64. Nagaro, H.; Ichikawa, H.; Takizawa, K.; Nagahashi, M.; Abe, S.; Hirose, Y.; Moro, K.; Miura, K.; Nakano, M.; Shimada, Y.; et al. Clinical Significance of Phosphorylated Sphingosine Kinase 1 Expression in Pancreatic Ductal Adenocarcinoma. *Anticancer Res.* **2023**, *43*, 3969–3977. [CrossRef] [PubMed]
65. Bonder, C.S.; Sun, W.Y.; Matthews, T.; Cassano, C.; Li, X.; Ramshaw, H.S.; Pitson, S.M.; Lopez, A.F.; Coates, P.T.; Proia, R.L.; et al. Sphingosine kinase regulates the rate of endothelial progenitor cell differentiation. *Blood* **2009**, *113*, 2108–2117. [CrossRef]
66. Meng, H.; Yuan, Y.; Lee, V.M. Loss of sphingosine kinase 1/S1P signaling impairs cell growth and survival of neurons and progenitor cells in the developing sensory ganglia. *PLoS ONE* **2011**, *6*, e27150. [CrossRef] [PubMed]
67. Mahajan-Thakur, S.; Bien-Moller, S.; Marx, S.; Schroeder, H.; Rauch, B.H. Sphingosine 1-phosphate (S1P) signaling in glioblastoma multiforme-A systematic review. *Int. J. Mol. Sci.* **2017**, *18*, 2448. [CrossRef] [PubMed]
68. Mukherjee, N.; Lu, Y.; Almeida, A.; Lambert, K.; Shiau, C.W.; Su, J.C.; Luo, Y.; Fujita, M.; Robinson, W.A.; Robinson, S.E.; et al. Use of a MCL-1 inhibitor alone to de-bulk melanoma and in combination to kill melanoma initiating cells. *Oncotarget* **2017**, *8*, 46801–46817. [CrossRef]
69. Luo, J.; Wang, P.; Wang, R.; Wang, J.; Liu, M.; Xiong, S.; Li, Y.; Cheng, B. The Notch pathway promotes the cancer stem cell characteristics of CD90+ cells in hepatocellular carcinoma. *Oncotarget* **2016**, *7*, 9525–9537. [CrossRef]
70. Hii, L.W.; Chung, F.F.; Mai, C.W.; Ng, P.Y.; Leong, C.O. Sphingosine Kinase 1 Signaling in Breast Cancer: A Potential Target to Tackle Breast Cancer Stem Cells. *Front. Mol. Biosci.* **2021**, *8*, 748470. [CrossRef]
71. Ketavarapu, V.; Ravikanth, V.; Sasikala, M.; Rao, G.V.; Devi, C.V.; Sripadi, P.; Bethu, M.S.; Amanchy, R.; Murthy, H.V.V.; Pandol, S.J.; et al. Integration of metabolites from meta-analysis with transcriptome reveals enhanced SPHK1 in PDAC with a background of pancreatitis. *BMC Cancer* **2022**, *22*, 792. [CrossRef]

72. Djokovic, N.; Djuric, A.; Ruzic, D.; Srdic-Rajic, T.; Nikolic, K. Correlating Basal Gene Expression across Chemical Sensitivity Data to Screen for Novel Synergistic Interactors of HDAC Inhibitors in Pancreatic Carcinoma. *Pharmaceuticals* **2023**, *16*, 294. [CrossRef] [PubMed]
73. Speirs, M.M.P.; Swensen, A.C.; Chan, T.Y.; Jones, P.M.; Holman, J.C.; Harris, M.B.; Maschek, J.A.; Cox, J.E.; Carson, R.H.; Hill, J.T.; et al. Imbalanced sphingolipid signaling is maintained as a core proponent of a cancerous phenotype in spite of metabolic pressure and epigenetic drift. *Oncotarget* **2019**, *10*, 449–479. [CrossRef] [PubMed]

Disclaimer/Publisher's Note: The statements, opinions and data contained in all publications are solely those of the individual author(s) and contributor(s) and not of MDPI and/or the editor(s). MDPI and/or the editor(s) disclaim responsibility for any injury to people or property resulting from any ideas, methods, instructions or products referred to in the content.

Article

Study on the Common Molecular Mechanism of Metabolic Acidosis and Myocardial Damage Complicated by Neonatal Pneumonia

Yifei Zhan [1,†], Huaiyan Wang [2,†], Zeying Wu [3,4,*] and Zhongda Zeng [1,*]

1. College of Environmental and Chemical Engineering, Dalian University, Dalian 116622, China; zhanyifei@chemdatasolution.com
2. Department of Neonatology, Changzhou Medical Center, Changzhou Maternity and Child Health Care Hospital, Nanjing Medical University, Changzhou 213000, China; huaiyanwang@njmu.edu.cn
3. State Key Laboratory of Analytical Chemistry for Life Science, School of Chemistry & Chemical Engineering, Nanjing University, Nanjing 210023, China
4. School of Chemical Engineering and Material Sciences, Changzhou Institute of Technology, Changzhou 213032, China
* Correspondence: wuzy@czu.cn (Z.W.); zeng@chemdatasolution.com (Z.Z.)
† These authors contributed equally to this work.

Abstract: Pneumonia is a common clinical disease in the neonatal period and poses a serious risk to infant health. Therefore, the understanding of molecular mechanisms is of great importance for the development of methods for the rapid and accurate identification, classification and staging, and even disease diagnosis and therapy of pneumonia. In this study, a nontargeted metabonomic method was developed and applied for the analysis of serum samples collected from 20 cases in the pneumonia control group (PN) and 20 and 10 cases of pneumonia patients with metabolic acidosis (MA) and myocardial damage (MD), respectively, with the help of ultrahigh-performance liquid chromatography–high-resolution mass spectrometry (UPLC–HRMS). The results showed that compared with the pneumonia group, 23 and 21 differential metabolites were identified in pneumonia with two complications. They showed high sensitivity and specificity, with the area under the curve (ROC) of the receiver operating characteristic curve (ROC) larger than 0.7 for each differential molecule. There were 14 metabolites and three metabolic pathways of sphingolipid metabolism, porphyrin and chlorophyll metabolism, and glycerophospholipid metabolism existing in both groups of PN and MA, and PN and MD, all involving significant changes in pathways closely related to amino acid metabolism disorders, abnormal cell apoptosis, and inflammatory responses. These findings of molecular mechanisms should help a lot to fully understand and even treat the complications of pneumonia in infants.

Keywords: UPLC–HRMS-based metabolomics; chemometrics; pneumonia; metabolic acidosis; myocardial damage

1. Introduction

Neonatal pneumonia is a common disease in very young infants. It is the leading cause of death in high under-5 mortality rate (U5MR) countries [1] and the major single killer of children outside the neonatal period [2]. Atypical clinical manifestations and the rapid onset and progress of neonatal pneumonia often result in associated complications such as respiratory failure, which presents a serious threat to children's health. Thus, early, accurate diagnosis and timely and effective treatment are particularly important.

However, due to the various causes of pneumonia infections, the symptoms of pneumonia are diverse and difficult to diagnose [3–5]. X-ray, CT examinations, and lung ultrasonography have been implemented for the diagnosis of pneumonia, but these methods have several disadvantages, such as the limited accuracy of imaging diagnosis, inconveniences, and high costs, resulting in a great challenge in current clinical practice [6,7]. Furthermore, there is also a lack of clinically reliable means for the prediction, classification, or effective screening of patients who require a higher level of care, making the appropriate triage of newborns with pneumonia rather problematic. Therefore, a thorough investigation of the molecular mechanisms of neonatal pneumonia is highly in demand, which will provide an in-depth understanding of the causes, identification and prevention approaches, and clinical management for patients.

Metabolomics, which bridges the gap between scientific interests and biological findings, is a strategy to understand and diagnose diseases, and reveal the mechanisms of disease occurrence and development. By obtaining the results of small molecule changes in metabolic phenotypes, it is possible to discover and explain the internal mechanisms that lead to diseases, and this is very important for early detection and treatment, leading to a decline in prevalence [8,9]. For example, Li et al. performed ultrahigh-performance liquid chromatography–tandem mass spectrometry analysis of metabolites in urine samples of healthy children and children with mycoplasma pneumoniae pneumonia in children (MPPC) [10]. In their study, acetyl phosphate and 2, 5-dioxovalerate were recognized for the first time as potential biomarkers for early diagnosis of MPPC. A similar approach was used by Del Borrello et al. to successfully reveal metabolic changes in community-acquired pneumonia (CAP) [11]. Three metabolites, sphingosine, lactate, and DHEA-S, were discovered to represent a panel of potential small molecule biomarkers for assessment of the severity of CAP [12]. The suitability of a new set of serum biomarkers, consisting of two proteins and three metabolites, for the identification of CAP and the recognition of severe pneumonia was proved by Wang et al. [13]. In these reports, the mechanisms were interpreted, and the development and evolution of diseases were explained based on the information obtained from the metabolomic analysis results.

The above studies have demonstrated the applicability of metabolomics for the pathogenic diagnosis of pneumonia and the assessment of pneumonia severity. However, the integrated and common molecular mechanisms of pneumonia with complications failed to be thoroughly investigated previously. Metabolomics-oriented research into the distinction between neonatal pneumonia and pneumonia complications is relatively rare, and clinical diagnosis, classification, and staging for pneumonia complications are also very limited.

In this work, a nontargeted metabolomics method using ultrahigh-performance liquid chromatography–high-resolution mass spectrometry (UPLC–HRMS) was established at first. The metabolites in serum samples collected from the control group suffering from pneumonia only (PN group), from patients suffering from pneumonia, metabolic acidosis, and other medical conditions except for myocardial damage (PN&MA group), and from patients suffering from pneumonia and myocardial damage (PN&MD group) were then fully investigated and compared with the PN group statistically. A comprehensive analysis of differences in the identified metabolites between PN, MA, and MD was performed for the discovery of key compounds involved in pneumonia complications. The common metabolite molecules were recognized and elucidated. The changes in these compounds and their metabolic pathways in neonatal pneumonia complications were further explored.

2. Materials and Methods

2.1. Sample Collection

This study was approved by the Ethics Committee of Changzhou Maternal and Child Health Hospital (CMCHH), Changzhou, China (approval number: 2019017). The blood samples of newborns were collected from December 2019 to September 2020 at CMCHH.

In this study, serum samples of 20 patients in PN group, 20 patients in PN&MA group, and 10 patients in PN&MD group were collected for nontargeted metabolomic analysis. The details of all the samples are shown in Figure 1. Most samples of patients with complications of MA or MD had other medical conditions. The details are given in the left column of the figure. The full names of all diseases with abbreviations are given in the Supplementary Material Table S1.

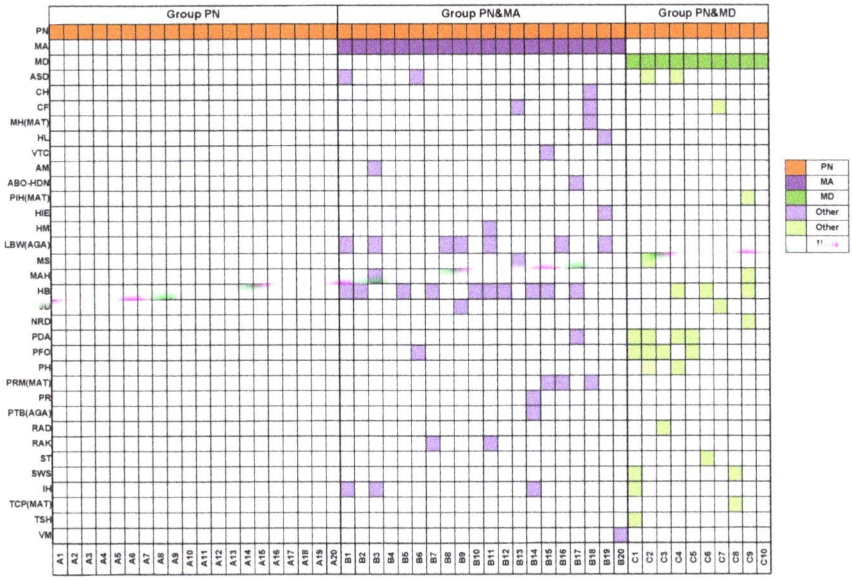

Figure 1. Clinical information of samples in groups PN, PN&MA, and PN&MD, respectively.

After collection, each blood sample was placed in a vacuum blood collection tube containing a procoagulant, left to stand for 1 h at room temperature, and centrifuged for 10 min at a speed of 3000 r/min. The supernatant was then placed in a centrifuge tube and stored in a refrigerator at −80 °C.

2.2. Materials

Methanol of chromatographic grade was purchased from Beijing Zhenxiang Company. All internal standards (IS) were obtained from the same company and prepared into a mixed standard solution with methanol. The final concentration of each standard in the mixed standard solution is given in Table 1.

Table 1. Each IS and its corresponding concentration in the prepared mixed standard solution.

No.	Compound Name	Concentration (μg/mL)
1	choline-d4	2.0
2	cannitine C2:0-d3	0.16
3	phenylalanine-d5	3.5
4	FFA16:0-d3	2.5
5	FFA18:0-d3	2.5
6	cannitine C10:0-d3	0.1
7	cannitine C8:0-d4	0.1
8	CA-d4	1.85
9	CDCA-d4	1.5
10	cannitine C16:0-d3	0.15
11	LPC 19:0	0.75
12	SM d30:1	0.75
13	glutamic acid-d3	0.15

2.3. Sample Preparation

The serum samples were prepared on ice. First, 50 μL sample was pipetted into a 1.5 mL centrifuge tube (Oxygen, Doral, FL, USA), 200 μL cold methanol was added, and it was shaken for 4 min at a speed of 1500 r/min on a Vortex Mixer T1 (Titan, West Springfield, MA, USA). The mixed solution was then left to stand for 10 min at a temperature of -20 °C prior to centrifugation for 15 min at 4 °C and 14,000 r/min. After this, 200 μL of supernatant was taken into a new centrifuge tube, while the remaining supernatant was used to prepare the quality control (QC) sample solution. The samples were then concentrated and stored at -40 °C. Prior to analysis, the lyophilized samples were redissolved in 100 μL methanol/water (80/20, v/v) solution, followed by shaking and centrifugation operations (1500 r/min). The supernatant was subject to analysis by UPLC–HRMS (Q-Exactive, Thermo Fisher, Waltham, MA, USA) in both positive and negative modes.

2.4. UPLC–HRMS Analysis

The UPLC–HRMS analysis was performed using a BEH C8 column (2.1 × 100 mm × 1.7 μm, Waters, Milford, MA, USA) in positive mode and an HSS T3 column (2.1 × 100 mm × 1.8 μm, Waters, USA) in negative mode. The flow rate was 0.35 mL/min on both columns. The mobile phases A and B were 0.1% formic acid in water and 0.1% methanol in acetonitrile, respectively. The gradient program started with 5% B and held for 1 min. Then, it was linearly changed to 100% B in 10 min and held for 2 min. The column temperature was set at 50 °C.

The ion source was operated with a spray voltage of 3.8 kV in positive mode and -3.0 kV in negative mode. The ion transfer capillary temperature was set at 320 °C. All the samples were analyzed in a nontargeted full scan acquisition mode from 70 to 1050 m/z at a resolution of 70,000. The MS^2 measurement was performed in independent data acquisition (IDA)-based auto-MS^2 mode, and the MS/MS fragments were acquired at a resolution of 17,500.

2.5. Data Processing and Statistical Analysis

The nontargeted raw data obtained from UPLC–HRMS analysis was first converted into mzML format using One-MAP/PTO software v2.8, which is freely available from www.5omics.com, accessed on 3 January 2023. It was further used to recognize and extract the MS characteristics of metabolites buried in the same sample and then to match the primary MS of the same substance in each sample to an Excel file for the generation of peak tables, which were the basis for the subsequent statistical analysis. The MS2 data were obtained from mgf files, in which the retention times and MS1 and MS2 information of MS features in peak tables were all included. Afterward, the peak lists in the table were preprocessed, such as filling in the missing values and sample normalization. The data quality was evaluated and calibrated by using One-MAP from www.5omics.com to improve the data quality. The peak table, together with the MS^2 data matched, was then imported into .mgf format for annotation of each MS feature and discovery of differential metabolites. The annotation was attained according to the qualitative characteristics of each compound built in the database of One-MAP.

The recognition of differential metabolites between different clinical groups was achieved with the help of a combination of univariate and multivariate analysis. For univariate statistical analysis, a volcano plot with information combination of fold change (FC) > 1.5 and $p < 0.05$ was used to effectively identify those differential components with statistically significant changes amongst different groups. For multivariate analysis, partial least squares discriminant analysis (PLS-DA) was applied to model the main potential variables and then screen the characteristic ions/metabolites that differed in the group with the values of variable importance in the projection (VIP) larger than 1.0. The relative levels of metabolite differences in each group were expressed as upward and downward changes in the multiplicity of change to visually represent the variation in differential metabolites between the experimental and control groups. Metabolic enrichment analysis was also performed on the basis of the aforementioned results of annotation and differential discovery of metabolites. The pathways to which these metabolites corresponded were determined, and the clinical interpretation and potentials were explained afterward. Both the groups of PN and PN&MA and PN and PN&MD were fully investigated following the strategies and methods introduced above.

3. Results

3.1. Metabolic Profiling Analysis

The typical total ion chromatograms (TICs) of serum samples in groups PN, PN&MA, and PN&MD obtained in positive mode are shown in Figure 2a–c, respectively. The corresponding TICs measured in negative mode are shown in Figure 2d–f, respectively. The horizontal axis in Figure 2 represents the retention time in minutes, while the vertical axis represents the peak intensity. It is obvious that all the samples were well separated under the experimental conditions described in Section 2.4 in both positive and negative modes.

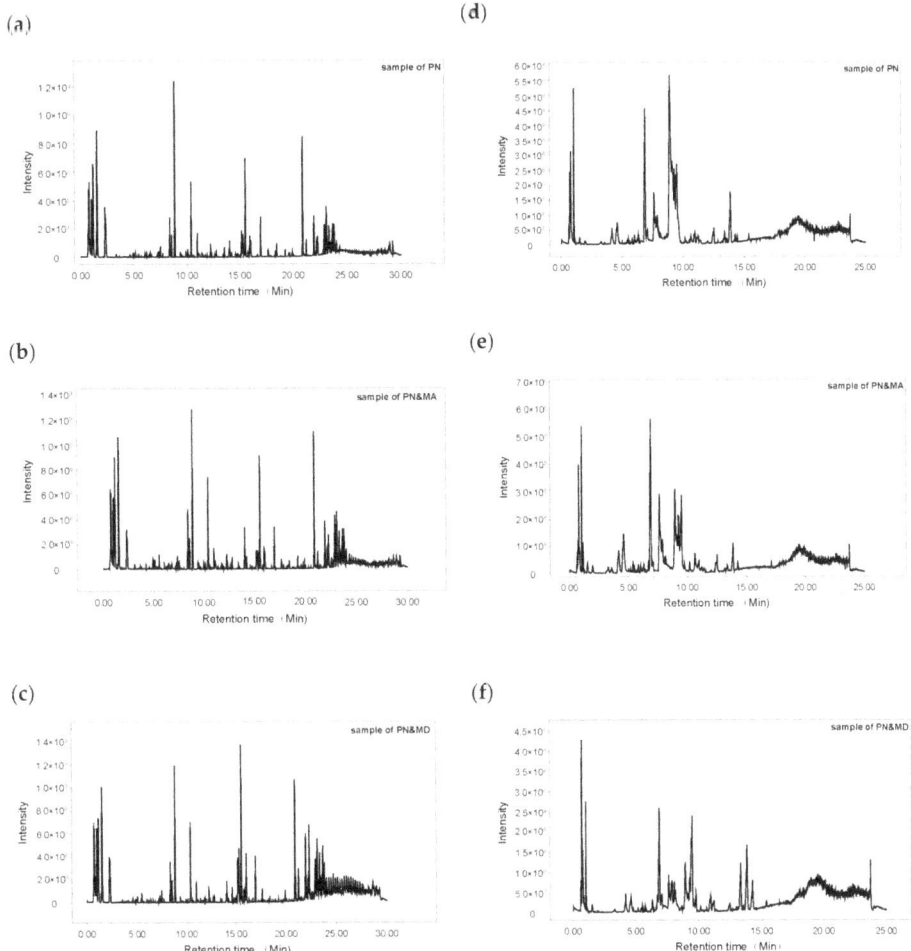

Figure 2. TICs of representative samples in groups PN (**a,d**), PN&MA (**b,e**), and PN&MD (**c,f**), respectively: (**a–c**) TICs measured in positive mode; (**d–f**) TICs measured in negative mode.

It is well known that both QC samples and IS compounds can provide important information for data quality evaluation and analysis in metabolomics studies and can be further used for data calibration to improve data quality if required. Figure 3a,b illustrate the distribution of relative standard deviations (RSDs) of mass spectral features in QC samples before data calibration in positive and negative modes, respectively. It can be observed that the data quality of the QC samples was relatively satisfactory. The percentage of MS features in QC samples with RSDs lower than 30% was greater than 75%. On the other hand, the PCA results of QC samples showed good consistency in both positive and negative modes, as given in Figure 3c,d. This indicated a high quality of UPLC–HRMS measurement.

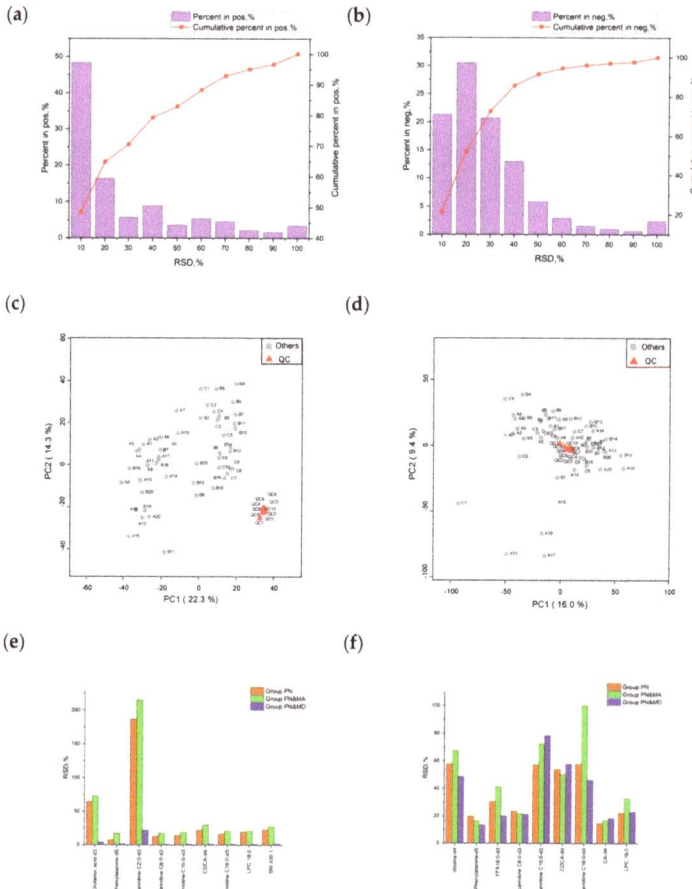

Figure 3. Quality evaluation based on QC samples and ISs. (**a**,**b**), (**c**,**d**), and (**e**,**f**) correspond to the results of RSD distribution of QC samples before data calibration, PCA evaluation results of QC samples, and RSD distribution of ISs in positive and negative mode, respectively.

The IS compounds can be used to maintain overall data quality during the experimental process, although some of the ISs may not be stable during the whole procedure, due to which the data quality was reduced. The ISs with RSD values below 30% in positive mode accounted for 84.6%, as given in Figure 3e. Among all the ISs, the RSD of cannitine C2:0-d3 and glutamic acid-d3 were relatively high. In negative mode, as shown in Figure 3f, the percentage of ISs with RSDs lower than 30% was 44.4%, which was not as good as in positive mode. This was probably due to the suitability of these components to be ionized by positive ion sources. Combining the results of the data quality analysis as described above, the accuracy and reproducibility of the experimental analysis proved that the proposed approach was feasible for the present study.

3.2. Discovery of Differential Metabolites and Further Analysis of Metabolic Pathway

The processes for nontargeted metabolomics data analysis were previously introduced in Section 2.5. Due to the large variance in the interfering diseases in the samples, some of them exhibited outliers in the analysis, but a trend of separation was still observed. The annotation results are given in the Supplementary Material Table S2. In Figure 4a,b, univariate volcano plots of univariate analysis were obtained with statistical p-values lower than 0.05 and FCs larger than 1.5 in both positive and negative modes, respectively. In positive mode, the results of PLS-DA modeling revealed a significant difference between the PN&MA group and PN controls, as shown in Figure 4c. It is obvious that samples collected from healthy children and patients were discriminated with 100% accuracy. Similarly, a consistent pattern was also observed between these two groups in negative mode, as given in Figure 4d. With the help of univariate and multivariate analysis, a total of 23 metabolites were discovered to be significantly different in the PN and PN&MA groups. The permutation test plots with number of running time equal to 200 ensured the credibility of the proposed model, which are shown in Figure 4e,f. These results of the distribution of R^2 and Q^2 values indicated that the model was reliable and could be used for predictive analysis.

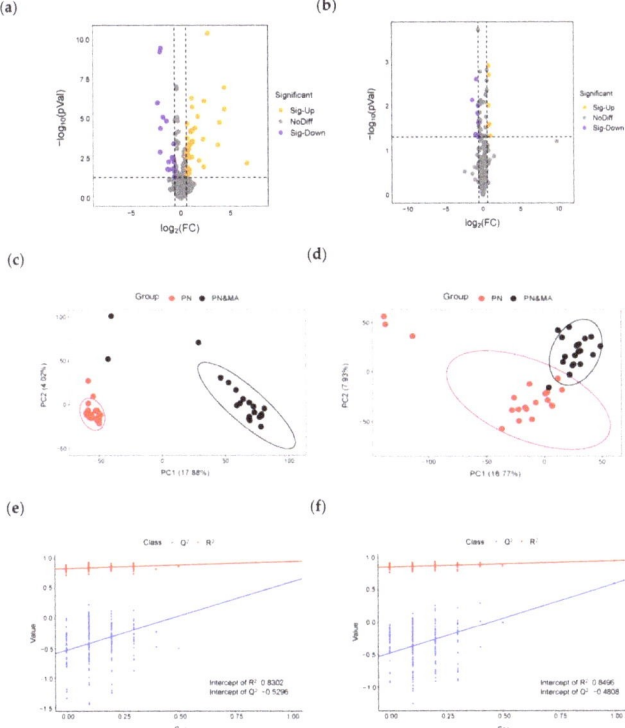

Figure 4. The results of univariate and multivariate analysis of PN and PN&MA groups. (a–f) were volcano plots, PLS-DA score plots, and permutation analysis obtained in positive and negative modes, respectively.

In addition, the specificity and sensitivity of the ROC model evaluation in positive and negative modes reached 100% and 100%, and 95% and 80%, respectively. Figure 5 shows the ROC curves of each differential molecule, with AUC values above 0.7, for discrimination of PN and PN&MA groups, which indicated a high sensitivity and a good specificity of the 23 differential molecules for discrimination analysis.

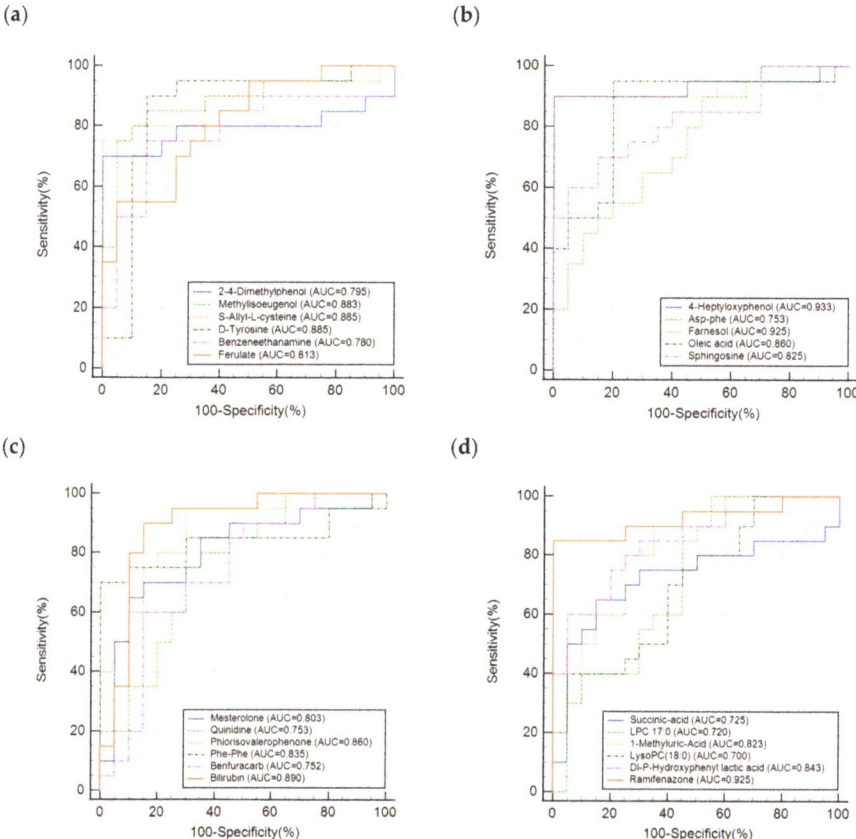

Figure 5. The results of ROC analysis of all the differential molecules obtained from PN and PN&MA groups. (a–d) correspond to the results obtained from different components, as shown in each figure.

Figure 6a,b are volcano plots in positive and negative modes obtained from univariate statistical analysis with p-values lower than 0.05 and FCs larger than 1.5. In terms of the processes for data analysis introduced above, the PLS-DA models of PN and PN&MD groups indicated a significant difference in positive mode, as shown in Figure 6c. An apparent separation between them in negative mode is presented in Figure 6d, and samples in the two groups are marked in red and black, respectively. With the results of PN and PN&MA given above, it can be concluded that the two types of samples can be discriminated with 100% accuracy. A total of 21 differential metabolites with statistically significant changes were discovered after analysis. The results of the permutation test with number of running time equal to 200 proved that the model was robust and reliable, as shown in Figure 6e,f. The specificity and sensitivity of the ROC model evaluation in both positive and negative modes reached 100%. Figure 7 shows the ROC results of all the differential molecules discovered in the PN and PN&MD groups, which are presented in Figure 7a–d with a total of 21 metabolite molecules, as introduced above. All the AUC values were above 0.75, which also indicated the high sensitivity and specificity of specific metabolite molecules for disease recognition.

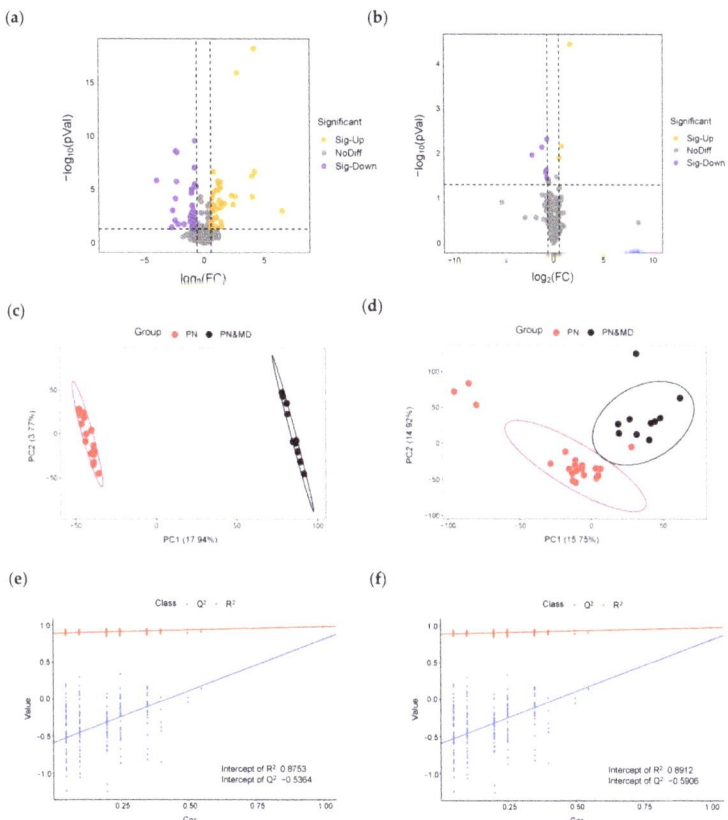

Figure 6. The results of univariate and multivariate analysis of PN and PN&MD groups. (a–f) are volcano plots, PLS-DA score plots, and permutation analysis obtained in positive and negative modes, respectively.

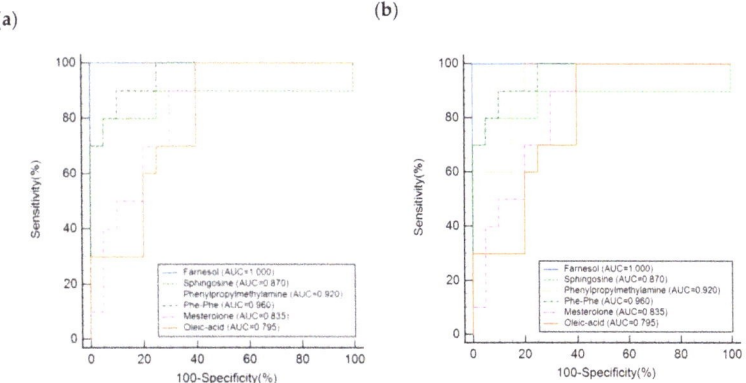

Figure 7. Cont.

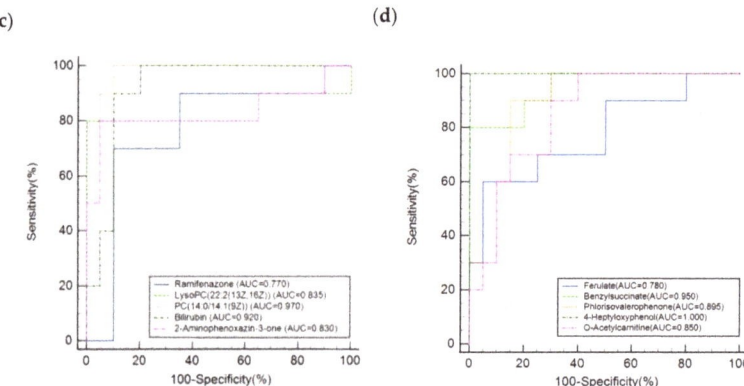

Figure 7. The results of ROC analysis of all the differential molecules obtained from PN and PN&MD groups. (**a**–**d**) correspond to the results obtained from different components, as shown in each figure.

3.3. Study of Common Differential Molecular Characteristics between PN&MA and PN&MD

A deeper commonality analysis of the differential metabolites found in the PN&MA and PN&MD groups was performed. A total of 14 overlapping differential metabolites in these two groups with the same up- and downregulation trends were recognized. The changes in concentrations of these substances may be strongly correlated with the severity of pneumonia, as shown in Figure 8a,b. Figure 8c,d are the differential metabolite network diagrams of the PN&MA and PN&MD groups. Among the common differential metabolite, farnesol, oleic acid, phlorisovalerophenone, methylisoeugenol, ferulate, 4-Heptyloxyphenol, bilirubin, sphingosine, ramifenazone, and Phe-Phe showed a significant correlation. Figure 8e,f are the enrichment pathway diagrams of differential metabolite molecules of the PN&MA and PN&MD groups, respectively. It can be seen that three common pathways were involved, including porphyrin and chlorophyll metabolism, sphingolipid metabolism, and glycerophospholipid metabolism.

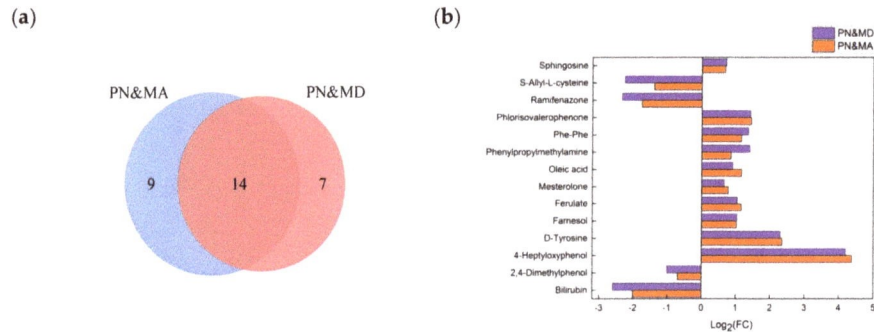

Figure 8. *Cont.*

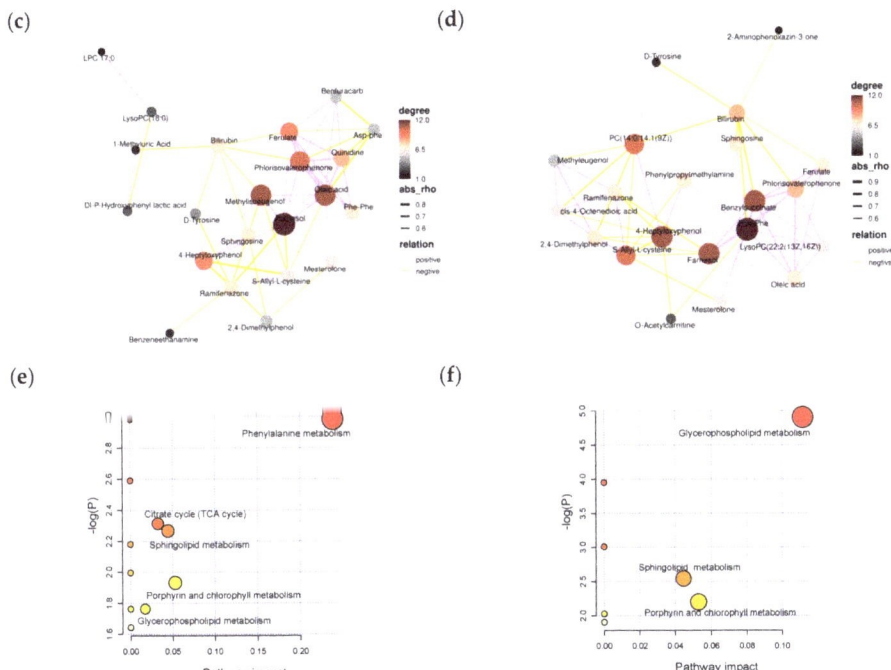

Figure 8. (a) A Venn diagram of differential metabolites found in PN and PN&MA groups and PN and PN&MD groups; (b) the up–down bar charts of the molecules found in (a); (c,d) the correlation network of differential molecules found in PN and PN&MA groups and PN and PN&MD groups; (e,f) the findings after metabolic pathway enrichment by using the differential molecules introduced in (c,d).

4. Discussion

This study identified 14 common differential metabolite molecules and three common pathways in the pneumonia group with metabolic acidosis and myocardial damage compared with the pneumonia group. These findings contribute to a better understanding of the molecular mechanisms underlying pneumonia with metabolic acidosis and myocardial damage. Furthermore, they provide guidance for the identification and validation of biomarkers for pneumonia complications and offer insights into the subsequent clinical application of hierarchical diagnosis and personalized treatment. However, it is important to note that the study has limitations such as the limited number of serum samples and a lack of specific differentiation between various types of pneumonia pathogens. Therefore, further study of large-scale population verification is necessary.

Bilirubin participates in porphyrin and chlorophyll metabolisms. The main source of bilirubin is aging red blood cells. The aging red blood cells are first damaged and transformed into biliverdin with heme from other sources by heme oxygenase. Afterward, biliverdin reacts with biliverdin reductase to give bilirubin, which is involved in porphyrin metabolism in the body [14]. The decrease in bilirubin content in the PN&MD group may be due to the poor function of the mononuclear phagocytic system in newborns. It is possible that aging red blood cells cannot be phagocytosed or that engulfed red blood cells were degraded at a slower rate, resulting in insufficient raw materials for bilirubin production. At the same time, the heme oxygenase activities on the microsomes of mononuclear macrophages reduced the oxidation rate of heme into biliverdin, which also led to a decrease in bilirubin production [15]. In addition, bilirubin itself is a strong antioxidant, and metabolic acidosis and myocardial damage in children may lead to an increase in oxygen free radicals in the body. Part of the bilirubin is consumed to exert its antioxidant function, and the content decreases [16].

Sphingosine, the main component of sphingolipids and a kind of sphingomyelin basic lipid, is involved in sphingolipid metabolism. Sphingomyelin and its metabolites are not only important constituents of cell membranes but also necessary regulators in a variety of signal transduction pathways, which play a significant role in many pathological processes [17,18]. Since sphingomyelin is widely interrelated in the body, its abnormal content will lead to a series of chain reactions leading to inflammation [19]. The content of sphingosine in the two groups of pneumonia complications increased more significantly than that in the PN group, indicating aggravation of an inflammatory reaction.

PC (14:0/14:1 (9Z)) and LysoPC (22:2 (13Z, 16Z)) are involved in glycerophospholipid metabolism. Glycerophospholipids are the most abundant phospholipids in the body and play a role in various physiological functions, such as inflammation and cell damage [20]. In various pathophysiological conditions, the ratio of free and albumin-bound lysophosphatidylcholine (LPC) can be profoundly altered by increasing the production of LPC or lowering plasma albumin levels [21,22], which means that to a certain extent, phospholipids act as important regulators in inflammatory reactions, and the changes in type and content of phospholipids can reflect the severity of inflammation [23]. LPC has many protective or anti-inflammatory effects, and LPC, at a higher level, can act as an anti-inflammatory molecule by producing vascular protective effects through prostacyclin or nitric oxide. During bacterial or viral infections, low levels of LPC can lead to immune disorders [24]. In the present study, the relative contents of substances PC (14:0/14:1 (9Z)) and LysoPC (22:2 (13Z, 16Z)) involved in this pathway increased, indicating that in the PN&MD and PN&MA groups, the inflammatory response was more severe compared with that in PN group. PC (14:0/14:1 (9Z)) and LysoPC (22:2 (13Z, 16Z)) may act as protective or anti-inflammatory compounds.

In addition, some important common differential substances, such as farnesol and oleic acid, have significant specificities, and previous reports have shown that they can regulate inflammatory responses and have beneficial effects on the immune response system in diseases, including edema, allergic asthma, and colon tumors [25]. A series of animal models have demonstrated that farnesol can eliminate tumor growth [26]. It has also been proved that farnesol exhibits potential pro/anti-inflammatory and anticancer effects in various diseases [27]. The increase in farnesol levels in both complication groups indicated that the body was regulating its own exacerbating inflammatory responses. Oleic acid affects cell membrane fluidity, receptors, and intracellular signaling pathways. It can directly regulate the synthesis and activity of antioxidant enzymes and has anti-inflammatory effects, which are reached through inhibiting proinflammatory cytokines and activating anti-inflammatory cytokines [28,29]. As shown in our studies, oleic acid was significantly upregulated in the complication groups, which also reflected the severe inflammatory response in the complication groups. Phe-Phe is a peptide composed of two phenylalanine molecules. Phenylalanine is an essential amino acid and a precursor of D-tyrosine, which can be used for protein synthesis and converted into nonessential amino acid tyrosine [30]. The levels of Phe-Phe and D-tyrosine significantly increased in both complication groups, indicating an abnormal disorder in amino acid metabolism. Changes in amino acid levels can reflect the immune response of the body to infection or tissue damage. Almost all tissue remodeling in the body involves the breakdown and synthesis of proteins, and abnormal amino acid metabolism can have adverse effects on the production of proteins [31]. In a previous study, Phe-Phe was found to be a marker of pancreatic cancer [32]. Therefore, the increase in Phe-Phe concentration may indirectly indicate an increased risk of cancer in the pneumonia complication groups compared with the PN group.

It was evident that pneumonia with metabolic acidosis or myocardial damage was two different types of complications compared with the pneumonia control group, but there were common molecules, and they acted in the same metabolic pathways. The common metabolite molecules revealed that the inflammatory response was more severe in the complication groups, and the risk of cancer tended to increase, while the increased inflammatory response led to a series of ripple reactions, such as abnormal apoptosis of body cells and energy metabolism, and disturbance of amino acid metabolism. These findings of common molecules and metabolic pathways between PN&MA and PN&MD provide a deep understanding of the complications of pneumonia disease, especially for newborns with relatively low immunity.

5. Conclusions

In this work, an untargeted UPLC–HRMS-based metabolomics approach was developed and applied for the study of neonatal pneumonia and pneumonia complicated by metabolic acidosis or myocardial damage. Serum samples were collected and analyzed to explore the important differential metabolites between healthy control and disease groups. The molecular mechanisms of actions that disrupt the pathways were discovered. Based on these findings, the common molecular mechanisms of the two complications of pneumonia were investigated. The results showed a significant decrease in bilirubin levels in the PN&MA and PN&MD groups, which implied an increased inflammatory response. Meanwhile, the levels of farnesol and sphingosine as anti-inflammatory substances increased, indicating the presence of the self-regulation of organisms in response to the exacerbation of the inflammatory response. The changes in the levels of these common substances and the metabolic pathways involved were mainly closely related to the series of immune responses caused by the exacerbation of inflammation levels. Comparing groups PN and PN&MA and PN and PN&MD, the relative content of common substances showed the same upward and downward trends with varying degrees, but the differences were not significant. The changes in the contents of these substances may represent changes in the severity of pneumonia complications in newborns. In the future, a targeted study will be conducted to take the effects of multiple pathologies into consideration for validation and clinical application, which aims to fully validate the investigation of the potential programming mechanisms of these differential molecules, as well as the effects and mechanisms of repair after drug treatment and/or prognostic assessment.

Supplementary Materials: The following supporting information can be downloaded at: https://www.mdpi.com/article/10.3390/metabo13111118/s1. Table S1: diseases abbreviations; Table S2: annotation results.

Author Contributions: Y.Z., sample preparation, experimental analysis, data processing, draft writing; H.W., clinical demand, sample collection, draft improvement; Z.W., results inspection, methodology validation, draft improvement; Z.Z., conceptualization, methodology, draft improvement. All authors have read and agreed to the published version of the manuscript.

Funding: This research was funded by a research project (No. LJKMZ20221836) from the Department of Education of Liaoning Province of China, the Natural Science Foundation (NSF) of the Jiangsu Higher Education Institutions of China (No. 21KJA610002), and China Tobacco Yunnan Industrial Co., Ltd. (No. 2022CP02, No. 2022539200340157).

Institutional Review Board Statement: The study was conducted in accordance with the Declaration of Helsinki, and approved by the Ethics Committee of Changzhou Maternal and Child Health Hospital (CMCHH), Changzhou, China (approval number: 2019017; date of approval: 5 July 2019).

Informed Consent Statement: Informed consent was obtained from all subjects involved in the study.

Data Availability Statement: The data presented in this study are available on request from the corresponding author. Data is not publicly available due to privacy.

Conflicts of Interest: The authors declare no conflict of interest.

References

1. Nascimento-Carvalho, C.M. Community-acquired pneumonia among children: The latest evidence for an updated man-agement. *J. Pediatr. (Rio J.)* **2020**, *96* (Suppl. 1), 29–38. [CrossRef]
2. le Roux, D.M.; Zar, H.J. Community-acquired pneumonia in children—A changing spectrum of disease. *Pediatr. Radiol. Oct.* **2017**, *47*, 1392–1398. [CrossRef]
3. Leung, A.K.C.; Wong, A.H.C.; Hon, K.L. Community-Acquired Pneumonia in Children. *Recent Pat. Inflamm. Allergy Drug Discov.* **2018**, *12*, 136–144. [CrossRef]
4. Zar, H.J.; Andronikou, S.; Nicol, M.P. Advances in the diagnosis of pneumonia in children. *BMJ* **2017**, *358*, j2739. [CrossRef]
5. Dean, P.; A Florin, T. Factors Associated With Pneumonia Severity in Children: A Systematic Review. *J. Pediatr. Infect. Dis. Soc.* **2018**, *7*, 323–334. [CrossRef]
6. Mujahid, M.; Rustam, F.; Álvarez, R.; Mazón, J.L.V.; Díez, I.D.l.T.; Ashraf, I. Pneumonia Classification from X-ray Images with Inception-V3 and Convolutional Neural Network. *Diagnostics* **2022**, *12*, 1280. [CrossRef]
7. Xin, H.; Li, J.; Hu, H.Y. Is Lung Ultrasound Useful for Diagnosing Pneumonia in Children? A Meta-Analysis and Systematic Review. *Ultrasound Q. Mar.* **2018**, *34*, 3–10. [CrossRef]
8. Schrimpe-Rutledge, A.C.; Codreanu, S.G.; Sherrod, S.D.; McLean, J.A. Untargeted Metabolomics Strategies—Challenges and Emerging Directions. *J. Am. Soc. Mass Spectrom.* **2016**, *27*, 1897–1905. [CrossRef]
9. Cui, L.; Lu, H.; Lee, Y.H. Challenges and emergent solutions for LC-MS/MS based untargeted metabolomics in diseases. *Mass Spectrom. Rev.* **2018**, *37*, 772–792. [CrossRef]
10. Li, J.; Fu, Y.; Jing, W.; Li, J.; Wang, X.; Chen, J.; Sun, S.; Yue, H.; Dai, Y. Biomarkers of *Mycoplasma pneumoniae* pneumonia in children by urine metabolomics based on Q Exactive liquid chromatography/tandem mass spectrometry. *Rapid Commun. Mass Spectrom.* **2022**, *36*, e9234. [CrossRef]
11. Del Borrello, G.; Stocchero, M.; Giordano, G.; Pirillo, P.; Zanconato, S.; Da Dalt, L.; Carraro, S.; Esposito, S.; Baraldi, E. New insights into pediatric community-acquired pneumonia gained from untargeted metabolomics: A preliminary study. *Pediatr. Pulmonol.* **2020**, *55*, 418–425. [CrossRef]
12. Ning, P.; Zheng, Y.; Luo, Q.; Liu, X.; Kang, Y.; Zhang, Y.; Zhang, R.; Xu, Y.; Yang, D.; Xi, W.; et al. Metabolic profiles in community-acquired pneumonia: Developing assessment tools for disease severity. *Crit. Care* **2018**, *22*, 130. [CrossRef]
13. Wang, Y.; Huang, X.; Li, F.; Jia, X.; Jia, N.; Fu, J.; Liu, S.; Zhang, J.; Ge, H.; Tai, J.; et al. Serum-integrated omics reveal the host response landscape for severe pediatric communi-ty-acquired pneumonia. *Crit. Care Mar.* **2023**, *27*, 79. [CrossRef]
14. McDonagh, A.F. Controversies in bilirubin biochemistry and their clinical relevance. *Semin. Fetal Neonatal Med.* **2010**, *15*, 141–147. [CrossRef]
15. Creeden, J.F.; Gordon, D.M.; Stec, D.E.; Hinds, T.D., Jr. Bilirubin as a metabolic hormone: The physiological relevance of low levels. *Am. J. Physiol.-Endocrinol. Metab.* **2021**, *320*, E191–E207. [CrossRef]
16. Soto Conti, C.P. Bilirubin: The toxic mechanisms of an antioxidant molecule. *Arch. Argent Pediatr.* **2021**, *119*, e18–e25.
17. Tan-Chen, S.; Guitton, J.; Bourron, O.; Le Stunff, H.; Hajduch, E. Sphingolipid Metabolism and Signaling in Skeletal Muscle: From Physiology to Physiopathology. *Front. Endocrinol.* **2020**, *11*, 491. [CrossRef]
18. Lai, Y.; Tian, Y.; You, X.; Du, J.; Huang, J. Effects of sphingolipid metabolism disorders on endothelial cells. *Lipids Health Dis.* **2022**, *21*, 101. [CrossRef]
19. Obinata, H.; Hla, T. Sphingosine 1-phosphate and inflammation. *Int. Immunol.* **2019**, *31*, 617–625. [CrossRef]
20. Wang, B.; Tontonoz, P. Phospholipid Remodeling in Physiology and Disease. *Annu. Rev. Physiol.* **2019**, *81*, 165–188. [CrossRef]
21. Liu, P.; Zhu, W.; Chen, C.; Yan, B.; Zhu, L.; Chen, X.; Peng, C. The mechanisms of lysophosphatidylcholine in the development of diseases. *Life Sci.* **2020**, *247*, 117443. [CrossRef]
22. Ren, J.; Lin, J.; Yu, L.; Yan, M. Lysophosphatidylcholine: Potential Target for the Treatment of Chronic Pain. *Int. J. Mol. Sci.* **2022**, *23*, 8274. [CrossRef]
23. Zhu, Q.; Wu, Y.; Mai, J.; Guo, G.; Meng, J.; Fang, X.; Chen, X.; Liu, C.; Zhong, S. Comprehensive Metabolic Profiling of Inflammation Indicated Key Roles of Glycerophospholipid and Arginine Metabolism in Coronary Artery Disease. *Front. Immunol.* **2022**, *13*, 829425. [CrossRef]
24. Kabarowski, J.H. G2A and LPC: Regulatory functions in immunity. *Prostaglandins Other Lipid Mediat.* **2009**, *89*, 73–81. [CrossRef]
25. Delmondes, G.D.; Santiago Lemos, I.C.; Dias, D.D.; Cunha, G.L.; Araújo, I.M.; Barbosa, R.; Coutinho, H.D.; Felipe, C.F.; Barbosa-Filho, J.M.; Lima, N.T.; et al. Pharmacological applications of farnesol ($C_{15}H_{26}O$): A patent review. *Expert Opin. Ther. Pat.* **2020**, *30*, 227–234. [CrossRef]
26. Sell, L.B.; Ramelow, C.C.; Kohl, H.M.; Hoffman, K.; Bains, J.K.; Doyle, W.J.; Strawn, K.D.; Hevrin, T.; Kirby, T.O.; Gibson, K.M.; et al. Farnesol induces protection against murine CNS inflammatory demyelination and modifies gut microbiome. *Clin. Immunol.* **2022**, *235*, 108766. [CrossRef]
27. Jung, Y.Y.; Hwang, S.T.; Sethi, G.; Fan, L.; Arfuso, F.; Ahn, K.S. Potential Anti-Inflammatory and Anti-Cancer Properties of Farnesol. *Molecules* **2018**, *23*, 2827. [CrossRef]
28. Palomer, X.; Pizarro-Delgado, J.; Barroso, E.; Vázquez-Carrera, M. Palmitic and Oleic Acid: The Yin and Yang of Fatty Acids in Type 2 Diabetes Mellitus. *Trends Endocrinol. Metab.* **2018**, *29*, 178–190. [CrossRef]

29. Sun, Y.; Wang, J.; Guo, X.; Zhu, N.; Niu, L.; Ding, X.; Xie, Z.; Chen, X.; Yang, F. Oleic Acid and Eicosapentaenoic Acid Reverse Palmitic Acid-induced Insulin Resistance in Human HepG2 Cells via the Reactive Oxygen Species/JUN Pathway. *Genom. Proteom. Bioinform.* **2021**, *19*, 754–771. [CrossRef]
30. Kopple, J.D. Phenylalanine and Tyrosine Metabolism in Chronic Kidney Failure. *J. Nutr.* **2007**, *137*, 1586S–1590S. [CrossRef]
31. Møller, N.; Meek, S.; Bigelow, M.; Andrews, J.; Nair, K.S. The kidney is an important site for in vivo phenylalanine-to-tyrosine conversion in adult humans: A metabolic role of the kidney. *Proc. Natl. Acad. Sci. USA* **2000**, *97*, 1242–1246. [CrossRef]
32. Du, Y.; Fan, P.; Zou, L.; Jiang, Y.; Gu, X.; Yu, J.; Zhang, C. Serum Metabolomics Study of Papillary Thyroid Carcinoma Based on HPLC-Q-TOF-MS/MS. *Front. Cell Dev. Biol.* **2021**, *9*, 593510. [CrossRef]

Disclaimer/Publisher's Note: The statements, opinions and data contained in all publications are solely those of the individual author(s) and contributor(s) and not of MDPI and/or the editor(s). MDPI and/or the editor(s) disclaim responsibility for any injury to people or property resulting from any ideas, methods, instructions or products referred to in the content.

Article

A Nucleotide Metabolism-Related Gene Signature for Risk Stratification and Prognosis Prediction in Hepatocellular Carcinoma Based on an Integrated Transcriptomics and Metabolomics Approach

Tianfu Wei [1,2,†], Jifeng Liu [1,2,†], Shurong Ma [1,2], Mimi Wang [3], Qihang Yuan [1,2], Anliang Huang [3], Zeming Wu [4], Dong Shang [1,2,3,*] and Peiyuan Yin [1,2,3,*]

[1] Clinical Laboratory of Integrative Medicine, First Affiliated Hospital of Dalian Medical University, Dalian 116000, China; tianfuwei0901@gmail.com (T.W.)
[2] Department of General Surgery, First Affiliated Hospital of Dalian Medical University, Dalian 116000, China
[3] Institute of Integrative Medicine, Dalian Medical University, Dalian 116000, China
[4] iPhenome Biotechnology (Yun Pu Kang) Inc., Dalian 116000, China
* Correspondence: shangdong@dmu.edu.cn (D.S.); yinpeiyuan@dmu.edu.cn (P.Y.)
† These authors contributed equally to this work.

Abstract: Hepatocellular carcinoma (HCC) is a leading cause of cancer-related mortality worldwide. The in-depth study of genes and metabolites related to nucleotide metabolism will provide new ideas for predicting the prognosis of HCC patients. This study integrated the transcriptome data of different cancer types to explore the characteristics and significance of nucleotide metabolism-related genes (NMGRs) in different cancer types. Then, we constructed a new HCC classifier and prognosis model based on HCC samples from TCGA and GEO, and detected the gene expression level in the model through molecular biology experiments. Finally, nucleotide metabolism-related products in serum of HCC patients were examined using untargeted metabolomics. A total of 97 NMGRs were obtained based on bioinformatics techniques. In addition, a clinical model that could accurately predict the prognostic outcome of HCC was constructed, which contained 11 NMGRs. The results of PCR experiments showed that the expression levels of these genes were basically consistent with the predicted trends. Meanwhile, the results of untargeted metabolomics also proved that there was a significant nucleotide metabolism disorder in the development of HCC. Our results provide a promising insight into nucleotide metabolism in HCC, as well as a tailored prognostic and chemotherapy sensitivity prediction tool for patients.

Keywords: nucleotide metabolism; hepatocellular carcinoma; prognosis signature; molecular classification; chemotherapy sensitivity; tumor immune microenvironment

1. Introduction

Hepatocellular carcinoma (HCC) is one of the dominant types of cancer all over the world [1]. HCC is the third leading cause of mortalities among all malignancies in the world [2–4]. In addition, effective prognostic indicators would be a boon for these patients. Thus, it is urgent to develop and verify new prognostic signals to predict the clinical prognosis of HCC patients at an early stage in order to improve the survival rate of patients.

Nucleotide is the basic building block of organisms, and it is an essential raw material for producing nucleic acid to sustain cell proliferation [5]. Nucleotide metabolism is in a state of dynamic equilibrium, which is important for maintaining normal physiological functions of cells [6,7]. Recently, researchers have affirmed that abnormal nucleotide metabolism enhances the growth of tumors and suppresses the normal immune responses in the tumor microenvironment [8]. For example, disrupting the homeostasis of the pools of nucleotides can produce mutations that influence antigen presentation and, ultimately, the immune response to the tumor [9,10]. Targeting nucleotide metabolism also provides new directions for the development of novel antitumor-specific drugs [11,12]. Therefore, focusing on the reprogramming of the nucleotide metabolism will provide new ideas for predicting prognostic outcomes in HCC patients. Moreover, the clinical relevance of nucleotide metabolism-related genes (NMRGs) in predicting outcomes and guiding chemotherapeutic strategies for patients with HCC remains unknown to the best of our knowledge. Thus, the development of the HCC risk stratification tool using NMRGs is promising.

In the present research, we will systematically evaluate the potential of NMRGs in predicting the prognosis of HCC patients using a bioinformatics approach and establish a risk score signal based on NMRGs to predict the clinical outcome of HCC patients. This model could be utilized in making clinical decisions and providing individualized care. To further validate the credibility of the model, we examined the expression of NMRGs in the model at the cellular level by molecular biology experiments. Ultimately, we used non-targeted metabolomics to detect the nucleotide metabolism-related products in serum samples of patients with HCC, further supporting our study from the metabolic point of view. We are optimistic that the findings of this investigation will avail a greater and new insight into the diagnosis and management of HCC. Additionally, it will be essential in availing a theoretical basis for upcoming nucleotide metabolism studies.

2. Materials and Methods

2.1. Data Collection and Processing

Firstly, 97 NMRGs were obtained based on the following dataset from the Molecular Signatures Database (MSigDB): REACTOME_METABOLISM_OF_NUCLEOTIDES. RNA-sequencing (RNA-seq) and the matched clinical characteristics were derived from the TCGA database. The samples that were obtained contained 373 and 49 HCC patient samples and normal samples, respectively. RNA-seq, along with clinical data obtained from the Gene Expression Omnibus (GEO) database (GSE14520), were used for external validation. Patients who did not have information on their survival were excluded from further analysis. To facilitate batch normalization, the "sva" package in R was employed. In addition, the TCGA database was another database that was utilized to acquire SNV, transcriptome profiles, CNV, methylation data, and pan-cancer transcriptomes' clinical features.

2.2. Pan-Cancer Analysis

Currently, inadequate research has been conducted to determine the link between nucleotide metabolism and malignancies. As a result, the differences in NMRGs in various malignancies are described inadequately. SNV, CNV, methylation, and mRNA expression data were examined and graphically illustrated as heatmaps to avail a pan-cancer summary of NMRGs. Moreover, a univariate Cox regression analysis between the mRNA expression and OS to probe into the value of NMRGs in the prognoses of patients with various malignancies was conducted using R version 4.0.3 and TBtools version 1.098 [13].

Single sample gene set enrichment analysis (ssGSEA) was used to calculate NMRG scores in every sample of each cancer to reveal the differential function of pathways regulated by NMRGs in various kinds of human tumors. Samples were categorized into two groups, one with the top 30% of NMRG scores and the other with the worst 30%. Gene set enrichment analysis (GSEA) was used to investigate the differences in pathway activity between the two groups based on the transcriptomes of the two groups.

2.3. Differentially Expressed Prognostic NMRG Identification

The "limma" packages were utilized to uncover the differentially expressed NMRGs between HCC and normal tissues (FDR < 0.05, fold change > 1.5). Next, 97 NMRGs screened out were put into univariate Cox regression analysis to acquire the genes with prognostic significance ($p < 0.05$). Afterward, the intersection of the two sets of genes was taken to obtain 32 NMRGs for subsequent analysis, as shown by the Venn diagram.

2.4. Non-Negative Matrix Factorization (NMF) Clustering Determination of NMRG Modification Subtypes

The HCC samples from the TCGA database were clustered by the NMF based on the expression data of 32 NMRGs. The range for the cluster count, k, was set from 2 to 10. The R package "NMF" calculated the common membership matrix's average contour width. On the basis of the dispersion, cophenetic, and silhouette metrics, the ideal cluster numbers were established. Afterward, the samples are split into two distinct molecular subtypes C1 and C2.

2.5. Gene Set Variation Analysis (GSVA) and NMRGs Different Expression Analysis

The NMRG scores of individual patients with HCC were computed by the "GSVA" package in R, which could serve as an indicator of nucleotide metabolism activities. Then, the "Wilcox.test" function in R and a T-test were employed to compare the difference in the scores and expression of NMRG between two clusters, respectively.

2.6. Differences in the Prognosis, Immune Checkpoint Genes, and Drug Sensitivity between Distinct NMRG-Based Clusters

The prognostic efficacy of clusters was assessed using Kaplan–Meier analyses, with the progression-free interval (PFI), disease-specific survival (DSS), disease-free interval (DFI), and overall survival (OS), as standards. Subsequently, the "Wilcox.test" function in R was adopted to explore the disparity between infiltration levels of typically immune checkpoint genes (ICGs). Additionally, we used pRRophetic [14], the R software that predicts the clinical chemotherapeutic response utilizing the expression levels of tumor genes, to calculate the semimaximum inhibitory concentration (IC50) of commonly used chemotherapeutic drugs in the HCC cohort. A Wilcoxon signed-rank test, on the other hand, determined if the difference in the IC50 between two clusters is statistically significant. A decreased semi-inhibitory mass concentration of the drug in malignant cells is always associated with a smaller IC50, indicating that the cancer cells are more susceptible to the medicine.

2.7. DEG Identification and Functional Analysis

DEGs between two clusters based on NMRGs were identified by the limma package, with the thresholds established as FDR < 0.05 and fold change > 1.5, which was further subjected to the Kyoto Encyclopedia of Genes and Genomes (KEGG) pathway and GO functional enrichment analyses using the R package "clusterProfiler".

2.8. Construction and Verification of a Prognostic Signature Based on NMRGs

The 32 differentially expressed and prognostically significant NMRGs obtained previously were incorporated in the least absolute shrinkage and selection operator (LASSO) Cox regression signature to develop the powerful prognostic signature. The risk score of each patient is calculated using the "Prediction" function in R, and then HCC patients in TCGA and GEO groups were classified into high- and low-risk groups as per the median risk score, and comparisons of their prognoses were done. To additionally test the viability of the risk score-based predictive signature in patients with HCC in the TCGA as well as GEO datasets, the principal component analysis (PCA) and the t-distributed stochastic neighbor embedding (t-SNE) analyses were done. Using the "survival ROC" R package version 4.0.3, time-dependent receiver operating characteristic (ROC) curves and AUC values were obtained to ascertain the specificity and sensitivity of the risk score.

2.9. Creating a Predictive Nomogram That Incorporates Clinical Characteristics and Risk Scores

The clinical data, which comprised age, gender, grade, and stage as well as the risk score of every patient in TCGA cohorts, were retrieved. The statistically significant indicators ($p < 0.05$) from the univariate Cox survival analysis of each indicator were then incorporated into the multivariate Cox survival analysis. These markers were regarded as independent prognostic variables ($p < 0.05$) in the multivariate Cox survival analysis. A nomogram was constructed utilizing the above clinical features and risk score. The nomogram's discriminating power and prediction accuracy were then assessed using calibration curves. The prediction performance was also assessed using the time-dependent ROC curve.

2.10. Reagents

Cell culture-related reagents such as Dulbecco's Modified Eagle Medium (DMEM), Minimum Essential Medium (MEM), and Roswell Park Memorial Institute 1640 Medium (RPMI-1640) were purchased from Gibco (Grand Island, NE, USA). PCR-related reagents were purchased from Accurate Biology (Changsha, China). Methanol, isopropanol, acetonitrile, formic acid, and ammonium acetate of mass spectrometry grade were supplied by Fisher Scientific (Fair Lawn, NJ, USA). Ammonium bicarbonate and methyl tert-butyl ether (MTBE) of mass spectrometry level were purchased from Sigma-Aldrich (St. Louis, MO, USA). Ultra-pure water (18.2 MΩ) was prepared by a Milli-Q water purification system (Merck KGaA, Darmstadt, Germany).

2.11. Cell Culture

The human HCC cell lines (HuH7, HepG2, and Hep3B2.1–7) were purchased from Procell Life Science & Technology (Wuhan, China). The L-02 cell line (human normal hepatocytes) was purchased from BeNa Culture Collection (Beijing, China). Briefly, HuH7 and L-02 were, respectively, grown in DMEM high glucose medium (Gibco, Grand Island, NE, USA) and RPMI-1640 medium, while HepG2 and Hep3B2.1–7 were incubated in MEM medium, all of which containing 10% fetal bovine serum and 1% penicillin-streptomycin solution. All the cells were incubated in a cell incubator under 37 °C with a concentration of 5% CO_2.

2.12. Real-Time Quantitative Polymerase Chain Reaction (qPCR)

The total RNA in the HuH7, HepG2, Hep3B2.1–7, and L-02 cell lines was extracted by the conventional Trizol method and the cDNA was obtained using a reverse transcription kit (Accurate Biology, Changsha, China). Furthermore, the expressed level of target gene was detected by using SYBR Green I fluorescent dye-based assay and β-actin was used as the internal reference gene. RNA level was analyzed and quantified by 2-ΔΔCt. The primer sequences of the genes were shown in Supplementary Table S1.

2.13. Participants and Criteria

Serum samples from HCC patients ($n = 26$) and healthy individuals ($n = 26$) were obtained from the biological sample bank of the First Affiliated Hospital of Dalian Medical University (collected from November 2016 to December 2019). In addition, the study has been approved by the Ethics Committee of the First Affiliated Hospital of Dalian Medical University (No. PJ-KS-KY-2021-129). Inclusion criteria for the HCC group included: (1) signed informed consent for collection and use of biological samples and aged ≥18 years; (2) the pathological diagnosis is HCC; (3) follow-up information is complete; (4) no other malignant tumors and no prior anti-tumor treatment was performed before surgery; (5) the biological sample is complete. Exclusion criteria include: (1) new adjuvant or chemical therapy before surgery; (2) accidental death during operation or postoperative relapse resulting in death within one month; (3) the follow-up information is incomplete or the biological sample is missing. Serum samples from the control group (CON group) were obtained from healthy individuals on physical examination and matched the sex and age composition of the HCC group.

2.14. Serum Sample Pretreatment and Non-Targeted Metabolomics Analysis

The pretreatment procedures of serum samples were divided into two parts, namely, extraction of polar small molecule metabolites and lipids. Briefly, to extract the polar metabolites, we added 150 µL of the serum sample to a 96-DeepWell plate followed by 600 µL of methanol solution. After the mixture was vortexed for 5 min, it was centrifuged at 5300 rpm for 20 min. The supernatants were divided into two 200 µL aliquots, and transferred to two individual 450 µL 96-well plates, and the liquid was lyophilized by a freeze dryer. Finally, the residual was redissolved prior to non-targeted metabolomics testing. Additionally, to extract the lipids, we added 20 µL of serum sample to a 1.5 mL microcentrifuge tube, followed by 120 µL of methanol solution and vortexed for 3 min. Then, 360 µL of methyl tertbutyl ether (MTBE) and 100 µL of ultra-pure water were sequentially added after oscillating for 3 min, and then the mixture was centrifuged at $13,000 \times g$ for 15 min. Similarly, the lipid layer was lyophilized and dissolved prior for the test. UltiMate 3000 ultra-high performance liquid chromatographic system and the Q-Exactive quadrupole-Orbitrap high resolution mass spectrometer (Thermo Fisher Scientific, Fair Lawn, NJ, USA) were used for non-targeted metabolomics analysis. For more information about metabolomics-related processes, please referred to the Supplementary Materials.

3. Results

3.1. Pan-Cancer Introduction with Respect to Differences in NMRGs

A chart displaying the research steps is provided in Figure 1. TCGA availed CNV, SNV, methylation, mRNA expression profiles, and survival data for 97 NMRGs in all kinds of malignancies for the pan-cancer study. We analyzed NMRG-related SNV data to ascertain the frequency as well as the variant types in every cancer subtype. As revealed in Supplementary Figure S1A, SKCM, UCEC, LUSC, LUAD, and STAD all had substantial SNV of NMRGs. The frequency of SNV of the NMRGs was 75.17% (2703 of 3596 tumors). Missense mutations were the predominant SNP type, according to the examination of variant types. The top five mutated genes, as determined by SNV percentage analyses, were CAD, DPYD, XDH, AK9, and AMPD1, with respective mutation percentages of 8%, 8%, 8%, 7%, and 6% (Supplementary Figure S1B). Moreover, to examine the genetic aberrations of NMRGs in malignancy, the percentage of CNV was evaluated and the findings revealed that, in general, CNV occurred at remarkable frequencies in a majority of cancer types (Figure 2A,B). In addition to CNV, aberrant DNA methylation of the promoter is linked to tumorigenesis [15]. The methylation of the promoter can modulate gene expression. We observed that most NMRGs in the 20 cancer types exhibited complex methylation patterns. However, TXNRD and ENTPD3 consistently showed hypermethylation in several tumors, while NME3, UPP2, and XDH showed the opposite (Figure 2C).

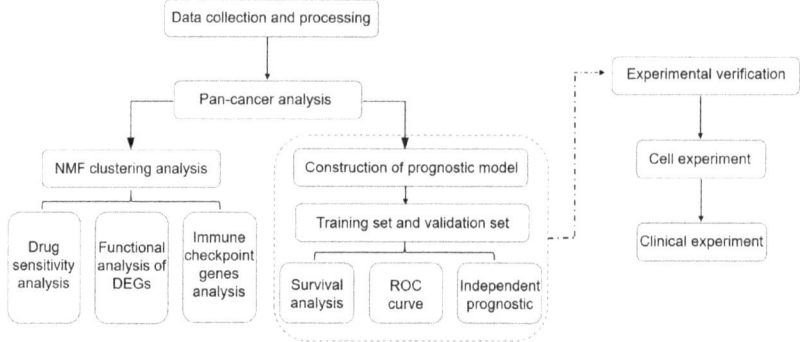

Figure 1. The investigation's flow chart.

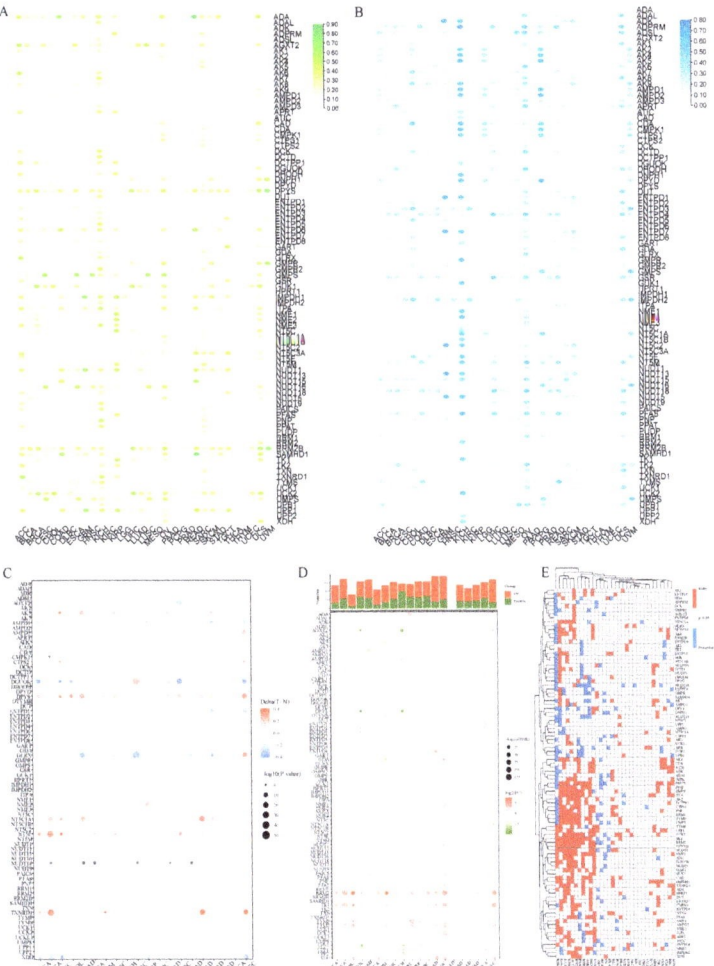

Figure 2. Panoramic view of nucleotide metabolism-related genes (NMRGs) in pan-cancer. (**A**,**B**) Histogram displays the frequency of copy number variation (CNV) for each NMRG in each tumor type ((**A**) amplification; (**B**) deletion). (**C**) Heatmap displays the differential methylation of NMRGs in cancers; hypermethylated and hypomethylated genes are denoted with red and blue, respectively (Wilcoxon rank-sum test). (**D**) Histogram (upper panel) and heatmap demonstrate the number of significant DEGs and the fold change and FDR of NMRGs, respectively, in each cancer. Substantially upregulated and downregulated genes are denoted with red and green, respectively. (**E**) NMRGs' survival profiles across cancers.

Besides genetic alterations, each cancer type's altered NMRG gene expression patterns were investigated using differential expression analysis between the malignant and nearby normal tissues. With the exception of pancreatic cancer tissues, we ascertained that most gene expression levels in cancer tissues varied in contrast with those in normal tissues. RRM2 and TK1 had remarkably increased expression levels in several cancers (Figure 2D). Afterward, utilizing univariate Cox regression of mRNA expression and OS, risky NMRGs with HR > 1 and p-Value < 0.05 as well as protective NMRGs with HR < 1 and p-Value < 0.05 were detected, as displayed in Figure 2E.

3.2. Identification of Differentially Expressed Prognostic NMRGs

RNA-seq data and clinical data of 49 normal samples and 373 HCC samples were retrieved from the TCGA database. A heatmap was developed with the aim of demonstrating the differentially expressed NMRGs between the normal and cancerous samples (Supplementary Figure S2A). A total of 69 out of 97 NMRGs were discovered to have differential expressions in normal and cancerous samples (Supplementary Table S2). Meanwhile, univariate Cox survival analysis was also done on NMRGs, of which 38 NMRGs were statistically significant (Supplementary Table S3). Finally, the intersection of the two sets of genes was taken to obtain 32 NMRGs for subsequent analysis (Supplementary Figure S2B).

3.3. NMF Clustering Identification of Molecular Typing Based on the NMRG

The NMF method selects the appropriate clustering number of two for the data, as per cophenetic, dispersion, and silhouette coefficients (Supplementary Figure S3, Figure 3A). The results of the following GSVA and KM analyses indicate that samples in C2 have higher NMRG scores and worse OS, DFI, PFI, and DSS, indicating the risky significance of NMRGs in HCC patients (Figure 3B–F). Supplementary Figure S4 shows the NMRGs that are differentially expressed in the two subgroups. Studies report adenosine block immune cell differentiation as well as maturation. It furthermore activates the expression of checkpoint molecules. We, therefore, compared the expression of ICGs between the two subtypes. Figure 3G shows all the statistically distinct ICGs, which are all expressed at higher levels in C2. In order to select appropriate administrating chemotherapeutic drugs for HCC patients, we performed chemotherapy sensitivity predictions between the two clusters. The results showed that Sorafenib, Metformin, Docetaxel, Dasatinib, Erlotinib, and Gefitinib are more suitable for C1 populations, while Gemcitabine, Doxorubicin, Cisplatin, Camptothecin, Bortezomib, and Etoposide are more suitable for C2 populations (Supplementary Figure S5).

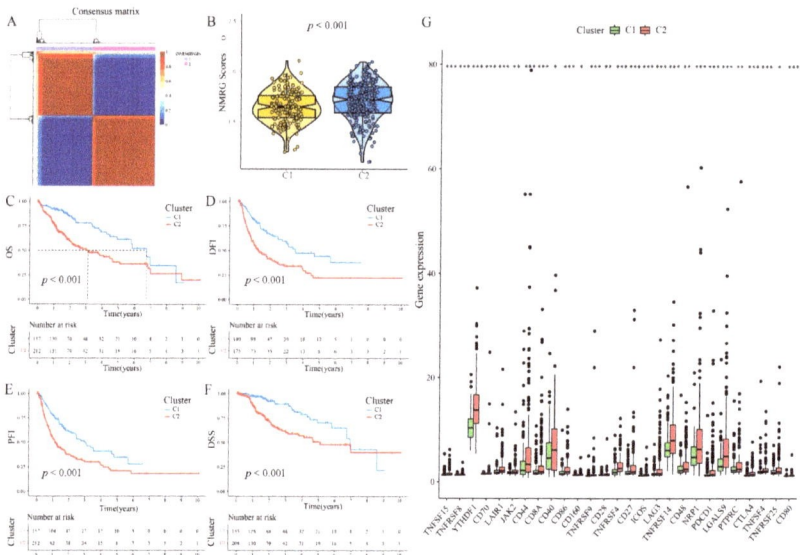

Figure 3. PanNMF clustering identification of two molecular subtypes with remarkably varied prognosis and expression of immune checkpoint genes. (**A**) The optimal clustering number of 2. (**B**) NMRG scores of the two subgroups are shown by violin plots. (**C–F**) Kaplan–Meier analyses (OS, DFI, PFI, and DSS) as regards two molecular subtypes. (**G**) Differential expression analysis of ICGs. * $p < 0.05$, ** $p < 0.01$, *** $p < 0.001$.

3.4. Functional Analysis for the NMRG Clusters

Then, in order to investigate probable mechanisms and biological functions at the gene level for the C1 and C2 groups, GO and KEGG pathway analyses were employed. Out of 995 DEGs that were subjected to screening (Supplementary Table S4), 356 and 639 genes were ascertained to be downregulated and upregulated in the C1 group, respectively (Figure 4A). The GO analysis affirmed that the genes were remarkably involved in the biological process of catabolic processing, inhibitor activity, and cell−substrate junction (Figure 4B). Meantime, the KEGG analysis revealed that these genes were also significantly related to various metabolic pathways, such as Tryptophan metabolism, Fatty acid degradation, Arginine and proline metabolism. (Figure 4C).

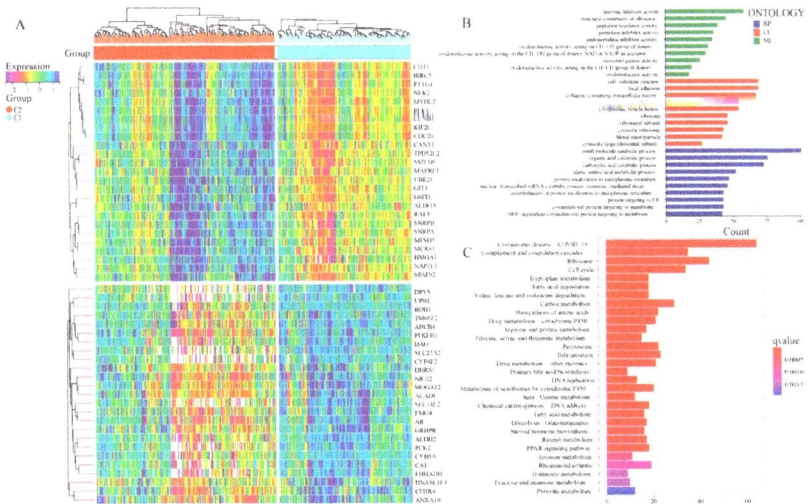

Figure 4. Functional analysis for the NMRG clusters. (**A**) Heatmap to display mRNA levels of DEGs in two NMRG clusters. (**B**) The analysis of GO enrichment for DEGs between two clusters. (**C**) The analysis of KEGG enrichment for DEGs between two clusters.

3.5. Determination and Verification of an NMRG-Based Prognostic Signature

To examine further the prognostic value of NMRGs, NMRG-based risk scores were created to anticipate HCC patients' survival. Upon conducting a LASSO regression and multivariate Cox analyses on the training cohort (Supplementary Figure S6), 11 genes (i.e., GMPS, UCK2, ENTPD2, PPAT, TXNRD1, RRM2, ATIC, ADSL, ADK, CDA, and DPYS) with prognostic values were uncovered from 32 NMRGs that had been previously obtained. Risk scores were subsequently determined for each HCC patient in the training cohort, and the training cohort sample was categorized into high- and low-risk subgroups based on the value of the median risk score (Figure 5A). Patients with greater risk scores had an increased likelihood of mortality, based on the risk score distributions and survival status. (Figure 5B). According to the PCA and t-SNE displayed in Figure 5C,D, patients belonging to the two risk groups may be distinguished with ease. Individuals that were in the high-risk subgroup had consistently reduced DSS, DFI, PFI, and OS values ($p < 0.05$), as shown in Figure 5E–H. Furthermore, the survival probability of the ROC curves of risk score-related AUC values were 0.798, 0.716, and 0.700 for 1, 3, and 5 years (Figure 5I), demonstrating that the risk score exerts a remarkable function in the prediction of the survival of HCC patients.

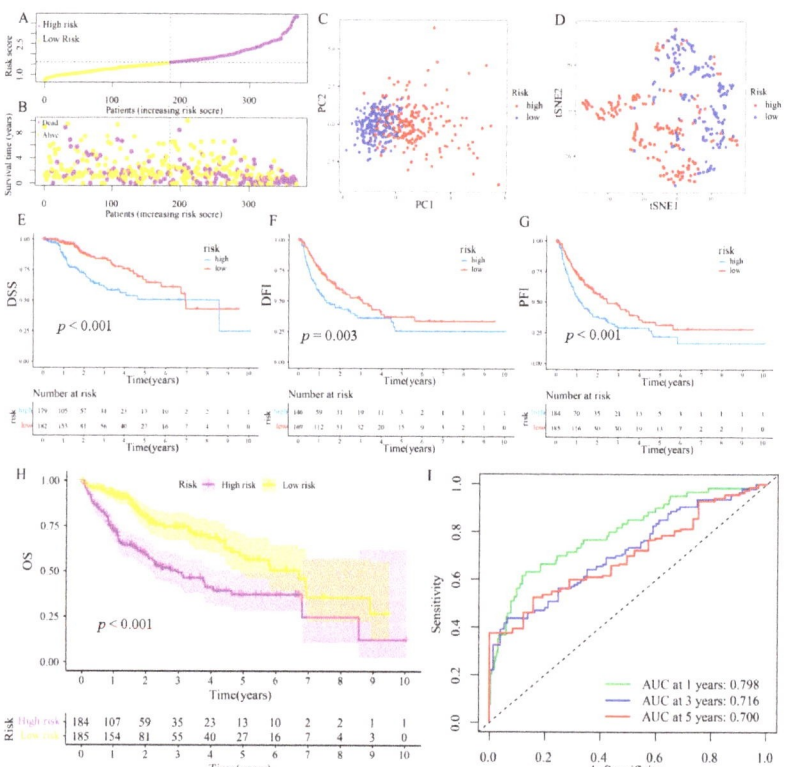

Figure 5. Construction of NMRG-related signature in the train cohort. (**A**) Various groups of the train cohort were created as per the median risk score. (**B**) Distributions of risk scores and the cohort's overall survival status. (**C**) Train cohort's PCA. (**D**) Train cohort's t-SNE. (**E–H**) Kaplan–Meier analyses (OS, DFI, PFI, and DSS) of the train cohort. (**I**) The train cohort's AUC values for ROC curves.

3.6. Predictive Efficiency of the Risk Signature Validation in the GEO Cohort

The GEO cohort (GSE14520) availed NMRG expression data on 225 HCC patients with complete survival data to confirm the replicability of the risk score in a different patient cohort. The GEO dataset was classified into high- and low-risk groups as per the median risk score of the training cohort (Supplementary Figure S7A). As displayed in Supplementary Figure S7B, the high-risk group was detected to have more death events, while the low-risk group demonstrated a remarkable probability of survival. PCA as well as t-SNE demonstrated that patients in the two risk groups were also distributed as per the two different groups (Supplementary Figure S7C,D). As demonstrated by the Kaplan–Meier curves for OS in Supplementary Figure S7E, patients in the high-risk group were discovered to exhibit a worse prognosis in contrast with the other risk group. Additionally, the high-risk group patients demonstrated a shorter survival time. A time-dependent ROC curve was examined to further determine the accuracy of the predictive risk signatures. Here, it was discovered that the AUC values of the signature in 1, 3, and 5 years were 0.611, 0.610, and 0.619, respectively (Supplementary Figure S7F).

3.7. Nomogram Development and Verification

To ascertain the link between immune function and the risk score, a heatmap was created. Statistically significant variations existed between the high as well as low-risk groups in the immune function of activated dendritic cells (aDCs), cytolytic activity, T cell regulation (Treg), Type I IFN Response, and Type II IFN Response in both the train and test cohorts (Supplementary Figure S8A,B). The univariate and multivariate Cox analyses evaluated the training cohort's clinical characteristics such as age, gender, grade, stage, and risk score. The findings of the univariate Cox and multivariate Cox regression analyses revealed that the training cohort's risk score and stage were independent prognostic predictors (Figure 6A,B). Afterward, the aforementioned factors were incorporated to generate a nomogram (Figure 6C). Furthermore, calibration curves were constructed to verify the anticipation power for the nomogram. The findings indicated an overall agreement between the nomogram's predicted survival rates and the actual survival rates (Figure 6D). The AUC values of the nomogram in 1, 3, and 5 years for HCC were 0.749, 0.732, and 0.719, respectively (Figure 6E).

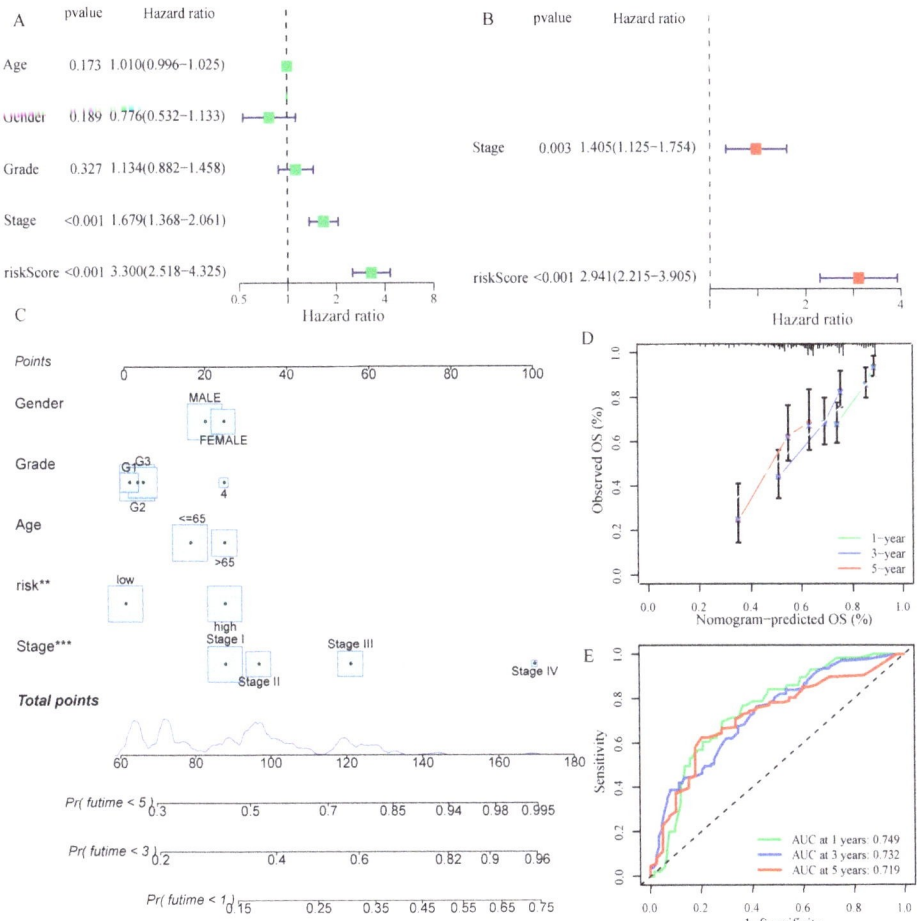

Figure 6. Nomogram (based on risk scores) development and verification. (**A**,**B**) Univariate and multivariate Cox regression analyses in the train cohort. (The green nodes in (**A**) indicate one-factor COX regression analysis, and the red nodes in (**B**) indicate multifactor COX regression analysis). (**C**) A nomogram of risk scores and clinical features. (The numbers in the overlapping part of (**C**) indicate the survival time (years)). (**D**) Calibration curves were utilized to validate the nomogram's 1-year, 3-year, and 5-year predictive ability. (**E**) The AUC values of the ROC curves for improved evaluation of the nomogram's prognostic ability. ** $p < 0.01$, *** $p < 0.001$.

3.8. The Expression of Hub Gene in Different HCC Cell Lines

To validate the bioinformatics predictions, we extracted the total RNA from different human HCC cell lines (HuH7, HepG2, and Hep3B2.1–7) and human normal hepatocyte line L-02. The mRNA level of the key genes, namely ADK, ADSL, ATIC, CDA, DPYS, ENTPD2, GMPS, PPAT, RRM2, TXNRD1, and UCK2, were determined. The results showed that the expression levels of ADK, ADSL, ATIC, DPYS, ENTPD2, TXNRD1, and UCK2 in at least one tumor cell line were consistent with the predictions (Figure 7A). We found that the expression levels of CDA, GMPS, and RRM2 in HCC patients were opposite to the predicted results (Figure 7B), which is an interesting phenomenon.

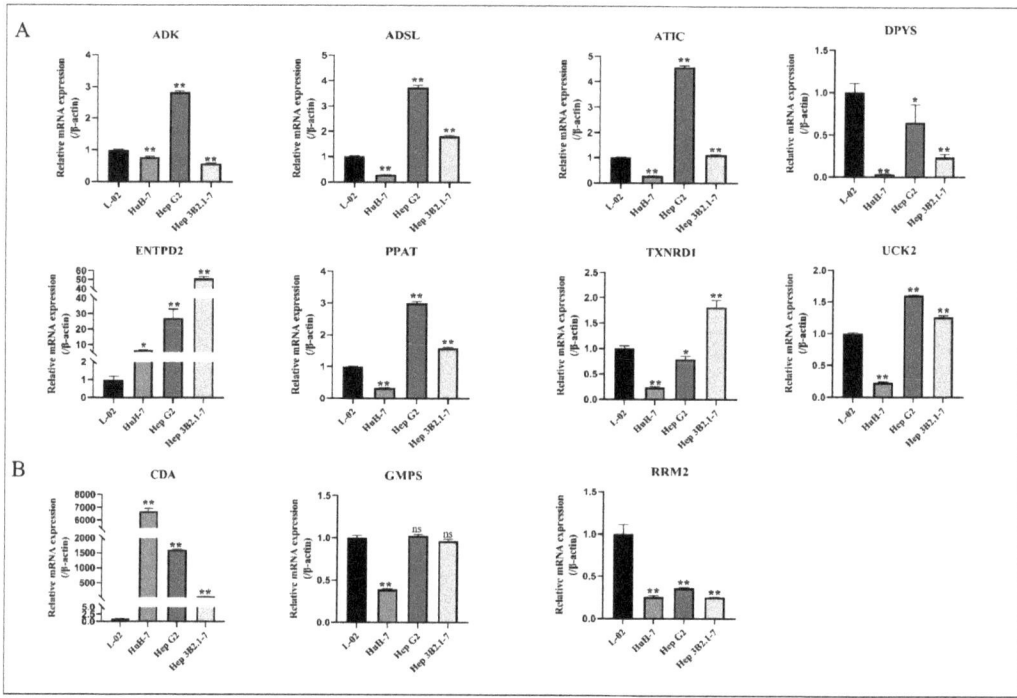

Figure 7. The differential expression of core genes in three hepatocellular carcinoma cell lines and normal hepatic epithelial cell lines based on RT-PCR. (**A**) Genes whose expression levels are consistent with the predicted results. (**B**) Genes whose expression levels are contrary to the predicted results. * $p < 0.05$, ** $p < 0.01$ versus L-02 group.

3.9. Metabolic Profiles of Hepatocellular Carcinoma and Differential Analysis of Nucleotide Metabolites

To observe the overall metabolic profiles in patients with hepatocellular carcinoma, we performed a non-targeted metabolomics analysis. A total of 26 serum samples from HCC patients obtained from the biological sample bank of the First Affiliated Hospital of Dalian Medical University were included in this study. In addition, we matched 26 serum samples from healthy control subjects according to the sex and age of the HCC patients. Baseline information for both groups is presented in Supplementary Table S5. The results of the OPLS-DA analysis showed a significant segregation in polar metabolites and lipids for both groups (Figure 8A). Next, volcanic maps were used to perform the differences between the two groups and the mean rate of change in intensity. The results are presented in Figure 8B.

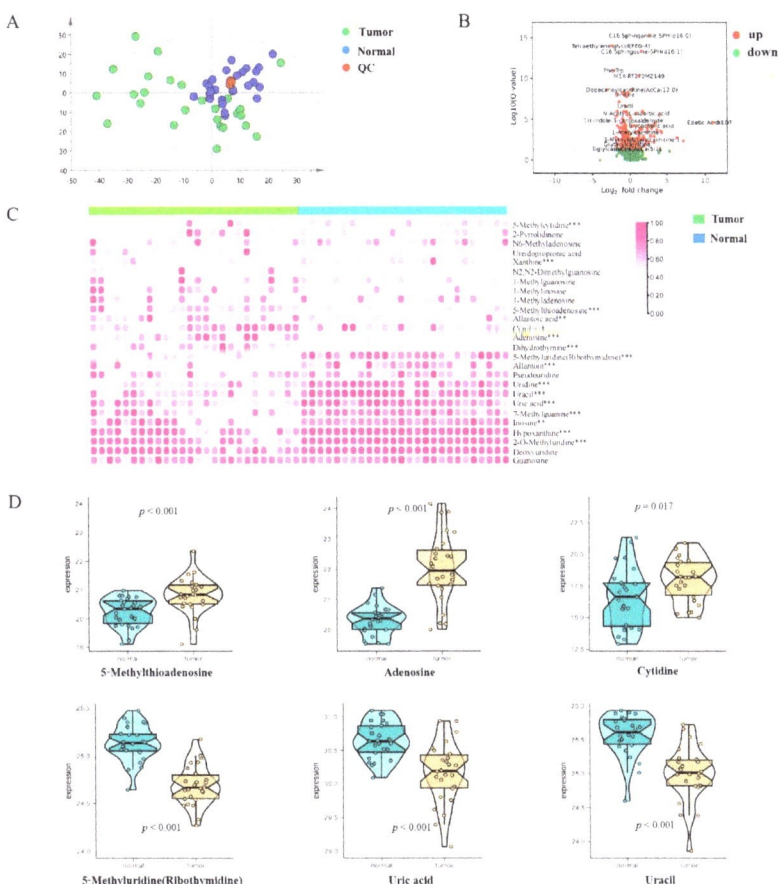

Figure 8. Characterization of the nucleotide metabolism landscape of hepatocellular carcinoma. (**A**,**B**) Overall metabolism of patients with hepatocellular carcinoma. (**C**,**D**) Nucleotide metabolism in patients with hepatocellular carcinoma. * $p < 0.05$, ** $p < 0.01$, *** $p < 0.001$ versus normal group.

To further explore the nucleotide metabolic profile in HCC patients, we compared the levels of nucleotide-related metabolites in those two groups of the samples and the results are presented as heat maps (Figure 8C). Specifically, a total of 26 products related to nucleotide metabolism were identified, of which 16 were significantly different, as follows: adenosine, dihydrothymine, cytidine, hypoxanthine, inosine, uric acid, xanthine, uridine, Uracil, Allantoin, 5-Methyluridine (Ribothymidine), 7-Methylguanine, 5-Methylcytidine, 5-MethylThioadenosine, Allantoic Acid, and 2-O-Methyluridine. We show some obviously different metabolites in Figure 8D. For the difference analysis of other metabolites, see Supplementary Figure S9.

4. Discussion

HCC is extremely aggressive, so it is clinically important to explore its effective prognostic indicators [16]. Recently, the traditional prognostic assessment system using clinicopathological parameters and staging has failed to meet the needs of precision medicine [17]. As sequencing technology has advanced, researchers have focused increasingly on disease molecular type and the quest for novel biomarkers to help with clinical diagnosis as well as treatment [18]. This approach not only enhances the standard prognostic assessment but also identifies a novel kind of pathogenesis. During the development of tumors, abnormal cancer metabolism takes place [19]. Recent research has demonstrated that aberrant nucleotide metabolism speeds up the progression of tumors while suppressing the tumor microenvironment's normal immune response [7,20]. The research on the link between nucleotide metabolism and the emergence of cancer is fast progressing, despite the paucity of pertinent experiments and studies. For malignancies treatment and prevention of recurrence as well as metastasis, the intervention, change, or modulation of molecular pathways connected to aberrant nucleotide metabolism in cancerous cells has emerged as a novel strategy and idea [8]. Thus, NMRG-based risk stratification of HCC is a promising strategy for prognosis assessment and individual management.

We sum up the differences in NMRGs across numerous cancers before studying the effect of aberrant nucleotide metabolism in HCC. The differences in NMRGs more or less happened and partial NMRGs had prognostic values in various malignancies. Additionally, it was evidently shown in several tumors that NMRGs had undergone genetic mutations and alterations. NMRGs were positively correlated with MYC targets, oxidative phosphorylation, mTORC1 signaling, E2F targets, and DNA repair in a majority of types of tumors. Nevertheless, they were negatively linked to UV response DN, myogenesis, and epithelial–mesenchymal transition. MYC orchestrates proliferation, apoptosis, differentiation, and metabolism and is frequently linked to poor prognosis and survival of patients. It plays a crucial function in practically every step of the neoplastic process [21]. Ectopic MYC expression in malignancies might simultaneously promote aerobic glycolysis and/or oxidative phosphorylation to supply adequate energy and anabolic substrates that are essential for the growth of cells and cell proliferation within the tumor microenvironment [22]. In cases of proliferative deregulation and in numerous different cancer types, mTOR signaling is triggered. Numerous mTOR pathway components have been documented to be dysregulated in malignancies including breast, colon, ovarian, kidney, and head and neck cancers [23]. Recent studies in HCC and pancreatic cancer suggest that E2F expression and/or increased E2F target expression in tumors have been linked to poor prognosis [24–26]. Genes involved in DNA repair responses exhibit a variety of mutations and abnormal expressions in cancer cells. These changes cause genomic instability and accelerate the processes of carcinogenesis and cancer progression [27]. Aberrant nucleotide metabolism may contribute to the development of cancer by regulating the above pathways.

Then, we filtered 97 NMRGs to obtain NMRGs that were differentially expressed in both cancerous and normal tissues and had prognostic significance. Thirty-two NMRGs were found for NMF clustering and a signature building. First, 32 NMRGs are applied to divide HCC samples into two molecular clusters with significantly distinct prognoses. C2 subtype is characterized by high NMRG scores and poor prognosis (PFI, DFI, DSS, and OS), indicating the risky significance of NMRGs in HCC patients. This result is consistent with the finding that the majority of NMRGs were HCC risk genes in the pan-cancer analysis. Considering that adenosine is able to induce the expression of checkpoint molecules, we compared the differences in ICG expression between the two subtypes. We discovered that ICGs are expressed at a high level in the C2 subtype, and these differentially expressed ICGs may be intrinsic to the differential prognosis of HCC and may be potential targets for treatment. Even though there are various therapeutic choices available for HCC patients, chemotherapy remains a primary treatment modality for those with advanced HCC. Nevertheless, the efficacy of chemotherapy is yet unreliable. Therefore, it is important to find a method to accurately anticipate HCC patients' chemotherapy responses. We then explored whether there were differences in the sensitivity of patients with two subtypes based on NMRG to commonly used chemotherapeutic agents. We found that the C1 subtype might benefit from Sorafenib, Metformin, Docetaxel, Dasatinib, Erlotinib, and Gefitinib; however, the C2 subtype might benefit from Gemcitabine, Doxorubicin, Cisplatin, Camptothecin, Bortezomib, and Etoposide. It demonstrates how NMRG-based clustering may be a huge help in accurately treating individuals with HCC.

In addition, we used the KEGG pathway enrichment analysis method to investigate the possible molecular biological mechanisms of C1 and C2 subtypes. The results showed that differential genes between subtypes were enriched in a variety of metabolic pathways, such as Tryptophan metabolism, Fatty acid degradation, Arginine and proline metabolism, Glycine, serine, and threonine metabolism, Primary bile acid biosynthesis, Fatty acid metabolism, Tyrosine metabolism, and Pyruvate metabolism among others. Dysregulation of these metabolic processes plays an important role in the development of HCC. Tryptophan catabolism has been reported to be involved in immune tolerance response and to promote response to other anticancer drugs [28]. Furthermore, altered lipid metabolism is increasingly recognized as a marker of tumor occurrence [29], and our enrichment analysis showed that fatty acid metabolic processes were indeed significantly altered. The above results give us an insight that the metabolic processes of the organism are interrelated and related to each other, while an abnormal nucleotide metabolism can lead to reprogramming situations of multiple metabolic processes, and finally jointly induce the occurrence of tumors. Therefore, focusing on the complex metabolic regulatory network may be a novel direction for predicting or treating tumors.

Additionally, to obtain a reliable signature with clinical significance, we screened 32 NMRGs by univariate Cox and LASSO regression analyses and tested the optimized candidate genes for signature development. After verification, a novel NMRG-related prognostic signature was created incorporating 11 genes (i.e., GMPS, UCK2, ENTPD2, PPAT, TXNRD1, RRM2, ATIC, ADSL, ADK, CDA, and DPYS).

Other research studies have examined these 11 genes in numerous cancer forms, some of which have also been examined in HCC. A glutamine amide is used by GMPS to generate the guanine nucleotide as part of the de novo purine biosynthesis process. Previous research has shown that GMPS was crucial to the development of ovarian cancer [30], HCC [31], myeloid [32], prostate cancer [33], etc. UCK2, which can catalyze the phosphorylation of uridine and cytidine to uridine monophosphate and cytidine monophosphate. UCK2 has been proven to enhance the migration and invasion of HCC cells [34], which was also identified to be a latent diagnostic as well as a prognostic indicator for lung cancer [35] and breast cancer [36]. ENTPD2 is regarded as a pivotal ectoenzyme engaged in extracellular ATP hydrolysis [37]. The upregulation of ENTPD2 is present in papillary thyroid carcinoma-derived cells [38], esophageal cancer cells [39], glioma cells [40], and liver cancer cells [41] in comparison to normal cells. While ENTPD2 overexpression was a poor predictor of prognosis for HCC, ENTPD2 inhibition was able to slow the progression of the tumor and improve the effectiveness and efficiency of immune checkpoint inhibitors [41]. PPAT catalyzes the initial committed step of de novo purine nucleotide biosynthesis [42,43], implying that targeting PPAT can serve as a successful cancer strategy [44]. Additionally, PPAT was discovered as a prognostic biomarker in HCC [45]. Modulation of TXNRD1 could influence the proliferation, invasion, and migration of carcinoma [46,47]. TXNRD1 is upregulated in breast cancer, head and neck cancer, and lung cancer, and its overexpression is linked to a bad prognosis [48,49]. By altering the redox balance in vitro, inactivation of TXNRD1 prevented HCC cells from proliferating and led to their apoptosis [50]. Several previous reports indicated that RRM2 functioned in the proliferation, invasion, and metastasis of malignant cells, and as a result, participated in several types of malignant tumors including HCC [51,52]. ATIC, a bifunctional protein enzyme, catalyzes the final two steps of the de novo purine biosynthesis pathway. Studies show that the overexpression of ATIC in HCC is associated with a shorter life expectancy and promotes the growth of HCC cells via controlling the AMPK-mTOR-S6 K1 signature [53]. ADSL, an essential enzyme for de novo purine biosynthesis, is thought to be a novel oncogene in prostate cancer and colorectal carcinoma [54,55]. ADK is a member of the ribokinases family and is an essential enzyme for the elimination of extracellular adenosine by phosphorylating it into 5′-adenosine monophosphate [56]. ADK can influence immune systems and aid in the development of cancer. In addition, lower ADK expression was linked to liver cancer relapse [57]. Gemcitabine became inactive as a result of the deamination of dFdC to dFdU caused by CDA [58]. According to several in vitro studies, overexpressing CDA resulted in gemcitabine resistance, whereas removing CDA restored gemcitabine sensitivity [59,60]. A zinc metalloenzyme, DPYS, which breaks down dihydropyrimidine, is expressed at a high level in tumors in contrast with the matching normal tissues [61]. According to studies, the DPYS subtype DPYSL3 was a potential biomarker for stomach cancer's malignant nature [62].

Utilizing the signature, HCC patients may be successfully classified into the high-risk subgroup with a worse prognosis as well as the low-risk subgroup in the train, test1, test2, and test3 cohorts with a better prognosis. The areas under the ROC curves affirmed that this signature has a good predictive value. Given the possible impact of the tumor immune function on cancer therapy, we evaluated the difference in immune function between two risk subgroups of HCC. The results showed Treg and aDCs were expressed at a high level in the high-risk group, whereas the opposite was true for IFN response and cytolytic activity. To explicitly exploit the signature's prognostic capability, the survival rate of HCC patients was quantitatively examined upon creating a nomogram based on risk score and other clinical features. ROC and calibration curves evaluated the nomogram's predictive potential, showing high accuracy. Finally, we verified the expression of these 11 genes through basic experiments.

However, some drawbacks are related to our research. All RNA sequence data and clinical information were from public databases, such as the TCGA and GEO databases. To develop the predictive significance of our prognostic signature, substantial prospective clinical research is needed. Lastly, the feature was developed using bioinformatics research and preliminary basic experimental analysis was performed, but further genetic functional research is needed to verify our findings.

5. Conclusions

In this study, we successfully obtained a clinical model that can accurately predict the prognosis of HCC patients by using bioinformatics-related analysis methods. The model contains 11 NMRGs, and its expression was verified in subsequent molecular biology experiments. Finally, the nucleotide-related metabolic profile under HCC was verified in patients based on non-targeted metabolomics data. It is expected that the current investigation might provide novel perspectives for clinical management and personalized treatment of HCC patients.

Supplementary Materials: The following supporting information can be downloaded at: https://www.mdpi.com/article/10.3390/metabo13111116/s1; Figure S1: Frequency of single nucleotide variations (SNVs) and various NMRG variant categories; Figure S2: Identifying NMRG-related prognostic DEGs in the TCGA dataset; Figure S3: Rank survey of NMF clustering; Figure S4: Heatmap to demonstrate mRNA levels of NMRGs in two clusters; Figure S5: The link between drug sensitivity and the NMRG clusters; Figure S6: Variable selection; Figure S7: Internal validation of NMRG-related; Figure S8: Relationship between NMRG-related signature and immune function; Figure S9: Supplement to the expression of nucleotide metabolism-related metabolites in patients with hepatocellular carcinoma; Table S1: Gene primer sequence; Table S2: The findings of DEGs between 373 HCC samples and 49 normal samples from the TCGA; Table S3: The findings of HCC samples from the TCGA database that underwent univariate Cox regression analysis; Table S4: The results of DEGs between two NMRGs clusters; Table S5: Baseline data of clinical samples.

Author Contributions: Conception and design of the study/experiments: D.S. and P.Y.; experimental implementation/data acquisition: T.W., Z.W. and M.W.; data analysis and interpretation: T.W., J.L., S.M., Q.Y., A.H. and Z.W.; drafting of the manuscript: T.W., J.L., S.M., D.S. and P.Y. All authors have read and agreed to the published version of the manuscript.

Funding: This work was supported by the key research and development project of Liaoning Province (No. 2018225054).

Institutional Review Board Statement: Human involved serum samples were obtained from the biological sample bank of the First Affiliated Hospital of Dalian Medical University (collected from

November 2016 to December 2019), all the samples were informed consent for scientific research. The study has been approved by the Ethics Committee of the First Affiliated Hospital of Dalian Medical University (No. PJ-KS-KY-2021-129).

Informed Consent Statement: Informed consent was obtained from all subjects involved in the study.

Data Availability Statement: The data presented in this study are available in the main article, further inquiries can be directed to the corresponding authors. The data are not publicly available due to privacy.

Conflicts of Interest: Z.W. and P.Y. are co-founders of iphenome (Yun Pu Kang) Biotechnology Inc. Z.W. is an employee of iphenome (Yun Pu Kang) Biotechnology Inc. The other authors declare no competing interests.

References

1. Llovet, J.M.; Zucman-Rossi, J.; Pikarsky, E.; Sangro, B.; Schwartz, M.; Sherman, M.; Gores, G. Hepatocellular carcinoma. *Nat. Rev. Dis. Primers* **2016**, *2*, 16018. [CrossRef] [PubMed]
2. Xiao, J.; Wang, F.; Wong, N.K.; He, J.; Zhang, R.; Sun, R.; Xu, Y.; Liu, Y.; Li, W.; Koike, K.; et al. Global liver disease burdens and research trends: Analysis from a Chinese perspective. *J. Hepatol.* **2019**, *71*, 212–221. [CrossRef] [PubMed]
3. Sarin, S.K.; Kumar, M.; Eslam, M.; George, J.; Al Mahtab, M.; Akbar, S.; Jia, J.; Tian, Q.; Aggarwal, R.; Muljono, D.H.; et al. Liver diseases in the Asia-Pacific region: A Lancet Gastroenterology & Hepatology Commission. *Lancet Gastroenterol. Hepatol.* **2020**, *5*, 167–228. [PubMed]
4. Xia, C.; Dong, X.; Li, H.; Cao, M.; Sun, D.; He, S.; Yang, F.; Yan, X.; Zhang, S.; Li, N.; et al. Cancer statistics in China and United States, 2022: Profiles, trends, and determinants. *Chin. Med. J.* **2022**, *135*, 584–590. [CrossRef]
5. Rathbone, M.P.; Middlemiss, P.J.; Kim, J.K.; Gysbers, J.W.; DeForge, S.P.; Smith, R.W.; Hughes, D.W. Adenosine and its nucleotides stimulate proliferation of chick astrocytes and human astrocytoma cells. *Neurosci. Res.* **1992**, *13*, 1–17. [CrossRef]
6. Vander Heiden, M.G.; DeBerardinis, R.J. Understanding the Intersections between Metabolism and Cancer Biology. *Cell* **2017**, *168*, 657–669. [CrossRef]
7. Pavlova, N.N.; Thompson, C.B. The Emerging Hallmarks of Cancer Metabolism. *Cell Metab.* **2016**, *23*, 27–47. [CrossRef]
8. Ma, J.; Zhong, M.; Xiong, Y.; Gao, Z.; Wu, Z.; Liu, Y.; Hong, X. Emerging roles of nucleotide metabolism in cancer development: Progress and prospect. *Aging* **2021**, *13*, 13349–13358. [CrossRef]
9. Lee, J.S.; Adler, L.; Karathia, H.; Carmel, N.; Rabinovich, S.; Auslander, N.; Keshet, R.; Stettner, N.; Silberman, A.; Agemy, L.; et al. Urea Cycle Dysregulation Generates Clinically Relevant Genomic and Biochemical Signatures. *Cell* **2018**, *174*, 1559–1570. [CrossRef]
10. Keshet, R.; Lee, J.S.; Adler, L.; Iraqi, M.; Ariav, Y.; Lim, L.; Lerner, S.; Rabinovich, S.; Oren, R.; Katzir, R.; et al. Targeting purine synthesis in ASS1-expressing tumors enhances the response to immune checkpoint inhibitors. *Nat. Cancer* **2020**, *1*, 894–908. [CrossRef]
11. Kepp, O.; Loos, F.; Liu, P.; Kroemer, G. Extracellular nucleosides and nucleotides as immunomodulators. *Immunol. Rev.* **2017**, *280*, 83–92. [CrossRef] [PubMed]
12. Kumar, V. Adenosine as an endogenous immunoregulator in cancer pathogenesis: Where to go. *Purinergic Signal* **2013**, *9*, 145–165. [CrossRef] [PubMed]
13. Chen, C.; Chen, H.; Zhang, Y.; Thomas, H.R.; Frank, M.H.; He, Y.; Xia, R. TBtools: An Integrative Toolkit Developed for Interactive Analyses of Big Biological Data. *Mol. Plant* **2020**, *13*, 1194–1202. [CrossRef] [PubMed]
14. Geeleher, P.; Cox, N.; Huang, R.S. pRRophetic: An R package for prediction of clinical chemotherapeutic response from tumor gene expression levels. *PLoS ONE* **2014**, *9*, e107468. [CrossRef]
15. Shen, H.; Laird, P.W. Interplay between the cancer genome and epigenome. *Cell* **2013**, *153*, 38–55. [CrossRef]
16. Ferlay, J.; Soerjomataram, I.; Dikshit, R.; Eser, S.; Mathers, C.; Rebelo, M.; Parkin, D.M.; Forman, D.; Bray, F. Cancer incidence and mortality worldwide: Sources, methods and major patterns in GLOBOCAN 2012. *Int. J. Cancer* **2015**, *136*, E359–E386. [CrossRef]
17. Engel, J.; Emeny, R.T.; Hölzel, D. Positive lymph nodes do not metastasize. *Cancer Metastasis Rev.* **2012**, *31*, 235–246. [CrossRef]
18. Wang, W.; Kandimalla, R.; Huang, H.; Zhu, L.; Li, Y.; Gao, F.; Goel, A.; Wang, X. Molecular subtyping of colorectal cancer: Recent progress, new challenges and emerging opportunities. *Semin. Cancer Biol.* **2019**, *55*, 37–52. [CrossRef]
19. Hanahan, D.; Weinberg, R.A. Hallmarks of cancer: The next generation. *Cell* **2011**, *144*, 646–674. [CrossRef]
20. Panther, E.; Corinti, S.; Idzko, M.; Herouy, Y.; Napp, M.; la Sala, A.; Girolomoni, G.; Norgauer, J. Adenosine affects expression of membrane molecules, cytokine and chemokine release, and the T-cell stimulatory capacity of human dendritic cells. *Blood* **2003**, *101*, 3985–3990. [CrossRef]
21. Chen, H.; Liu, H.; Qing, G. Targeting oncogenic Myc as a strategy for cancer treatment. *Signal Transduct. Target Ther.* **2018**, *3*, 5. [CrossRef] [PubMed]
22. Dang, C.V.; Le, A.; Gao, P. MYC-induced cancer cell energy metabolism and therapeutic opportunities. *Clin. Cancer Res.* **2009**, *15*, 6479–6483. [CrossRef] [PubMed]

23. Pópulo, H.; Lopes, J.M.; Soares, P. The mTOR signalling pathway in human cancer. *Int. J. Mol. Sci.* **2012**, *13*, 1886–1918. [CrossRef] [PubMed]
24. Kent, L.N.; Rakijas, J.B.; Pandit, S.K.; Westendorp, B.; Chen, H.Z.; Huntington, J.T.; Tang, X.; Bae, S.; Srivastava, A.; Senapati, S.; et al. E2f8 mediates tumor suppression in postnatal liver development. *J. Clin. Investig.* **2016**, *126*, 2955–2969. [CrossRef] [PubMed]
25. Kent, L.N.; Bae, S.; Tsai, S.Y.; Tang, X.; Srivastava, A.; Koivisto, C.; Martin, C.K.; Ridolfi, E.; Miller, G.C.; Zorko, S.M.; et al. Dosage-dependent copy number gains in E2f1 and E2f3 drive hepatocellular carcinoma. *J. Clin. Investig.* **2017**, *127*, 830–842. [CrossRef] [PubMed]
26. Lan, W.; Bian, B.; Xia, Y.; Dou, S.; Gayet, O.; Bigonnet, M.; Santofimia-Castaño, P.; Cong, M.; Peng, L.; Dusetti, N.; et al. E2F signature is predictive for the pancreatic adenocarcinoma clinical outcome and sensitivity to E2F inhibitors, but not for the response to cytotoxic-based treatments. *Sci. Rep.* **2018**, *8*, 8330. [CrossRef]
27. Motegi, A.; Masutani, M.; Yoshioka, K.I.; Bessho, T. Aberrations in DNA repair pathways in cancer and therapeutic significances. *Semin. Cancer Biol.* **2019**, *58*, 29–46. [CrossRef]
28. Brochez, L.; Chevolet, I.; Kruse, V. The rationale of indoleamine 2,3-dioxygenase inhibition for cancer therapy. *Eur. J. Cancer* **2017**, *76*, 167–182. [CrossRef]
29. Glaysher, J. Lipid metabolism and cancer. *Curr. Opin. Lipidol.* **2013**, *24*, 530–531. [CrossRef]
30. Wang, P.; Zhang, Z.; Ma, Y.; Lu, J.; Zhao, H.; Wang, S.; Tan, J.; Li, B. Prognostic values of GMPS, PR, CD40, and p21 in ovarian cancer. *PeerJ* **2019**, *7*, e6301. [CrossRef]
31. Yin, L.; He, N.; Chen, C.; Zhang, N.; Lin, Y.; Xia, Q. Identification of novel blood-based HCC-specific diagnostic biomarkers for human hepatocellular carcinoma. *Artif. Cells Nanomed. Biotechnol.* **2019**, *47*, 1908–1916. [CrossRef] [PubMed]
32. Chen, X.; Burkhardt, D.B.; Hartman, A.A.; Hu, X.; Eastman, A.E.; Sun, C.; Wang, X.; Zhong, M.; Krishnaswamy, S.; Guo, S. MLL-AF9 initiates transformation from fast-proliferating myeloid progenitors. *Nat. Commun.* **2019**, *10*, 5767. [CrossRef] [PubMed]
33. Wang, Q.; Guan, Y.F.; Hancock, S.E.; Wahi, K.; van Geldermalsen, M.; Zhang, B.K.; Pang, A.; Nagarajah, R.; Mak, B.; Freidman, N.; et al. Inhibition of guanosine monophosphate synthetase (GMPS) blocks glutamine metabolism and prostate cancer growth. *J. Pathol.* **2021**, *254*, 135–146. [CrossRef] [PubMed]
34. Zhou, Q.; Jiang, H.; Zhang, J.; Yu, W.; Zhou, Z.; Huang, P.; Wang, J.; Xiao, Z. Uridine-cytidine kinase 2 promotes metastasis of hepatocellular carcinoma cells via the Stat3 pathway. *Cancer Manag. Res.* **2018**, *10*, 6339–6355. [CrossRef]
35. Wu, Y.; Jamal, M.; Xie, T.; Sun, J.; Song, T.; Yin, Q.; Li, J.; Pan, S.; Zeng, X.; Xie, S.; et al. Uridine-cytidine kinase 2 (UCK2): A potential diagnostic and prognostic biomarker for lung cancer. *Cancer Sci.* **2019**, *110*, 2734–2747. [CrossRef]
36. Shen, G.; He, P.; Mao, Y.; Li, P.; Luh, F.; Ding, G.; Liu, X.; Yen, Y. Overexpression of Uridine-Cytidine Kinase 2 Correlates with Breast Cancer Progression and Poor Prognosis. *J. Breast Cancer* **2017**, *20*, 132–141. [CrossRef]
37. Lua, I.; Li, Y.; Zagory, J.A.; Wang, K.S.; French, S.W.; Sévigny, J.; Asahina, K. Characterization of hepatic stellate cells, portal fibroblasts, and mesothelial cells in normal and fibrotic livers. *J. Hepatol.* **2016**, *64*, 1137–1146. [CrossRef]
38. Bertoni, A.; de Campos, R.P.; Tsao, M.; Braganhol, E.; Furlanetto, T.W.; Wink, M.R. Extracellular ATP is Differentially Metabolized on Papillary Thyroid Carcinoma Cells Surface in Comparison to Normal Cells. *Cancer Microenviron.* **2018**, *11*, 61–70. [CrossRef]
39. Santos, A.A., Jr.; Cappellari, A.R.; de Marchi, F.O.; Gehring, M.P.; Zaparte, A.; Brandão, C.A.; Lopes, T.G.; da Silva, V.D.; Pinto, L.; Savio, L.; et al. Potential role of P2X7R in esophageal squamous cell carcinoma proliferation. *Purinergic Signal.* **2017**, *13*, 279–292. [CrossRef]
40. Braganhol, E.; Zanin, R.F.; Bernardi, A.; Bergamin, L.S.; Cappellari, A.R.; Campesato, L.F.; Morrone, F.B.; Campos, M.M.; Calixto, J.B.; Edelweiss, M.I.; et al. Overexpression of NTPDase2 in gliomas promotes systemic inflammation and pulmonary injury. *Purinergic Signal.* **2012**, *8*, 235–243. [CrossRef]
41. Chiu, D.K.; Tse, A.P.; Xu, I.M.; Di Cui, J.; Lai, R.K.; Li, L.L.; Koh, H.Y.; Tsang, F.H.; Wei, L.L.; Wong, C.M.; et al. Hypoxia inducible factor HIF-1 promotes myeloid-derived suppressor cells accumulation through ENTPD2/CD39L1 in hepatocellular carcinoma. *Nat. Commun.* **2017**, *8*, 517. [CrossRef] [PubMed]
42. Iwahana, H.; Oka, J.; Mizusawa, N.; Kudo, E.; Ii, S.; Yoshimoto, K.; Holmes, E.W.; Itakura, M. Molecular cloning of human amidophosphoribosyltransferase. *Biochem. Biophys. Res. Commun.* **1993**, *190*, 192–200. [CrossRef] [PubMed]
43. Yamaoka, T.; Kondo, M.; Honda, S.; Iwahana, H.; Moritani, M.; Ii, S.; Yoshimoto, K.; Itakura, M. Amidophosphoribosyltransferase limits the rate of cell growth-linked de novo purine biosynthesis in the presence of constant capacity of salvage purine biosynthesis. *J. Biol. Chem.* **1997**, *272*, 17719–17725. [CrossRef] [PubMed]
44. Bibi, N.; Parveen, Z.; Nawaz, M.S.; Kamal, M.A. In Silico Structure Modeling and Molecular Docking Analysis of Phosphoribosyl Pyrophosphate Amidotransferase (PPAT) with Antifolate Inhibitors. *Curr. Cancer Drug Targets* **2019**, *19*, 408–416. [CrossRef] [PubMed]
45. Hu, X.; Bao, M.; Huang, J.; Zhou, L.; Zheng, S. Identification and Validation of Novel Biomarkers for Diagnosis and Prognosis of Hepatocellular Carcinoma. *Front. Oncol.* **2020**, *10*, 541479. [CrossRef] [PubMed]
46. Huang, S.; Zhu, X.; Ke, Y.; Xiao, D.; Liang, C.; Chen, J.; Chang, Y. LncRNA FTX inhibition restrains osteosarcoma proliferation and migration via modulating miR-320a/TXNRD1. *Cancer Biol. Ther.* **2020**, *21*, 379–387. [CrossRef]
47. Hua, S.; Quan, Y.; Zhan, M.; Liao, H.; Li, Y.; Lu, L. miR-125b-5p inhibits cell proliferation, migration, and invasion in hepatocellular carcinoma via targeting TXNRD1. *Cancer Cell Int.* **2019**, *19*, 203. [CrossRef]

48. Bhatia, M.; McGrath, K.L.; Di Trapani, G.; Charoentong, P.; Shah, F.; King, M.M.; Clarke, F.M.; Tonissen, K.F. The thioredoxin system in breast cancer cell invasion and migration. *Redox Biol.* **2016**, *8*, 68–78. [CrossRef]
49. Leone, A.; Roca, M.S.; Ciardiello, C.; Costantini, S.; Budillon, A. Oxidative Stress Gene Expression Profile Correlates with Cancer Patient Poor Prognosis: Identification of Crucial Pathways Might Select Novel Therapeutic Approaches. *Oxid. Med. Cell. Longev.* **2017**, *2017*, 2597581. [CrossRef]
50. Lee, D.; Xu, I.M.; Chiu, D.K.; Leibold, J.; Tse, A.P.; Bao, M.H.; Yuen, V.W.; Chan, C.Y.; Lai, R.K.; Chin, D.W.; et al. Induction of Oxidative Stress through Inhibition of Thioredoxin Reductase 1 Is an Effective Therapeutic Approach for Hepatocellular Carcinoma. *Hepatology* **2019**, *69*, 1768–1786. [CrossRef]
51. Wang, N.; Zhan, T.; Ke, T.; Huang, X.; Ke, D.; Wang, Q.; Li, H. Increased expression of RRM2 by human papillomavirus E7 oncoprotein promotes angiogenesis in cervical cancer. *Br. J. Cancer* **2014**, *110*, 1034–1044. [CrossRef] [PubMed]
52. Tian, H.; Ge, C.; Li, H.; Zhao, F.; Hou, H.; Chen, T.; Jiang, G.; Xie, H.; Cui, Y.; Yao, M.; et al. Ribonucleotide reductase M2B inhibits cell migration and spreading by early growth response protein 1-mediated phosphatase and tensin homolog/Akt1 pathway in hepatocellular carcinoma. *Hepatology* **2014**, *59*, 1459–1470. [CrossRef] [PubMed]
53. Li, M.; Jin, C.; Xu, M.; Zhou, L.; Li, D.; Yin, Y. Bifunctional enzyme ATIC promotes propagation of hepatocellular carcinoma by regulating AMPK-mTOR-S6 K1 signaling. *Cell Commun. Signal.* **2017**, *15*, 52. [CrossRef] [PubMed]
54. Liao, J.; Song, Q.; Li, J.; Du, K.; Chen, Y.; Zou, C.; Mo, Z. Carcinogenic effect of adenylosuccinate lyase (ADSL) in prostate cancer development and progression through the cell cycle pathway. *Cancer Cell Int.* **2021**, *21*, 46. [CrossRef]
55. Taha-Mehlitz, S.; Bianco, G.; Coto-Llerena, M.; Kancherla, V.; Bantug, G.R.; Gallon, J.; Ercan, C.; Panebianco, F.; Eppenberger-Castori, S.; von Strauss, M.; et al. Adenylosuccinate lyase is oncogenic in colorectal cancer by causing mitochondrial dysfunction and independent activation of NRF2 and mTOR-MYC-axis. *Theranostics* **2021**, *11*, 4011–4029. [CrossRef] [PubMed]
56. Park, J.; Gupta, R.S. Adenosine kinase and ribokinase--the RK family of proteins. *Cell. Mol. Life Sci.* **2008**, *65*, 2875–2896. [CrossRef]
57. Zhulai, G.; Oleinik, E.; Shibaev, M.; Ignatev, K. Adenosine-Metabolizing Enzymes, Adenosine Kinase and Adenosine Deaminase, in Cancer. *Biomolecules* **2022**, *12*, 418. [CrossRef]
58. Shipley, L.A.; Brown, T.J.; Cornpropst, J.D.; Hamilton, M.; Daniels, W.D.; Culp, H.W. Metabolism and disposition of gemcitabine, and oncolytic deoxycytidine analog, in mice, rats, and dogs. *Drug Metab. Dispos.* **1992**, *20*, 849–855.
59. Weizman, N.; Krelin, Y.; Shabtay-Orbach, A.; Amit, M.; Binenbaum, Y.; Wong, R.J.; Gil, Z. Macrophages mediate gemcitabine resistance of pancreatic adenocarcinoma by upregulating cytidine deaminase. *Oncogene* **2014**, *33*, 3812–3819. [CrossRef]
60. Maréchal, R.; Bachet, J.B.; Mackey, J.R.; Dalban, C.; Demetter, P.; Graham, K.; Couvelard, A.; Svrcek, M.; Bardier-Dupas, A.; Hammel, P.; et al. Levels of gemcitabine transport and metabolism proteins predict survival times of patients treated with gemcitabine for pancreatic adenocarcinoma. *Gastroenterology* **2012**, *143*, 664–674. [CrossRef]
61. Basbous, J.; Aze, A.; Chaloin, L.; Lebdy, R.; Hodroj, D.; Ribeyre, C.; Larroque, M.; Shepard, C.; Kim, B.; Pruvost, A.; et al. Dihydropyrimidinase protects from DNA replication stress caused by cytotoxic metabolites. *Nucleic Acids Res.* **2020**, *48*, 1886–1904. [CrossRef] [PubMed]
62. Kanda, M.; Nomoto, S.; Oya, H.; Shimizu, D.; Takami, H.; Hibino, S.; Hashimoto, R.; Kobayashi, D.; Tanaka, C.; Yamada, S.; et al. Dihydropyrimidinase-like 3 facilitates malignant behavior of gastric cancer. *J. Exp. Clin. Cancer Res.* **2014**, *33*, 66. [CrossRef] [PubMed]

Disclaimer/Publisher's Note: The statements, opinions and data contained in all publications are solely those of the individual author(s) and contributor(s) and not of MDPI and/or the editor(s). MDPI and/or the editor(s) disclaim responsibility for any injury to people or property resulting from any ideas, methods, instructions or products referred to in the content.

Article

A New Biomarker Profiling Strategy for Gut Microbiome Research: Valid Association of Metabolites to Metabolism of Microbiota Detected by Non-Targeted Metabolomics in Human Urine

Sijia Zheng [1,2], Lina Zhou [1], Miriam Hoene [3], Andreas Peter [3,4,5], Andreas L. Birkenfeld [4,5,6], Cora Weigert [3,4,5], Xinyu Liu [1], Xinjie Zhao [1], Guowang Xu [1,*] and Rainer Lehmann [3,4,5,*]

[1] CAS Key Laboratory of Separation Science for Analytical Chemistry, Dalian Institute of Chemical Physics, Chinese Academy of Sciences, Dalian 116023, China; zhengsj@dicp.ac.cn (S.Z.); zhouln@dicp.ac.cn (L.Z.); liuxy2012@dicp.ac.cn (X.L.); xj_zhao@dicp.ac.cn (X.Z.)
[2] University of Chinese Academy of Sciences, Beijing 100049, China
[3] Institute for Clinical Chemistry and Pathobiochemistry, Department for Diagnostic Laboratory Medicine, University Hospital Tübingen, 72076 Tuebingen, Germany; miriam.hoene@med.uni-tuebingen.de (M.H.); andreas.peter@med.uni-tuebingen.de (A.P.); cora.weigert@med.uni-tuebingen.de (C.W.)
[4] Institute for Diabetes Research and Metabolic Diseases of the Helmholtz Zentrum München at the University of Tübingen, 72076 Tübingen, Germany; andreas.birkenfeld@med.uni-tuebingen.de
[5] German Center for Diabetes Research (DZD), 90451 Neuherberg, Germany
[6] Internal Medicine 4, University Hospital Tuebingen, 72076 Tuebingen, Germany
* Correspondence: xugw@dicp.ac.cn (G.X.); rainer.lehmann@med.uni-tuebingen.de (R.L.); Tel.: +86-411-84379530 (G.X.); +49-7071-2983193 (R.L.)

Abstract: The gut microbiome is of tremendous relevance to human health and disease, so it is a hot topic of omics-driven biomedical research. However, a valid identification of gut microbiota-associated molecules in human blood or urine is difficult to achieve. We hypothesize that bowel evacuation is an easy-to-use approach to reveal such metabolites. A non-targeted and modifying group-assisted metabolomics approach (covering 40 types of modifications) was applied to investigate urine samples collected in two independent experiments at various time points before and after laxative use. Fasting over the same time period served as the control condition. As a result, depletion of the fecal microbiome significantly affected the levels of 331 metabolite ions in urine, including 100 modified metabolites. Dominating modifications were glucuronidations, carboxylations, sulfations, adenine conjugations, butyrylations, malonylations, and acetylations. A total of 32 compounds, including common, but also unexpected fecal microbiota-associated metabolites, were annotated. The applied strategy has potential to generate a microbiome-associated metabolite map (M3) of urine from healthy humans, and presumably also other body fluids. Comparative analyses of M3 vs. disease-related metabolite profiles, or therapy-dependent changes may open promising perspectives for human gut microbiome research and diagnostics beyond analyzing feces.

Keywords: microbiome; gut flora; metabolomics; metabolites; urine; diagnosis; profiling; gut microbiota

1. Introduction

The gut microbiota very closely interacts with its human host and influences human health [1]. A continuously increasing number of reports show an important role of the gut microbiome in disease development, but also for recovery from diseases, for remission, as well as for disease prevention [2–5]. Consequently, the luminal (fecal) and mucosal gut microbiota has been intensively investigated in animal models and humans in a comprehensive manner, applying various omics approaches [6]. These studies are first and foremost performed in feces [6,7]. As a result, a tremendous increase in knowledge has been achieved, for instance, about nutritional effects on the microbiome, the pathophysiological consequences of a disbalance of bacterial phyla (e.g., in metabolic diseases), and the role of distinct bacterial species in health and diseases [4,7–11].

These investigations were paralleled by efforts to detect and identify compounds involved in the crosstalk between the gut microbiota and human cells, tissues, or organs [12–15]. Comprehensive investigations of these metabolites have just started [16–18]. Most of the reported metabolites were studied in feces and blood, and just a few in urine [19–23].

Urine is a non-invasively collected sample material. It stands in direct connectivity to blood, since metabolites from the gut microbiome passing the intestinal wall can be transported via the splanchnic bed and the mesenteric veins to the liver, and then partially filtered blood passes through the hepatic veins into the systemic circulation, including the kidneys. Consequently, it could be a useful and easy-to-collect biospecimen to study gut microbiome-associated metabolites. However, the unequivocal linking of a detected metabolite in a body fluid, like blood or urine, to the gut microbiota is quite challenging.

Aiming to contribute to the efforts of closing this gap in knowledge, we hypothesized that a bowel evacuation would change the levels of metabolites associated with the fecal microbiota, thereby enabling the detection of these compounds in human urine. To test this hypothesis, in the current study, we investigated urine samples collected before and after a bowel evacuation, applying a comprehensive, non-targeted, and modifying group-assisted metabolomics approach.

2. Materials and Methods

2.1. Study Design

The first experiment was performed over 10 days, collecting urine samples at 40 time points before and after a bowel evacuation in an individual self-experiment (n = 1), as well as during the same period after starting again to consume food. Bowel evacuation, as preparation for a colonoscopy as a healthcare check, was achieved using CitraFleet® (sodium picosulfate, sodium citrate) and Tirgon® (Bisacodyl), both from Recordati Pharma (Ulm, Germany). Propofol anesthesia was applied during the colonoscopy. A procedure-related 48 h fasting period was included. Therefore, as a control, a similar experiment was performed, including 48 h of fasting but without bowel evacuation. This experiment ran for 9 days and 30 urine samples were collected. Both urine sample sets consisted of the 1st and 2nd morning urine, as well as spot urine, since urine samples were collected at various time points during the day and night. In a subsequent experiment, 6 healthy volunteers (age: 25–56 years; two females and four males) performed the same bowel evacuation with a total fasting period of 24 h before refeeding (12 h bowel evacuation and a preceding fasting period of 12 h). Urine samples were collected at 4 time points. The first urine sample was taken immediately after waking up at 7.00 am after a 12 h overnight fasting period (1st morning urine). The second sample was the 2nd morning urine, collected at 8.00 am, directly before the start of the bowel evacuation. The third and fourth samples were collected 10 h and 12 h after bowel evacuation, respectively. Furthermore, three out of the six volunteers (age: 25–56 years; one female and two males) performed in addition a 24 h "fasting-only" experiment, as a control. All urine samples were stored at −80 °C until further processing and analyses. The study was conducted according to the Declaration of Helsinki of 1964 and its later amendments. The ethics committee of the University of Tuebingen approved the protocol (188/2017BO2). All volunteers provided written informed consent before the start of the study.

2.2. Sample Preparation

Urine was thawed on ice, vortexed, and an internal standard (IS) mix was added (v/v, 1:10) containing the following 13 ISs: carnitine C2:0-d3, carnitine C6:0-d3, carnitine C10:0-d3, leucine-d3, phenylalanine-d5, tryptophan-d5, cholic acid-2,2,4,4-d4, chenodeoxycholic acid-2,2,4,4-d4, leucine enkephalin, indoxyl sulfate-[$^{13}C_6$], L-valine-d8, sodium-2-hydroxybutyrate-2,3,3-d3, and L-4-hydroxyphenyl-d4-alanine (details in Table S1). After the addition of the IS-mix, the samples were vortexed again, centrifuged at 18,000× g (4 °C for 10 min), and subsequently 100 μL of the supernatant was evaporated. For mass spectrometric analysis, samples were dissolved in 300 μL water:acetonitrile (v/v, 95:5).

2.3. Metabolites Profiling by Liquid-Chromatography Mass Spectrometry (LC-MS)

Non-targeted profiling was performed with an Ultra Performance Liquid Chromatography system (UPLC, Waters Corporation, Manchester, UK) coupled to a Triple TOF 5600+ mass spectrometer (AB SCIEX, Framingham, MA, USA). Chromatographic separation was performed using ACQUITY HSS T3 column (2.1 × 100 mm, 1.8 μm, Waters, Milford, MA, USA). The mobile phases were water (A) and acetonitrile (B) acidified by 0.1% formic acid, respectively. The flow rate was 0.35 mL/min and the total run time was 26 min. The elution gradient initiated with 5% B for 1 min, linearly increased to 50% B at 18 min, then increased to 100% B after 0.5 min, maintained for 4 min, then went back to 5% B after 0.5 min and maintained for 3 min for post-equilibrium. The injection volume was 5 μL. Column temperature was set at 40 °C.

For the MS instrument, full MS-ddMS^2 mode was used with mass ranges of m/z 50–1000 Da and 30–1000 Da, respectively. Accumulation times for full scan and ddMS^2 acquisition modes were 0.25 s and 0.03 s, respectively. Cycle time was 0.75 s. Declustering potential was set at 90. Both electrospray positive ion (ESI$^+$) and negative ion (ESI$^-$) modes were used. Electrospray voltages were set at 4.5 kV for ESI$^+$ mode and −4.0 kV for ESI$^-$ mode; ionspray temperature was set at 500 °C; ion source gas1 was 50 psi; ion source gas2 was 50 psi; and curtain gas was 35 psi. For dd-MS^2, collision energies of 15 V, 30 V, and 45 V were applied and MS/MS fragmentation patterns of the 15 most intense ions in full scan were acquired. Every sixth sample was followed by a quality control (QC) analysis of pooled urine.

2.4. Data Processing

Peak detection and alignment were conducted by MarkerView software 1.2.1(AB SCIEX, Framingham, MA, USA). The parameters for peak detection were as follows: minimum spectral peak width of 10 ppm, minimum retention time (t_R) peak width of 5 scans, and noise threshold of 1000. For peak alignment, a t_R tolerance of 0.5 min and mass tolerance of 10 ppm were used. After applying "modified 80% rule" to remove missing values and then removing isotope ions [24], the intensity of each peak was normalized to an appropriate IS. Only peaks with relative standard deviation (RSD) of responses in QC samples less than 30% were kept for subsequent statistical analysis. Creatinine concentrations were further used to normalize the responses of features.

Before statistical analysis, the relative peak response of each metabolite at time point 1 was set to 100%. Then we compared the relative responses at time point 3 (10 h after bowel evacuation) in the bowel evacuation group (n = 6) versus the only fasting group (n = 3) using the two-tailed unpaired t test. $p < 0.05$ indicated that the difference was statistically significant. The heatmap of differential metabolites was obtained by Multi-Experiment viewer 4.9.0.

2.5. Metabolite Annotation

Metabolite annotation was firstly carried out with the OSI/SMMS software 2.4.1. In brief, the accurate MS and MS/MS spectra of ion features in analyzed samples were searched against an in-house database containing comprehensive qualitative information of more than 2000 reference chemical standards [25]. For modification-type determination of other gut microbiota-associated ion features unannotated, we used MS2Analyzer software to search for characteristic neutral losses of 40 typical metabolite modifications in human urine in the extracted MS^2 spectra by setting parameters of the m/z window as 0.005 Da and the intensity threshold as 0.1 [26].

3. Results

3.1. Distinct Metabolites Are Decreased Subsequent to Laxative-Induced Bowel Evacuation

Samples from the self-experiment were analyzed by non-targeted metabolomics to test our hypothesis of the detection of fecal microbiota-associated metabolites in human urine by a comparison of samples collected before and after a laxative-induced bowel evacuation. Exemplarily, Figure 1A shows at 40 different time points during a 10-day period the time courses of the levels of phenylacetylglutamine, hippuric acid, p-cresol glucuronide, glutamine, and glutamate. A persistent decrease after the bowel evacuation until the start of refeeding is clearly visible. To exclude the possibility that the detected decrease was caused by fasting, since the procedure of bowel evacuation entailed a 48 h fasting period, the same male individual performed a second self-experiment lasting 9 days, which included a "fasting-only" phase of 48 h (Figure 1D). Differences in the time courses of metabolite signal intensities between the bowel evacuation and the exclusively fasting experiment are obvious (Figure 1). Anesthesia by propofol was included in the period of the bowel evacuation, because a colonoscopy was executed in the scope of a health check. Therefore, a possible propofol effect on metabolite levels could not be excluded at this time point.

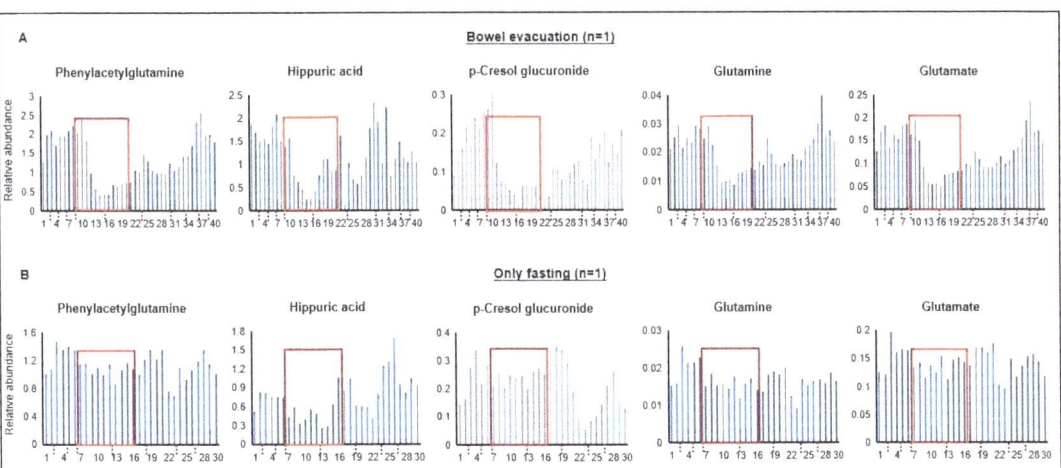

Figure 1. (**A**) Time courses of levels of exemplarily selected metabolites in human urine over a 10-day period before and during a laxative-induced bowel evacuation, and after starting refeeding. The red rectangles mark the 48 h period without food consumption and the dash dotted lines on the x-axes separate the different days. In total, 40 urine samples were collected (1st and 2nd morning urine, as well as spot urine) throughout the whole day (sample numbers are provided on the x-axes). The experiment was conducted as self-experiment from one male individual. (**B**) Control experiment, i.e., 9-day time courses of the levels of the same metabolites before and during a 48 h fasting period, and after starting refeeding. The fasting period is marked by red rectangle. In total, 30 urine samples were collected all day from the same individual. The x-axes show the different days and the y-axes the relative peak responses in arbitrary units.

Based on this single individuum experiment, we concluded that the experimental design was suitable. In addition, possible time points for sample collection after bowel evacuation ($\geq$10 h) could be extracted from the achieved data for subsequent studies.

3.2. Confirmation of the Findings of Luminal (Fecal) Microbiota-Associated Metabolites in Human Urine

Next, we aimed to (a) confirm the preceding findings, (b) exclude possible propofol effects on the findings, and (c) shorten the fasting period to adjust the sample collection to the usual procedure during a common health check colonoscopy. This laxative-induced bowel evacuation experiment was performed by six volunteers but without subsequent colonoscopy, i.e., without propofol anesthesia. The time from the start of bowel evacuation until refeeding was reduced to 12 h and the total fasting time was reduced to 24 h. Urine samples were collected at four time points. A scheme illustrating the experimental design and sample collection time points is provided in Figure 2A. Furthermore, a "fasting-only" experiment with identical sample collection time points was performed by three volunteers as a control experiment. Figure 2B shows the confirmation of all findings achieved in the n = 1 experiment (Figure 1). Based on these findings we could not only validate our preceding results, but could also exclude effects of propofol, as well as confirm the time frame of 10–12 h after bowel evacuation for sample collection as well-suited.

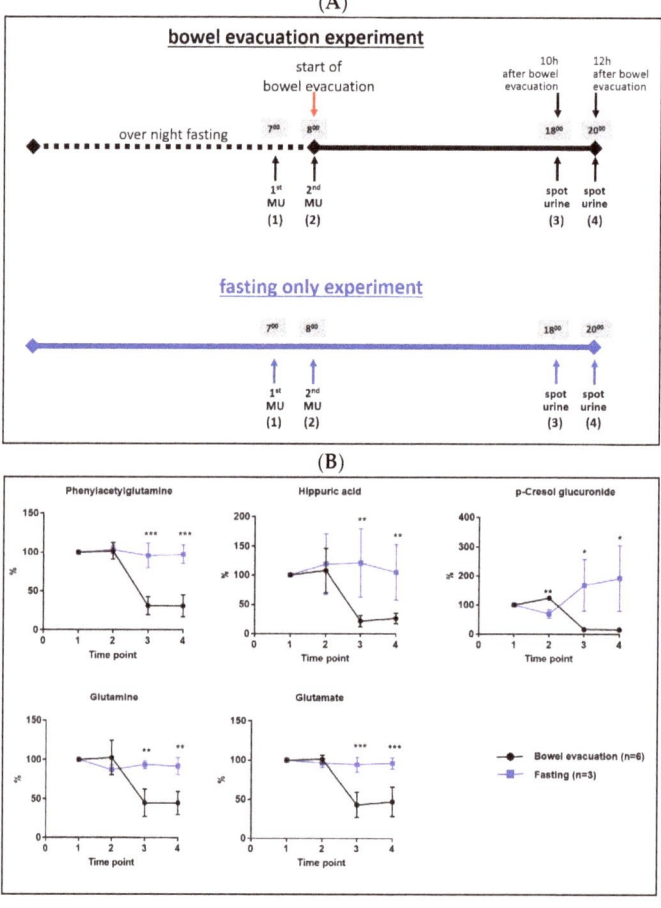

Figure 2. (**A**) Scheme of the experimental design and sample collection time points. (**B**) Metabolite levels in human urine collected at four time points before and after a bowel evacuation including 24 h fasting period (n = 6, black lines), and only fasting for 24 h (n = 3, blue lines). Time point 1: 1st morning urine; time point 2: 2nd morning urine, collected directly before the start of the bowel evacuation; time point 3: collected 10 h after bowel evacuation; time point 4: collected 12 h after bowel evacuation. Bars represent mean ± SD; the student's *t*-test between groups: * $p < 0.05$, ** $p < 0.01$, *** $p < 0.001$.

3.3. A Considerable Number of Metabolites in Human Urine Are Associated to the Luminal (Fecal) Microbiome

Next, we evaluated the data of all covered metabolite ion masses detected by non-targeted metabolomics analysis. After LC-MS data pretreatment, 7501 features remained. Around 4% (331 metabolite ion masses) were significantly altered by bowel evacuation and were therefore labeled as associated with the fecal microbiota. The majority of these metabolites were decreased after bowel evacuation, suggesting a production or transformation of these compounds by luminal gut microbes (Figure 3A). Profiling 40 different kinds of modifications led to the detection of 94 modified metabolites among these 310 fecal microbiota-associated metabolites that decreased after bowel evacuation (details in Table S2). Glucuronidation dominated these modifications (36%), followed by carboxylation (26%), sulfation (5%), adenine conjugation (4%), butyrylation (4%), malonylation (3%), acetylation (2%), and other modification types (20%). Interestingly, 21 fecal microbiome-associated metabolites, including six modified metabolite signals, were increased in comparison to the "fasting-only" control experiment (Figure 3B and Table S2), which may imply that gut microbes contribute to the suppression of their levels.

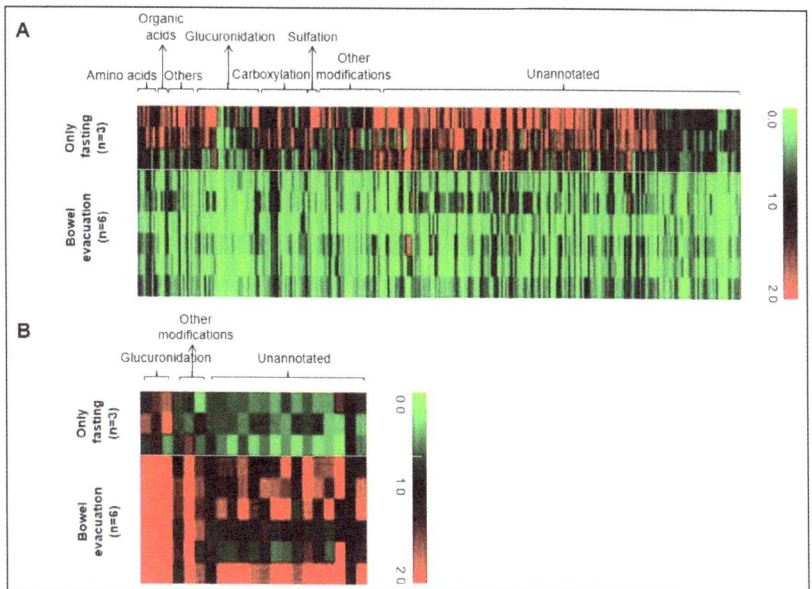

Figure 3. (**A**) Heat map of 310 metabolites in human urine showing significantly decreased metabolite levels after laxative-induced bowel evacuation (n = 6) in comparison to only fasting (n = 3). (**B**) Significantly increased levels of 21 metabolites after bowel evacuation (n = 6) in comparison to only fasting (n = 3). A significant difference was defined as $p < 0.05$ in a two-tailed unpaired t test comparing relative responses at time point 3 (10 h after bowel evacuation versus the only fasting group at the same time point). In the heat map, each urinary metabolite is represented by a single column. Rows represent different individuals. Black is the intensity at time point 1, green labels show decreased signal intensities, and red labels show increased signal intensities.

Among the 331 luminal microbiota-associated metabolite features, 32 were structurally elucidated and 6 were confirmed by standard compounds. Table 1 provides a list of annotated fecal microbiota-associated metabolites detected in human urine.

Table 1. Annotated fecal microbiota-associated metabolites. Identity is either based on confirmation with a standard compound or putative annotation based on exact mass (MS^1) and fragmentation patterns (MS^2).

No.	Metabolites	Annotation Base	Category	Selected Microbiota-Related References
1	Phenylalanine	a, b, c	Amino acid	[8]
2	Glutamine	a, b	Amino acid	[12,27]
3	Glutamate	a, b	Amino acid	[12,27]
4	Methionine	a, b	Amino acid	[8]
5	Tryptophan	a, b	Amino acid	[8]
6	N-Acetyltryptophan	a, b	Amino acid	[8]
7	5-Hydroxytryptophan	a, b	Amino acid	[8]
8	N-Acetyltyrosine	a, b	Amino acid	[8]
9	N-(3-Indolylacetyl)-L-alanine	a, b	Amino acid	[8]
10	N-cyclohexyltaurine	a, b	Amino acid	
11	Phenylacetylglutamine	a, b, c	Amino acid	[13,28]
12	Hippuric acid	a, b, c	Organic acid	[29,30]
13	Hydroxyhippuric acid	a, b	Organic acid	[8]
14	Hydroxyphenyl lactic acid	a, b	Organic acid	[8]
15	5-Hydroxyindole-3-acetic acid	a, b	Organic acid	[8]
16	Aminobutyric acid	a, b, c	Organic acid	[8]
17	Dimethyluric acid	a, b, c	Organic acid	[8]
18	Aminooctanoic acid	a, b	Organic acid	[8]
19	p-Cresol glucuronide	a, b	Organic acid	[16,31,32]
20	Dimethylxanthine	a, b	Nucleoside	[33]
21	Orotidine	a, b	Nucleoside	[8]
22	8-Hydroxy-2-deoxyguanosine	a, b	Nucleoside	
23	Decanoylcarnitine	a, b, c	Others	[8]
24	Tyrosol	a, b	Others	[34]
25	Hydroxybenzyl alcohol	a, b	Others	
26	2-Methyl-1,2,3,4-tetrahydro-6,7-isoquinolinediol	a, b	Others	
27	4-Hydroxyquinoline	a, b	Others	[8]
28	Hydroxybenzaldehyde	a, b	Others	[35]
29	Dihydroxyacetone	a, b	Others	[36]
30	Acetamidophenyl glucuronide	a, b	Others	[32]
31	3-Methyloxindole	a, b	Others	[8]
32	Phenylacetamide	a, b	Others	[8]

a: exact mass, b: MS/MS spectra, c: confirmed by a standard.

4. Discussion

In our study, we followed the hypothesis that gut microbiota-associated metabolites can be profiled by metabolomics investigations of urine samples. As a proof of concept, we compared the urinary metabolome before and after a bowel evacuation. We speculated that the massive reduction in the luminal (fecal) microbiota should affect the levels of associated metabolites. In Figures 1–3, the association of a considerable number of metabolites with the fecal gut microbiota was demonstrated, first in a self-experiment of one male individual and then confirmed in a subsequent experiment. Furthermore, the conditions were optimized and adjusted in a way that the sample collection matched the regular procedure during a colonoscopy performed as a healthcare check, meaning two sampling time points, one just before bowel evacuation and the other 10–12 h thereafter. This could open promising perspectives for gut microbiome research studies, particularly for the generation of a fecal microbiome-associated metabolite map (M3) in urine from healthy individuals as a first step, which then could build a base for comparisons of fecal gut microbiome-associated urinary metabolome profiles in disease-related contexts in the future.

The gut microbiota consists of two fractions, namely the luminal (fecal) microbiota and mucosal microbiota. The bacterial composition, overall abundance, and diversity of luminal and mucosal microbiota vary along the longitudinal axis of the gut based on differences in environmental parameters like pH, pO2, osmolality, or mucus type. Also, the microbiota mass varies. In comparison to their sparse distribution in the small intestine, the colon is densely colonized [37] and fecal matter is the most intensely studied sample material in gut microbiome research. Recently, it was reported that fecal samples provided a good approximation of the luminal microbiome [38]. Most likely, the laxative-induced bowel evacuation applied in our study affected or reduced most of the fecal microbiome of the colon. This suggests that our findings describe foremost gut microbiome-associated metabolites in urine related to the luminal microbiota, although it was recently reported that the analysis of feces also provides a good approximation of the average gut mucosal microbiome [38].

It is well-known that gut microbiota are suitable for producing and releasing compounds which subsequently affect either positively or negatively the health state of their human host [1–4]. Furthermore, gut microbiota can introduce modifications in metabolites by numerous enzymatic reactions, thereby changing the molecular structure, chemical properties, and as a consequence, frequently also their functions. Comprehensive profiling with a relatively high certainty that the detected biomarkers originate from the gut microbiome is only suitable in feces, but not in blood or other body fluids. Consequently, until now, a considerable number of those microbiota-derived metabolites in human body fluids as well as their functions for human health or disease remain unknown. On the other hand, gut microbiota-associated metabolites may reach the liver via the splanchnic bed and the mesenteric veins and then partially filtered blood passes through the hepatic veins into the systemic circulation, including the kidneys. In the kidneys, small molecules like metabolites are filtrated into the primary urine, reach the bladder, and can finally be collected in urine samples.

Phenylacetylglutamine (PAGN), one example of the fecal gut microbiota-associated metabolites detected in urine in our study, was very recently described as a product of the interaction of intestinal microbiota and human metabolism [13]. For decades, PAGN has been recognized as a side product of phenylalanine catabolism formed in liver and renal tissues of humans and primates from phenylacetic acid [39,40]. In 2017, Dodd and colleagues showed in bacterial cultures that the PAGN precursor phenylacetic acid can also be produced by bacterial fermentation of phenylalanine [28]. The prerequisite in vivo in humans for this fermentation step is that dietary phenylalanine reaches the large intestine. Phenylalanine is then metabolized by gut microbiota to phenylpyruvic acid and subsequently to phenylacetic acid, which is taken up into the portal system [1]. Recently, PAGN has been reported to be associated with cardiovascular diseases (CVD) and incident major adverse cardiovascular events (myocardial infarction, stroke, or death) [13] as well as heart failure [41].

Hippuric acid, one of the most abundant organic acids in human urine, is derived from two different metabolic pathways; both are interplays between gut microbiota and the liver [29]. On the one hand, it can originate from phenylalanine metabolized by gut microbiota to phenylpropionic acid, which is then re-oxidized to hippuric acid involving medium-chain acyl-CoA dehydrogenase [14]. On the other hand, dietary polyphenols from fruits and vegetables like epicatechins or chlorogenic acid are metabolized by the gut microbiota to benzoic acid, which is subsequently taken up into the splanchnic bed and transported to the human liver [30]. In the liver, hippuric acid is formed by the conjugation of glycine to benzoic acid.

The production of cresol from tyrosine has been recently attributed to four intestinal strains with high cresol production activity belonging to *Coriobacteriaceae* or to *Clostridium* cluster XI or XIVa, and 55 bacterial strains were described with cresol-producing potential [31]. The cresol metabolite p-cresol glucuronide as well as the above-mentioned phenylacetylglutamine showed in plasma stronger associations with several species from the *Clostridiales* order of the corresponding gut microbiota [16].

An unexpected interesting finding of our study was the detected decrease in glutamine and glutamate after bowel evacuation, both well-known amino acids in human metabolic pathways, generated by various tissues and cells in the human body. Recently, it was demonstrated that androgen modulated circulating glutamine and that the glutamine/glutamate (Gln/Glu) ratio partially depends on the gut microbiome [3,42], and the association of metabolic disorders and diabetes with blood levels of Glu, Gln has been reported [3]. The gut microbiome can modulate brain function and behaviors through the microbiota–gut–brain axis by affecting the Glu and Gln levels, which has recently been described for example in schizophrenia [12]. In mice, probiotic treatment with *Lactobacillus rhamnosus* (JB-1) showed a significant increase in brain Glu and Gln levels [27]. Future studies to clarify the pathophysiological role of the gut microbiota-associated glutamate and glutamine metabolism for mental diseases as well as possible probiotic therapeutic options are needed.

We also detected a considerable number of modified metabolites in urine within the group of fecal gut microbiota-associated metabolites (Table S2), dominated by glucuronidations, carboxylations, and sulfations. Glucuronidation of molecules is mainly known as a detoxification reaction of endogenous and exogenous compounds in the liver; however, as shown by our data and in the literature, glucuronidation can also be gut microbiota-associated [32,43]. Recently, applying a new specific profiling strategy for carboxylations, including a derivatization step, 261 gut microbiome-associated modified metabolites were detected [44]. In various metabolite classes, carboxylated compounds have been described, like fatty acids, bile acids, N-acyl amino acids, benzoheterocyclic acids, or aromatic acids [44]. Sulfated metabolites are a group of modified metabolites derived from gut microbiota–human co-metabolism, which have also been reported in the context of disease development. Recently a new enzyme-assisted metabolic profiling approach reported the discovery of 206 sulfated metabolites in human feces and urine, which was three times more than the content of the commonly used Human Metabolome Database [22]. In a subsequent study, the authors showed that a polyphenol-rich diet led to an increase in the levels of 236 sulfated metabolites [45]. Interestingly, although a standardized polyphenol-rich diet was consumed, the authors observed a broad interindividual variability in the generation of these modified metabolites, which led them to speculate about high- and low-sulfate metabolizers [45]. Metabolites modified by gut microbiota could be an interesting additional class of metabolites in biomedical research, but they often remain until now unknown in metabolomics datasets, since they are still underrepresented in all common big databases.

A potential weakness with respect to the data presented here is that in our approach with laxative use, only the luminal (fecal) microbiome in the gut is massively reduced, but no total eradication of the gut microbiome (luminal and mucosal) was achieved, unless antibiotics were used. Consequently, gut microbiota-associated metabolites of distinct species or maybe also phyla may not be detectable with our strategy. However, it was recently reported that fecal samples provide a good approximation of the luminal as well as of the average gut mucosal microbiome [38]. Furthermore, since urine was investigated as a non-invasively collected sample material, the covered fecal microbiota-associated metabolites were limited to water-soluble compounds, which were either per se more or less polar or which were modified before urinary excretion by the liver, e.g., by conjugation of glucuronides. Hence, water-insoluble, apolar gut microbiome-associated metabolites, which can only be detected by the investigation of feces, or in body fluids by invasive sampling (e.g., in blood samples), were not covered. Concerning sex differences in the gut microbiome, which are under intense discussion and investigation [46], we cannot draw any conclusions on the metabolite level based on our data.

5. Conclusions

Beginning with a single-individual self-experiment for hypothesis testing, which covered 40 sample collection time points spreading over 10 days, we revealed gut microbiota association of metabolites in urine after the depletion of the fecal microbiome by a laxative-induced bowel evacuation. These findings were confirmed in a subsequent study performed in six individuals with four sample collection time points by the detection of numerous fecal microbiota-associated metabolites, including modified metabolites. We conclude that our strategy is suitable to profile luminal or fecal microbiome-associated metabolites in urine and consequently detect, e.g., disease-related differences in human gut microbiomes, as well as therapy-dependent changes. Overall, our strategy opens new perspectives for comprehensive human microbiome studies in biomedical research and beyond.

Supplementary Materials: The following supporting information can be downloaded at: https://www.mdpi.com/article/10.3390/metabo13101061/s1, Table S1: Detailed information of internal standards; Table S2: Detailed information about specific modifying groups of gut microbiota-associated metabolite features.

Author Contributions: Conceptualization, R.L.; methodology and validation, S.Z., L.Z., X.L., X.Z., G.X. and R.L.; investigation, S.Z., L.Z., M.H., G.X. and R.L.; writing—original draft preparation, S.Z. and R.L.; writing—review and editing, S.Z., A.P., C.W., A.L.B., X.L., G.X. and R.L.; visualization, S.Z.; supervision and funding acquisition, G.X. and R.L. All authors have read and agreed to the published version of the manuscript.

Funding: This work was supported by the DFG/NSFC Sino-German mobility program (M-0257) to G.X. and R.L. and the Chinese Academy of Sciences (CAS)-President's International Fellowship Initiative (Grant No. 2019VBA0038) to R.L.

Institutional Review Board Statement: The study was conducted according to the Declaration of Helsinki of 1964 and its later amendments, and the ethics committee of the University of Tuebingen approved the protocol (188/2017BO2).

Informed Consent Statement: All volunteers provided written informed consent before beginning the study.

Data Availability Statement: The data presented in this study are available on request from the corresponding author. The data are not publicly available due to privacy restrictions.

Acknowledgments: We acknowledge support from the Open Access Publication Fund of the University of Tübingen.

Conflicts of Interest: The authors declare no conflict of interest. The funders had no role in the design of the study; in the collection, analyses, or interpretation of data; in the writing of the manuscript; or in the decision to publish the results.

References

1. Gentile, C.L.; Weir, T.L. The gut microbiota at the intersection of diet and human health. *Science* **2018**, *362*, 776–780. [CrossRef] [PubMed]
2. Selber-Hnatiw, S.; Sultana, T.; Tse, W.; Abdollahi, N.; Abdullah, S.; Al Rahbani, J.; Alazar, D.; Alrumhein, N.J.; Aprikian, S.; Arshad, R.; et al. Metabolic networks of the human gut microbiota. *Microbiology* **2020**, *166*, 96–119. [CrossRef] [PubMed]
3. Liu, R.; Hong, J.; Xu, X.; Feng, Q.; Zhang, D.; Gu, Y.; Shi, J.; Zhao, S.; Liu, W.; Wang, X.; et al. Gut microbiome and serum metabolome alterations in obesity and after weight-loss intervention. *Nat. Med.* **2017**, *23*, 859–868. [CrossRef] [PubMed]
4. Fan, Y.; Pedersen, O. Gut microbiota in human metabolic health and disease. *Nat. Rev. Microbiol.* **2021**, *19*, 55–71. [CrossRef] [PubMed]
5. Hills, R.D., Jr.; Pontefract, B.A.; Mishcon, H.R.; Black, C.A.; Sutton, S.C.; Theberge, C.R. Gut Microbiome: Profound Implications for Diet and Disease. *Nutrients* **2019**, *11*, 1613. [CrossRef]
6. Whon, T.W.; Shin, N.R.; Kim, J.Y.; Roh, S.W. Omics in gut microbiome analysis. *J. Microbiol.* **2021**, *59*, 292–297. [CrossRef]
7. Manor, O.; Dai, C.L.; Kornilov, S.A.; Smith, B.; Price, N.D.; Lovejoy, J.C.; Gibbons, S.M.; Magis, A.T. Health and disease markers correlate with gut microbiome composition across thousands of people. *Nat. Commun.* **2020**, *11*, 5206. [CrossRef]
8. Han, S.; Van Treuren, W.; Fischer, C.R.; Merrill, B.D.; DeFelice, B.C.; Sanchez, J.M.; Higginbottom, S.K.; Guthrie, L.; Fall, L.A.; Dodd, D.; et al. A metabolomics pipeline for the mechanistic interrogation of the gut microbiome. *Nature* **2021**, *595*, 415–420. [CrossRef]
9. Wu, H.; Tremaroli, V.; Schmidt, C.; Lundqvist, A.; Olsson, L.M.; Kramer, M.; Gummesson, A.; Perkins, R.; Bergstrom, G.; Backhed, F. The Gut Microbiota in Prediabetes and Diabetes: A Population-Based Cross-Sectional Study. *Cell Metab.* **2020**, *32*, 379–390.e3. [CrossRef]

10. David, L.A.; Maurice, C.F.; Carmody, R.N.; Gootenberg, D.B.; Button, J.E.; Wolfe, B.E.; Ling, A.V.; Devlin, A.S.; Varma, Y.; Fischbach, M.A.; et al. Diet rapidly and reproducibly alters the human gut microbiome. *Nature* **2014**, *505*, 559–563. [CrossRef]
11. Li, M.; Wang, B.; Zhang, M.; Rantalainen, M.; Wang, S.; Zhou, H.; Zhang, Y.; Shen, J.; Pang, X.; Zhang, M.; et al. Symbiotic gut microbes modulate human metabolic phenotypes. *Proc. Natl. Acad. Sci. USA* **2008**, *105*, 2117–2122. [CrossRef] [PubMed]
12. Zheng, P.; Zeng, B.; Liu, M.; Chen, J.; Pan, J.; Han, Y.; Liu, Y.; Cheng, K.; Zhou, C.; Wang, H.; et al. The gut microbiome from patients with schizophrenia modulates the glutamate-glutamine-GABA cycle and schizophrenia-relevant behaviors in mice. *Sci. Adv.* **2019**, *5*, eaau8317. [CrossRef] [PubMed]
13. Nemet, I.; Saha, P.P.; Gupta, N.; Zhu, W.; Romano, K.A.; Skye, S.M.; Cajka, T.; Mohan, M.L.; Li, L.; Wu, Y.; et al. A Cardiovascular Disease-Linked Gut Microbial Metabolite Acts via Adrenergic Receptors. *Cell* **2020**, *180*, 862–877.e22. [CrossRef] [PubMed]
14. Pruss, K.M.; Chen, H.; Liu, Y.; Van Treuren, W.; Higginbottom, S.K.; Jarman, J.B.; Fischer, C.R.; Mak, J.; Wong, B.; Cowan, T.M.; et al. Host-microbe co-metabolism via MCAD generates circulating metabolites including hippuric acid. *Nat. Commun.* **2023**, *14*, 512. [CrossRef]
15. Kikuchi, K.; Saigusa, D.; Kanemitsu, Y.; Matsumoto, Y.; Thanai, P.; Suzuki, N.; Mise, K.; Yamaguchi, H.; Nakamura, T.; Asaji, K.; et al. Gut microbiome-derived phenyl sulfate contributes to albuminuria in diabetic kidney disease. *Nat. Commun.* **2019**, *10*, 1835. [CrossRef]
16. Dekkers, K.F.; Sayols-Baixeras, S.; Baldanzi, G.; Nowak, C.; Hammar, U.; Nguyen, D.; Varotsis, G.; Brunkwall, L.; Nielsen, N.; Eklund, A.C.; et al. An online atlas of human plasma metabolite signatures of gut microbiome composition. *Nat. Commun.* **2022**, *13*, 5370. [CrossRef]
17. Zierer, J.; Jackson, M.A.; Kastenmuller, G.; Mangino, M.; Long, T.; Telenti, A.; Mohney, R.P.; Small, K.S.; Bell, J.T.; Steves, C.J.; et al. The fecal metabolome as a functional readout of the gut microbiome. *Nat. Genet.* **2018**, *50*, 790–795. [CrossRef]
18. Hu, J.; Ding, J.; Li, X.; Li, J.; Zheng, T.; Xie, L.; Li, C.; Tang, Y.; Guo, K.; Huang, J.; et al. Distinct signatures of gut microbiota and metabolites in different types of diabetes: A population-based cross-sectional study. *EClinicalMedicine* **2023**, *62*, 102132. [CrossRef]
19. Hryhorczuk, L.M.; Novak, E.A.; Gershon, S. Gut flora and urinary phenylacetic acid. *Science* **1984**, *226*, 996. [CrossRef]
20. Goodwin, B.L.; Ruthven, C.R.; Sandler, M. Gut flora and the origin of some urinary aromatic phenolic compounds. *Biochem. Pharmacol.* **1994**, *47*, 2294–2297. [CrossRef]
21. Li, R.J.; Jie, Z.Y.; Feng, Q.; Fang, R.L.; Li, F.; Gao, Y.; Xia, H.H.; Zhong, H.Z.; Tong, B.; Madsen, L.; et al. Network of Interactions between Gut Microbiome, Host Biomarkers, and Urine Metabolome in Carotid Atherosclerosis. *Front. Cell Infect. Microbiol.* **2021**, *11*, 708088. [CrossRef] [PubMed]
22. Ballet, C.; Correia, M.S.P.; Conway, L.P.; Locher, T.L.; Lehmann, L.C.; Garg, N.; Vujasinovic, M.; Deindl, S.; Lohr, J.M.; Globisch, D. New enzymatic and mass spectrometric methodology for the selective investigation of gut microbiota-derived metabolites. *Chem. Sci.* **2018**, *9*, 6233–6239. [CrossRef] [PubMed]
23. Jain, A.; Li, X.H.; Chen, W.N. An untargeted fecal and urine metabolomics analysis of the interplay between the gut microbiome, diet and human metabolism in Indian and Chinese adults. *Sci. Rep.* **2019**, *9*, 9191. [CrossRef] [PubMed]
24. Yang, J.; Zhao, X.; Lu, X.; Lin, X.; Xu, G. A data preprocessing strategy for metabolomics to reduce the mask effect in data analysis. *Front. Mol. Biosci.* **2015**, *2*, 4. [CrossRef]
25. Zhao, X.; Zeng, Z.; Chen, A.; Lu, X.; Zhao, C.; Hu, C.; Zhou, L.; Liu, X.; Wang, X.; Hou, X.; et al. Comprehensive Strategy to Construct In-House Database for Accurate and Batch Identification of Small Molecular Metabolites. *Anal. Chem.* **2018**, *90*, 7635–7643. [CrossRef]
26. Zheng, S.; Zhang, X.; Li, Z.; Hoene, M.; Fritsche, L.; Zheng, F.; Li, Q.; Fritsche, A.; Peter, A.; Lehmann, R.; et al. Systematic Modifying Group-Assisted Strategy Expanding Coverage of Metabolite Annotation in Liquid Chromatography-Mass Spectrometry-Based Nontargeted Metabolomics Studies. *Anal. Chem.* **2021**, *93*, 10916–10924. [CrossRef]
27. Janik, R.; Thomason, L.A.M.; Stanisz, A.M.; Forsythe, P.; Bienenstock, J.; Stanisz, G.J. Magnetic resonance spectroscopy reveals oral Lactobacillus promotion of increases in brain GABA, N-acetyl aspartate and glutamate. *Neuroimage* **2016**, *125*, 988–995. [CrossRef]
28. Dodd, D.; Spitzer, M.H.; Van Treuren, W.; Merrill, B.D.; Hryckowian, A.J.; Higginbottom, S.K.; Le, A.; Cowan, T.M.; Nolan, G.P.; Fischbach, M.A.; et al. A gut bacterial pathway metabolizes aromatic amino acids into nine circulating metabolites. *Nature* **2017**, *551*, 648–652. [CrossRef]
29. Ticinesi, A.; Guerra, A.; Nouvenne, A.; Meschi, T.; Maggi, S. Disentangling the Complexity of Nutrition, Frailty and Gut Microbial Pathways during Aging: A Focus on Hippuric Acid. *Nutrients* **2023**, *15*, 1138. [CrossRef]
30. Penczynski, K.J.; Krupp, D.; Bring, A.; Bolzenius, K.; Remer, T.; Buyken, A.E. Relative validation of 24-h urinary hippuric acid excretion as a biomarker for dietary flavonoid intake from fruit and vegetables in healthy adolescents. *Eur. J. Nutr.* **2017**, *56*, 757–766. [CrossRef]
31. Saito, Y.; Sato, T.; Nomoto, K.; Tsuji, H. Identification of phenol- and p-cresol-producing intestinal bacteria by using media supplemented with tyrosine and its metabolites. *FEMS Microbiol. Ecol.* **2018**, *94*, fiy125. [CrossRef] [PubMed]
32. Pellock, S.J.; Redinbo, M.R. Glucuronides in the gut: Sugar-driven symbioses between microbe and host. *J. Biol. Chem.* **2017**, *292*, 8569–8576. [CrossRef] [PubMed]
33. Yue, S.; Zhao, D.; Peng, C.; Tan, C.; Wang, Q.; Gong, J. Effects of theabrownin on serum metabolites and gut microbiome in rats with a high-sugar diet. *Food Funct.* **2019**, *10*, 7063–7080. [CrossRef] [PubMed]

34. Mosele, J.I.; Martin-Pelaez, S.; Macia, A.; Farras, M.; Valls, R.M.; Catalan, U.; Motilva, M.J. Faecal microbial metabolism of olive oil phenolic compounds: In vitro and in vivo approaches. *Mol. Nutr. Food Res.* **2014**, *58*, 1809–1819. [CrossRef] [PubMed]
35. Li, Y.; Sui, L.; Zhao, H.; Zhang, W.; Gao, L.; Hu, W.; Song, M.; Liu, X.; Kong, F.; Gong, Y.; et al. Differences in the Establishment of Gut Microbiota and Metabolome Characteristics between Balb/c and C57BL/6J Mice after Proton Irradiation. *Front. Microbiol.* **2022**, *13*, 874702. [CrossRef] [PubMed]
36. Jamshidi, N.; Nigam, S.K. Drug transporters OAT1 and OAT3 have specific effects on multiple organs and gut microbiome as revealed by contextualized metabolic network reconstructions. *Sci. Rep.* **2022**, *12*, 18308. [CrossRef]
37. McCallum, G.; Tropini, C. The gut microbiota and its biogeography. *Nat. Rev. Microbiol.* **2023**; ahead of print. PMID: 37740073. [CrossRef]
38. Vaga, S.; Lee, S.; Ji, B.; Andreasson, A.; Talley, N.J.; Agreus, L.; Bidkhori, G.; Kovatcheva-Datchary, P.; Park, J.; Lee, D.; et al. Compositional and functional differences of the mucosal microbiota along the intestine of healthy individuals. *Sci. Rep.* **2020**, *10*, 14977. [CrossRef]
39. Moldave, K.; Meister, A. Synthesis of phenylacetylglutamine by human tissue. *J. Biol. Chem.* **1957**, *229*, 463–476. [CrossRef]
40. Yang, D.; Brunengraber, H. Glutamate, a window on liver intermediary metabolism. *J. Nutr.* **2000**, *130*, 991S–994S. [CrossRef]
41. Romano, K.A.; Nemet, I.; Prasad Saha, P.; Haghikia, A.; Li, X.S.; Mohan, M.L.; Lovano, B.; Castel, L.; Witkowski, M.; Buffa, J.A.; et al. Gut Microbiota-Generated Phenylacetylglutamine and Heart Failure. *Circ. Heart Fail.* **2023**, *16*, e009972. [CrossRef] [PubMed]
42. Gao, A.; Su, J.; Liu, R.; Zhao, S.; Li, W.; Xu, X.; Li, D.; Shi, J.; Gu, B.; Zhang, J.; et al. Sexual dimorphism in glucose metabolism is shaped by androgen-driven gut microbiome. *Nat. Commun.* **2021**, *12*, 7080. [CrossRef] [PubMed]
43. Wikoff, W.R.; Anfora, A.T.; Liu, J.; Schultz, P.G.; Lesley, S.A.; Peters, E.C.; Siuzdak, G. Metabolomics analysis reveals large effects of gut microflora on mammalian blood metabolites. *Proc. Natl. Acad. Sci. USA* **2009**, *106*, 3698–3703. [CrossRef] [PubMed]
44. Wang, Y.Z.; Chen, Y.Y.; Wu, X.Z.; Bai, P.R.; An, N.; Liu, X.L.; Zhu, Q.F.; Feng, Y.Q. Uncovering the Carboxylated Metabolome in Gut Microbiota-Host Co-metabolism: A Chemical Derivatization-Molecular Networking Approach. *Anal. Chem.* **2023**, *95*, 11550–11557. [CrossRef] [PubMed]
45. Correia, M.S.P.; Jain, A.; Alotaibi, W.; Young Tie Yang, P.; Rodriguez-Mateos, A.; Globisch, D. Comparative dietary sulfated metabolome analysis reveals unknown metabolic interactions of the gut microbiome and the human host. *Free Radic. Biol. Med.* **2020**, *160*, 745–754. [CrossRef]
46. Valeri, F.; Endres, K. How biological sex of the host shapes its gut microbiota. *Front. Neuroendocrinol.* **2021**, *61*, 100912. [CrossRef]

Disclaimer/Publisher's Note: The statements, opinions and data contained in all publications are solely those of the individual author(s) and contributor(s) and not of MDPI and/or the editor(s). MDPI and/or the editor(s) disclaim responsibility for any injury to people or property resulting from any ideas, methods, instructions or products referred to in the content.

Article

Gas Chromatography–Mass Spectrometry Reveals Stage-Specific Metabolic Signatures of Ankylosing Spondylitis

Yixuan Guo [1,†], Shuangshuang Wei [1,†], Mengdi Yin [1], Dandan Cao [1], Yiling Li [1], Chengping Wen [1,2,*] and Jia Zhou [1,2,*]

[1] Institute of Basic Research in Clinical Medicine, College of Basic Medical Sciences, Zhejiang Chinese Medical University, Hangzhou 310053, China; guoyixuan1725@gmail.com (Y.G.); w91315@gmail.com (S.W.); 202111114811060@zcmu.edu.cn (M.Y.); 202211110211002@zcmu.edu.cn (D.C.); 202211110211021@zcmu.edu.cn (Y.L.)

[2] Key Laboratory of Chinese Medicine Rheumatology of Zhejiang Province, Zhejiang Chinese Medical University, Hangzhou 310053, China

* Correspondence: wencp@zcmu.edu.cn (C.W.); zhoujia@zcmu.edu.cn (J.Z.)

† These authors contributed equally to this work.

Abstract: Ankylosing spondylitis (AS) is a type of chronic rheumatic immune disease, and the crucial point of AS treatment is identifying the correct stage of the disease. However, there is a lack of effective diagnostic methods for AS staging. The primary objective of this study was to perform an untargeted metabolomic approach in AS patients in an effort to reveal metabolic differences between patients in remission and acute stages. Serum samples from 40 controls and 57 AS patients were analyzed via gas chromatography–mass spectrometry (GC–MS). Twenty-four kinds of differential metabolites were identified between the healthy controls and AS patients, mainly involving valine/leucine/isoleucine biosynthesis and degradation, phenylalanine/tyrosine/tryptophan biosynthesis, glutathione metabolism, etc. Furthermore, the levels of fatty acids (linoleate, dodecanoate, hexadecanoate, and octadecanoate), amino acids (serine and pyroglutamate), 2-hydroxybutanoate, glucose, etc., were lower in patients in the acute stage than those in the remission stage, which may be associated with the aggravated inflammatory response and elevated oxidative stress in the acute stage. Multiple stage-specific metabolites were significantly correlated with inflammatory indicators (CRP and ESR). In addition, the combination of serum 2-hydroxybutanoate and hexadecanoate plays a significant role in the diagnosis of AS stages. These metabolomics-based findings provide new perspectives for AS staging, treatment, and pathogenesis studies.

Keywords: ankylosing spondylitis; metabolomics; gas chromatography–mass spectrometry; acute and remission stages

1. Introduction

Ankylosing spondylitis (AS) is a systemic disease dominated by chronic inflammation of the axial joints, which mainly involve the sacroiliac joint, spine bone protrusion, paraspinal soft tissue, and peripheral joints, and it is often accompanied by varying degrees of extra-articular manifestations including ocular and gastrointestinal systems [1,2]. The pathogenesis of ankylosing spondylitis involves abnormalities of the immune system and chronic inflammation, and AS has a genetic predisposition with a strong association with the HLA–B27 gene [3,4].

AS is a chronic, incurable disease called undead cancer. Its clinical treatment focuses on suppressing inflammation, improving symptoms, preventing bone destruction, and decelerating disease progression [5]. Particularly, AS patients in acute and remission stages require different treatment strategies. During the acute stage, pain and inflammation may worsen, and non-steroidal anti-inflammatory drugs (NSAIDs) can help relieve pain and reduce inflammation [6,7]. Adequate rest is also necessary to prevent further joint damage. The remission stage, on the other hand, focuses on maintaining joint mobility and stability. Through appropriate exercises and treatments, stiffness and deformities of the joints can be slowed, and the risk of further complications can be reduced. Therefore, accurate diagnosis and prediction of the acute and remission stages of AS contributes to the development of personalized treatment plans, thereby effectively slowing the progression of the disease, improving clinical symptoms, and enhancing the quality of life for patients. Presently, the judgement of AS stage primarily relies on clinical manifestations (morning stiffness, pain, inflammation, etc.) and imaging examinations (X-rays, MRI, or CT) [8,9]. The erythrocyte sedimentation rate (ESR) and C-reactive protein (CRP) levels are important in the evaluation of disease progression to some extent, but these inflammatory indicators are currently only used as a reference due to their low specificity [10].

Metabolomics is a simultaneous qualitative and quantitative analysis of all metabolites in a specific biological sample under defined conditions, aiming to gain insight into the dynamics of metabolic networks in organisms, and thus, to reveal changes in disease progression and physiological states [11,12]. Metabolomics has gained significant attention in the field of disease research as it provides valuable insights into the metabolic alterations associated with the disease [13–15]. Targeted and untargeted metabolomics and lipidomics provide strong support for the diagnosis, prognosis, and treatment of gout, diabetes, SLE, cancer, etc. [16–19]. Several metabolomics studies have been conducted to investigate the metabolic profile of AS patients utilizing different analytical techniques such as mass spectrometry and nuclear magnetic resonance spectroscopy, aiming to identify potential biomarkers and unravel the underlying metabolic pathways involved in AS [20,21]. Perturbations in various metabolites such as tryptophan, lysine, proline, serine, and alanine have been discovered in AS patients [22–27], which provides information for exploring the pathogenesis of AS.

The pathological states of patients are different at various disease stages, and these differences can be manifested through alterations in metabolite composition and metabolic pathways. Therefore, investigating shifts in the humoral metabolome can help characterize the disease stages. However, metabolic changes associated with different stages of AS patients have not yet been investigated. The primary objective of this study was to perform a GC–MS-based untargeted metabolomic strategy in AS patients in an effort to identify metabolic differences between patients in remission and acute stages. Discriminant models were constructed to help differentiate AS patients at different stages based on the acquired metabolic data. Furthermore, we sought to reveal potential metabolic biomarkers to enable a more precise discrimination of AS stages. The findings may be useful in the clinical management of AS by providing valuable information for the diagnosis, treatment, and monitoring of patients.

2. Materials and Methods

2.1. Study Participants

A total of 57 patients (including 17 patients in the acute stage) in accordance with the New York Criteria for AS revised by the American College of Rheumatology in 1984 were included in the study. Meanwhile, an age- and gender-matched healthy control population of 40 cases was also enrolled. Subjects with severe cardiovascular, hepatic, and other organic pathologies and psychiatric abnormalities, as well as pregnant or lactating women were excluded. The study protocol was approved by the Ethics Committee of the First Affiliated Hospital of Zhejiang Chinese Medical University (2021-KLJ-010-03), and all patients signed an informed consent form.

2.2. Sample Preparation and Metabolomics Analysis

Fasting elbow venous blood of each subject was collected in the morning, followed by centrifugation, and the upper serum was stored at $-80\ °C$ for further testing. Before analysis, serum samples were thawed and vortex-mixed, and an aliquot of 50 µL of serum was spiked with 200 µL of acetonitrile (all operations were carried out on ice). After vortex-mixing, the samples were centrifuged at 4 °C for 10 min at 12,000 g to remove precipitated proteins, and the supernatant were freeze-dried under vacuum. Afterward, freeze-dried extracts were dissolved in 50 µL of methoxyamine pyridine solution (20 mg/mL), and the reaction mixtures were heated in a water bath at 40 °C for 90 min; then, 50 µL of MSTFA (N-trimethylsilyl-N-methyl trifluoroacetamide) was added to the samples for silylation reaction, followed by heating at 40 °C for 60 min. All samples to be analyzed were mixed in equal volumes to create quality control (QC) samples, and QC samples were prepared according to the aforementioned protocol.

All samples were analyzed on an Agilent 7890A gas chromatograph coupled with a 5975C MSD system (Agilent Technologies Inc., Santa Clara, CA, USA), and DB–5MS capillary column (30 m × 0.25 mm × 0.25 µm) was applied for separation. The carrier gas was helium (99.9996%) at a flow rate of 1.2 mL/min. The initial column temperature was kept at 70 °C for 3 min, programmed to 300 °C at a rate of 5 °C/min, and kept constant at 300 °C for 5 min. The injection volume was 1 µL, and the split ratio was 5:1. The injection port temperature and the transfer line temperature were set to 300 °C and 280 °C, respectively.

Mass spectrometry parameters were as follows: the ion source temperature was 230 °C, the scan range was 33~600 amu, and the solvent delay was 4.8 min. All samples were analyzed in a random order and QC samples were analyzed at every seven test samples to monitor the instrument stability.

2.3. Statistical Analysis

The QC sample is a homogeneous mix of all samples to be tested and contains the majority of the metabolite information in samples, so it was used as a template to establish a quantitative table of metabolites. The metabolic profile of the QC sample was firstly subjected to peak identification and overlapping peak resolution using AMDIS 2.62 (NIST, Boulder, CO, USA) software to obtain a quantitative table containing retention time and mass-to-charge ratio information of each metabolite. Then, the original data of all samples were imported into the workstation to integrate according to the above established table, and the peak area of each metabolite was obtained and combined into a two-dimensional data matrix.

After peak area normalization, data from all subjects were subjected to unit variance (UV) scaling and principal component analysis (PCA) using SIMCA-P 14.0 (Umetrics AB, Umea, Sweden) to obtain an overview of the differences in serum metabolic phenotypes between AS patients and healthy controls. The data were then filtered via orthogonal signal correction (OSC) to remove variables irrelevant to classification, and partial least squares discriminant analysis (PLS–DA) was utilized to discover specific metabolic patterns in AS patients, further differentiating between AS patients in acute and remission stages. Permutation test (200 permutations) and CV–ANOVA were performed to check the validity of OSC PLS–DA models. VIP (variable importance on projection) is an important indicator used to screen for differential metabolites in multivariate analyses, and variables with large VIP values contribute to model classification to a greater extent. The threshold of VIP values should be selected and adjusted according to the research requirements and data characteristics. In general, a VIP value greater than 1 is a standard threshold commonly used in the field to identify the most important variables in PLS–DA models. SPSS 21.0 (International Business Machines Corp., Armonk, NY, USA) was employed to conduct t-test and ANOVA. Here, serum metabolites with VIP > 1.0 in OSC PLS–DA and significance test p value < 0.05 between AS patients and controls or patients in different AS stages were screened as important candidate metabolites pending further structural annotation. Identification of metabolites was achieved by matching the mass spectra with commercial mass spectral libraries using NIST MS Search 2.0 (National Institute of Standards and Technology, Gaithersburg, MD, USA). The candidate results with similarity degree higher than 80% were further verified with standard compounds. The identified metabolites were introduced into MetaboAnalyst 4.0 (https://www.metaboanalyst.ca) for enrichment analysis to investigate which pathways were disturbed in AS patients [28].

A diagnostic model for discriminating acute and remission AS stages was established through binary logistic regression analysis SPSS 21.0 (International Business Machines Corp., Armonk, NY, USA) based on the AS stage-specific metabolites, and the diagnostic efficacy of each metabolite or different metabolites combination were evaluated by drawing receiver operating characteristic (ROC) curves. Spearman correlation analysis was performed to evaluate the correlation of serum differential metabolites with ESR and CRP in patients in acute and remission stages.

3. Results

3.1. Characterization of the Subjects

The demographic and clinical information of the selected subjects is shown in Table 1. The AS patient group and the control group were compatible in terms of age, gender and BMI, without significant differences; the levels of CRP, ESR, and WBC (white blood cells) in the AS group were higher than those in the control group, which was consistent with the clinical characteristics of AS. In addition, HLA–B27 was positive in 78.90% of AS patients.

Table 1. The clinical information of AS patients and healthy controls.

Characteristics	Control (n = 40)	AS (n = 57)	p
Male/Female,	20/20	28/29	0.932
Age, y	36.08 ± 1.53	39.37 ± 1.61	0.158
BMI	22.22 ± 0.46	22.68 ± 0.53	0.534
CRP (mg/L)	1.90 ± 0.38	4.18 ± 0.85	0.036
ESR (mm/h)	9.32 ± 0.67	15.02 ± 2.08	0.032
WBC (10^9/L)	5.90 ± 0.19	6.66 ± 0.20	0.010
HLA–B27 (+/−)	2/38	45/12	<0.001

AS, ankylosing spondylitis; BMI, body mass index; CRP, C-reactive protein; ESR, erythrocyte sedimentation rate; WBC, white blood cell; HLA–B27, hydrophile–lipophile balance–27. Data are shown as mean ± SEM.

3.2. Characteristics of the Serum Metabolomics Analysis

In this study, the typical total ion current (TIC) chromatograms of AS patients and healthy controls obtained via GC–MS are shown in Figure S1A. A total of 219 metabolites were identified, integrated, and statistically analyzed. In order to examine whether the analytical method was stable and reproducible, one QC was added to every seven samples in the analytical sequence for a total of fifteen QC samples. The relative standard deviation (RSD) of the peak area of each metabolite in the QCs was calculated. It was found that 191 metabolites had RSD less than 30%, and the cumulative peak area was 98.3% of the total peak area (Figure S1B). Moreover, all QC samples were found to be within a range of 2-fold SD based on the PCA analysis in Figure S1C. The results demonstrated that the sample pretreatment and instrumental analysis processes had sufficient stability and reproducibility, and the acquired metabolomics data were reliable.

3.3. Serum Metabolic Differences between AS Patients and Controls

In the PCA analysis, the contribution of each principal component (PC) is shown in the scree plot (Figure 1A), which shows that approximately 61.5% of the variance in the raw data is explained by 11 PCs. As shown in the score plots (Figure 1B,C), there is a tendency for separation between AS patients and healthy controls, although there is some overlap. To identify the differences in metabolic profiles between AS patients and controls, OSC PLS–DA was carried out (Figure 1D). R2Y and Q2Y are used to evaluate the fitting effect of the OSC PLS–DA model. Here, both parameters are close to 1 (R2Y = 0.987, Q2Y = 0.965), which indicates that the model has good classification and prediction ability and can clearly distinguish between the two groups (Figure 1D). In the permutation test, all Q2 values are lower than the original points on the right, the intercept of the regression line at Q2 is less than zero, and all R2 values are lower than the original points on the right (Figure S2A). Moreover, the p value of the CV–ANOVA is less than 0.001. The above results indicate that the OSC PLS–DA model is not overfitted. The V plot presenting the VIP values and correlation coefficients of OSC PLS–DA model is shown in Figure 1E. The metabolites with VIP > 1.0 are located at both ends of "V", where metabolites on the right side of the y-axis are positively correlated with AS patients and metabolites on the left side of the y-axis are negatively correlated with AS patients.

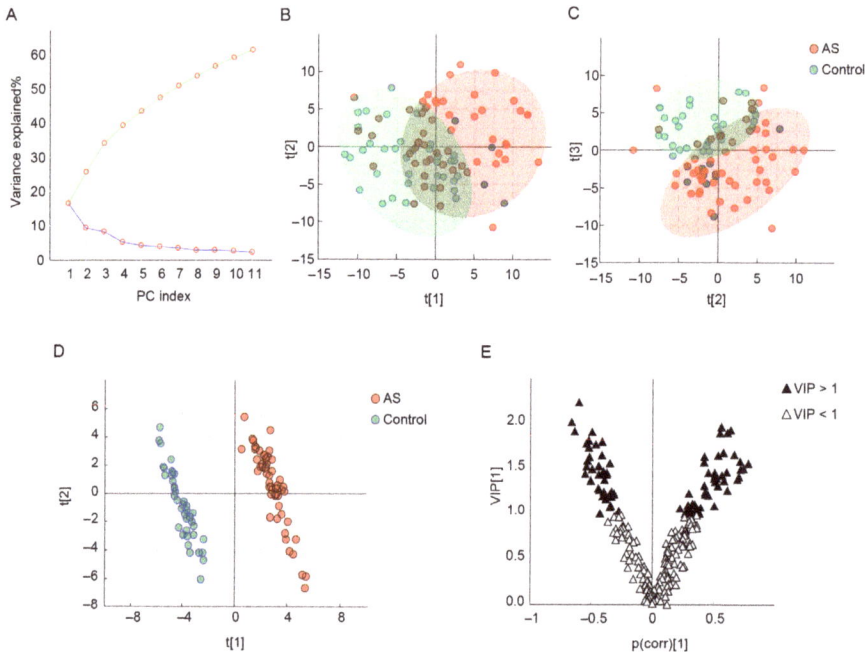

Figure 1. PCA and OSC PLS–DA of AS patients and healthy controls based on serum metabolic profiles. (**A**) PCA scree plot, (**B**) PCA score plot of PC1 and PC2, (**C**) PCA score plot of PC2 and PC3, (**D**) OSC PLS–DA score plot, and (**E**) V plot.

The metabolites with VIP > 1.0 and $p < 0.05$ were considered to be metabolic characteristics of AS patients, and the structures of 24 metabolites have been identified, including amino acids and their derivatives (methionine, serine, threonine, valine, phenylalanine, pyroglutamate, tryptophan, proline, leucine, glycine, isoleucine, tyrosine, alanine, 2-aminobutyrate, and ethanolamine), lipids and their derivatives (hexadecanoate and glycerol-3-phosphate), carbohydrates and their derivatives (glucose, ribofuranose, propylene glycol, 1,5-anhydrosorbitol, and 1,3-propanediol), creatinine, and cholesterol (Figure 2A,B). These metabolites were mainly enriched in valine/leucine/isoleucine biosynthesis and degradation (threonine, leucine, isoleucine, and valine), phenylalanine/tyrosine/tryptophan biosynthesis (phenylalanine, tryptophan, and tyrosine), glutathione metabolism (glycine, alanine, and pyroglutamate), and so on (Figure 2C).

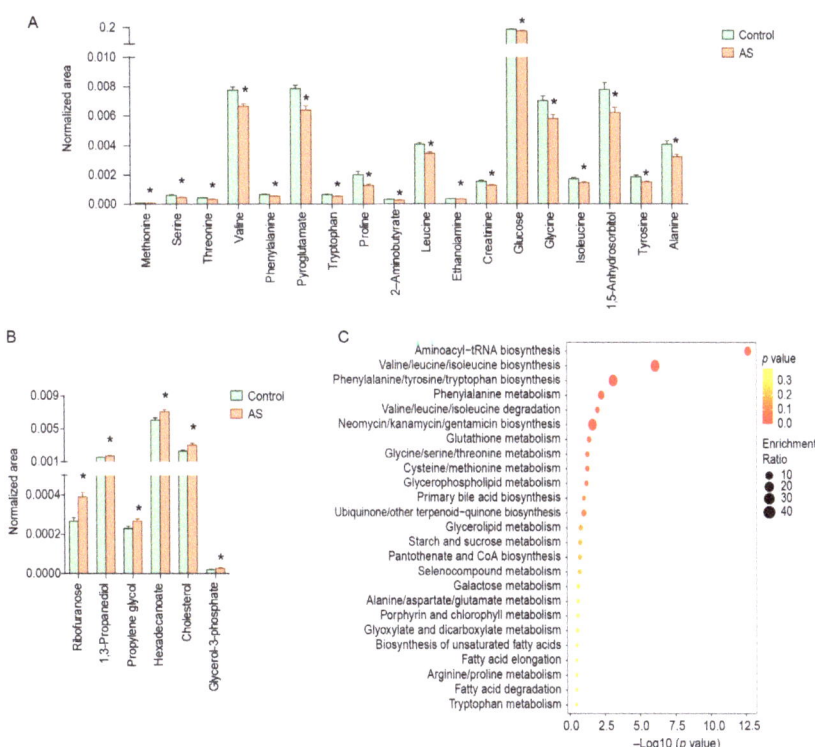

Figure 2. Differential metabolites and enrichment pathways in serum of AS patients and healthy controls. (**A**) Down-regulated metabolites in serum of AS patients, (**B**) up-regulated metabolites in serum of AS patients, (**C**) enriched pathways. * represents $p < 0.05$ between controls and AS patients.

3.4. Serum Metabolic Differences between Acute and Remission Stages of AS Patients

Based on the clinical symptoms and inflammatory indicators of AS, the AS patients were differentiated into patients in the acute stage (17 cases) and patients in the remission stage (40 cases); there was no statistically significant difference between the two groups in terms of age, gender, and BMI, and the CRP and ESR indices of the patients in the acute stage were significantly higher than those of the patients in the remission stage (Figure 3A). Differences in metabolite levels between controls and acute- and remission-stage AS patients were compared using ANOVA (Figure 3B), and OSC PLS–DA analysis was also performed to distinguish between the different stages of AS patients (Figure 3C). The model parameters (R2Y = 0.985, Q2Y = 0.938) are close to 1, indicating that the OSC PLS–DA model fits the data well. The permutation test (Figure S2B) and CV–ANOVA ($p < 0.001$) support the goodness of fit of the model. Subsequently, the metabolites with VIP > 1.0 were marked in the OSC PLS–DA V plot, which may play a very important role in classification (Figure 3D).

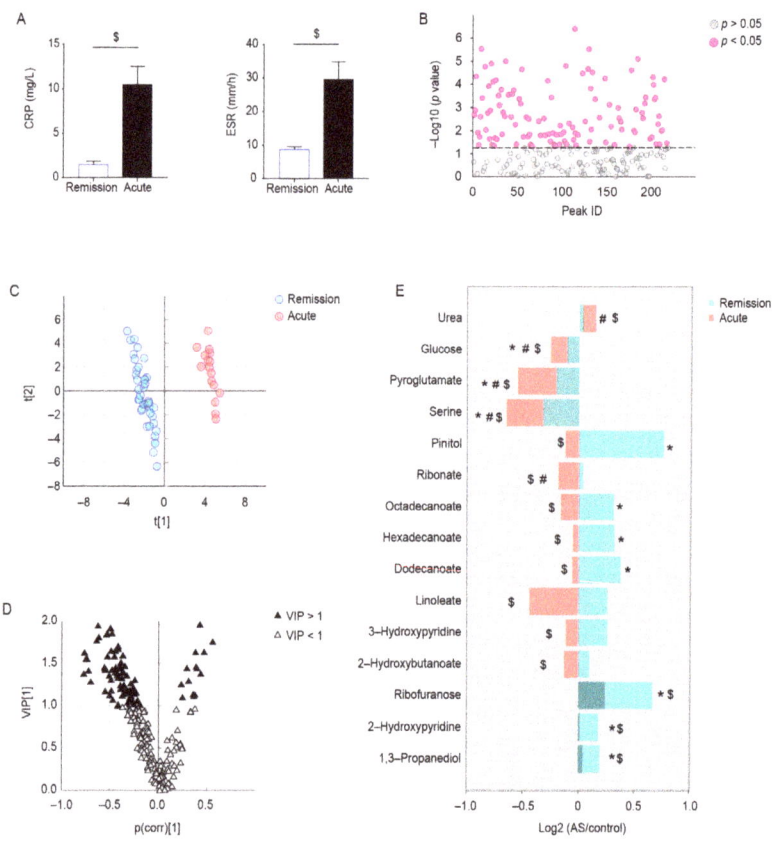

Figure 3. Metabolic differences between acute- and remission-stage patients with AS. (**A**) Differences in CRP and ESR in AS patients during different stages, (**B**) ANOVA of serum metabolites between controls, AS patients in the remission stage, and AS patients in the acute stage. (**C**) OSC PLS–DA score plot of AS patients in remission versus acute stages, (**D**) OSC PLS–DA V plot. (**E**) Trends of serum metabolites in AS patients during acute and remission stages. * represents $p < 0.05$ between controls and AS patients in the remission stage, # represents $p < 0.05$ between controls and AS patients in the acute stage, and $ represents $p < 0.05$ between AS patients in acute and remission stages.

Ultimately, 44 stage-specific metabolites were screened out on the basis of VIP and p value. Of these, 15/44 were identified, and the trends of these metabolites in the control group and patients during acute and remission stages are presented in Figure 3E. Serum levels of seven metabolites, 1,3-propanediol, 2-hydroxypyridine, ribofuranose, dodecanoate, hexadecanoate, octadecenoate, and pinitol, were higher in the AS patients than in controls during the remission stage, whereas they were reduced to control levels during the acute stage. Three metabolites (serine, pyroglutamate, and glucose) appeared to be reduced in AS patients during the remission stage compared with the controls, and the decrease was more pronounced during the acute stage. In addition, two metabolites were unchanged in AS patients during the remission stage and significantly changed during the acute stage (ribonate decreased and urea increased); and the levels of 2-hydroxybutanoate, 3-hydroxypyridine, and linoleate were significantly lower in the acute stage than in the remission stage.

Spearman's correlation analysis of serum differential metabolites with CRP and ESR in the acute and remission stages of AS patients was carried out (Figure 4A,B). There was a significant correlation between CRP levels and metabolites such as 2-hydroxypyridine, ribofuranose, 2-hydroxybutanoate, 3-hydroxypyridine, dodecanoate, hexadecanoate, linoleate, octadecanoate, ribonate, and pyroglutamate, and metabolites including ribofuranose, linoleate, ribonate, serine, and pyroglutamate were negatively correlated with ESR. Particularly, ribofuranose, linoleate, ribonate, and pyroglutamate had a strong negative correlation with both CRP and ESR, indicating that the changes in levels might be associated with the progression of AS.

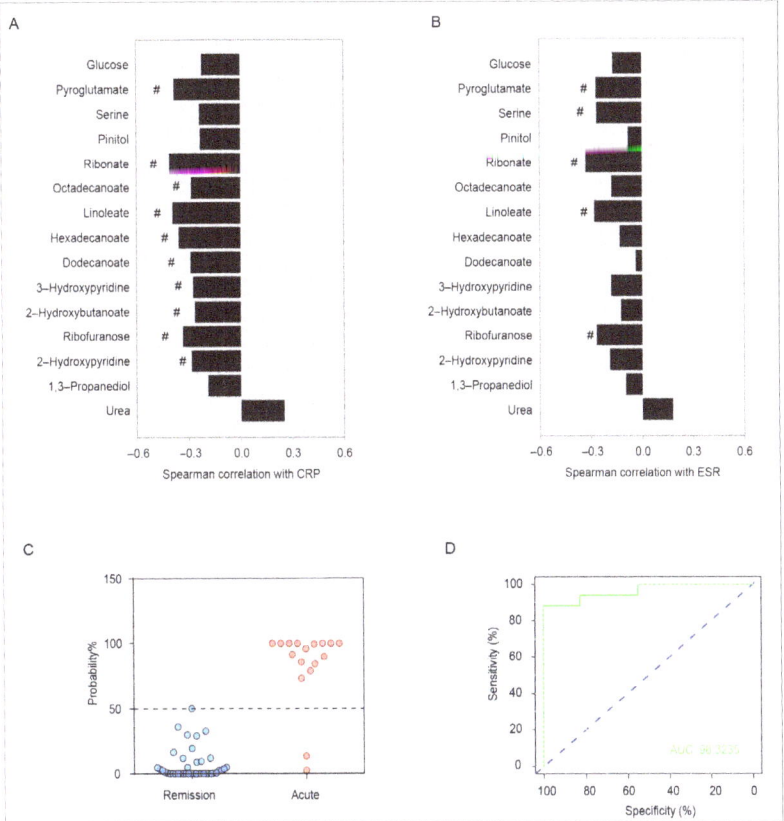

Figure 4. (**A**) Spearman correlation coefficient of serum differential metabolites and CRP index in AS patients during the remission and acute stages. # indicates a statistically significant correlation. (**B**) Spearman correlation coefficient of serum differential metabolites and ESR index in AS patients during the remission and acute stages. # indicates a statistically significant correlation. (**C**) Predictive probability of acute AS patients and remission patients based on serum differential metabolites. Individuals below the blue dashed line are predicted to be in remission stage, and those above the blue dashed line are predicted to be in the acute stage. (**D**) ROC analysis for the discrimination of AS patients during different stages using the combination of serum 2-hydroxybutanoate and hexadecanoate levels. The blue dashed line is the reference line for performance, and the ROC curve above the diagonal means the performance of the model is relatively good.

Clinical manifestations, imaging examinations, and inflammatory indexes are important indicators of AS progression to some extent, but with low sensitivity and specificity for AS staging. In this study, some significantly different metabolites in the serum of AS patients during the acute and remission stages were discovered, which may have potential for the staging of AS. Through the binary logistic regression analysis, a model was established to distinguish the acute stage and remission stage of AS patients by the serum levels of 2-hydroxybutanoate and hexadecanoate. The model can effectively distinguish the different stages of AS (Figure 4C). ROC curve analysis showed that the combined application of serum 2-hydroxybutanoate and hexadecanoate had a good diagnostic effect for AS staging (AUC = 0.963, Figure 4D).

4. Discussion

AS is a chronic inflammatory systemic disease with still unknown etiology, and recent studies have recognized that it might be correlated with infection, autoimmune, and genetic factors [2,6,29]. Correct staging is very important for the treatment of AS. In the acute stage, the main goal is to relieve pain and inflammation in patients with AS as soon as possible, while in the remission stage, the focus is on maintaining joint mobility and stability and slowing the progression of the disease and the likelihood of flare-ups. However, there is a lack of effective diagnostic methods for staging AS. In this study, we found that valine/leucine/isoleucine biosynthesis and degradation (threonine, leucine, isoleucine, and valine), phenylalanine/tyrosine/tryptophan biosynthesis (phenylalanine, tryptophan, and tyrosine), and glutathione metabolism (glycine, alanine, and pyroglutamate) were altered in AS patients compared with healthy individuals. Furthermore, 15 metabolites showed a significant difference between AS patients in the acute and remission stages.

A large number of inflammatory factors can be generated during the course of AS. Some studies have reported that high levels of serum-free fatty acids could induce chronic low-grade inflammation [30]; concurrently, inflammatory mediators can interfere with lipid metabolism as well [31,32]. In this study, we found that the levels of dodecanoate, hexadecanoate, and octadecanoate were elevated in the remission stage of AS patients, whereas a decrease in fatty acids was observed in AS patients in the acute stage, which could be attributed to the intensified energy demand stemming from a more robust inflammatory response during the acute stage, and thus more serum fatty acids entering the mitochondrial β-oxidation to provide the necessary energy [33,34]. Importantly, the serum levels of fatty acids, including linoleate, dodecanoate, hexadecanoate, and octadecanoate, exhibited significant correlations with the inflammatory marker CRP, suggesting a potential linkage between the development of AS and the alterations in fatty acids.

Glucose catabolism is a major source of energy [35], and glycerol-3-phosphate is mainly derived from glucose metabolism. Decreased serum glucose and 1,5-anhydrosorbitol, along with elevated glycerol-3-phosphate in AS patients, suggest an accelerated glucose metabolism, possibly linked to an increased energy demand due to chronic inflammation [36]. 2-Hydroxybutanoate can also enter the energy metabolism pathway and participate in the production of adenosine triphosphate [37]. Compared with the remission stage of AS, patients in the acute stage experience heightened inflammatory responses and greater energy requirements, leading to further consumption of glucose and 2-hydroxybutanoate to provide additional energy for the body.

Multiple amino acid metabolic pathways and related metabolites showed abnormalities in AS patients. The biosynthesis and degradation of valine/leucine/isoleucine primarily influence the levels of branched-chain amino acids (BCAAs) [38], such as leucine, isoleucine, and valine. BCAAs play an important role in the metabolism and immune regulation of the body [39]. Several studies have indicated an association between BCAA levels and the inflammatory state [40]. In inflammatory diseases or infections, BCAAs can be converted into energy in the liver, thereby maintaining the energy balance of the body [41] and leading to the consumption of BCAAs during the inflammatory state. Here, the altered BCAA levels in AS patients, characterized by significant reductions in valine, leucine, and isoleucine, suggest that AS patients are experiencing sustained energy depletion.

Lymphocyte abnormality is one of the most important factors in the development of AS [42]. Amino acids are the basic substances that constitute proteins, which are closely related to various aspects of the immune system, including the development of immune organs, the differentiation and proliferation of immune cells, the secretion of cytokines, the production of antibodies, and so on [43–45]. Disruptions in the biosynthesis of phenylalanine/tyrosine/tryptophan and decreased tryptophan were observed in AS patients. Previous studies have demonstrated that tryptophan depletion and the accumulation of tryptophan metabolites can impede T cell function, including suppressing T cell proliferation, impairing T cell survival, and promoting T cell apoptosis [46,47]. Threonine and methionine are major amino acids comprising immunoglobulins, and methionine has a protective effect on lymphocytes. The decrease in threonine and methionine may inhibit the synthesis and secretion of immunoglobulins, affecting humoral immunity [43,48]. The change in these amino acids is closely associated with the immune dysfunction in AS patients.

In AS patients, the glutathione metabolism was significantly changed. With the increased level of oxidative stress in AS patients, the demand for antioxidants, such as glutathione, increases to counteract the excessive production of reactive oxygen species (ROS) [49,50]. Amino acids like glycine and alanine play crucial roles in the synthesis of glutathione [51,52]. Our research has revealed a reduction in these amino acids in AS patients, which could be associated with an increased demand for glutathione synthesis.

Notably, serine and pyroglutamate were consistently decreased in the acute stage of AS compared with the remission stage. Serine is one of the building blocks of neutrophil serine proteases (NSPs), which play a significant role in inflammatory responses and immune reactions during acute infections [53,54]. Pyroglutamate is considered an endogenous antioxidant, which may help neutralize intracellular ROS, thus protecting cells from damage caused by oxidative stress [55,56]. The sustained reduction in serine and pyroglutamate during the acute stage of AS implies that AS is characterized by inflammatory activation and oxidative damage during the acute stage.

The main limitations of the study are as follows: first, only a portion of the metabolites associated with AS stages were structurally identified, and the unidentified metabolites may have important biological functions, which need to be further confirmed in combination with other analytical techniques; second, we did not set up an independent validation set to evaluate the diagnostic efficacy of the potential biomarkers in AS staging in clinical practice.

5. Conclusions

Taken together, a GC–MS-based metabolomics analysis was performed to shed light on significant changes in AS patients. In particular, it was demonstrated that patients in the acute stage and remission stage showed different metabolic characteristics, especially in terms of the changes in fatty acids (linoleate, dodecanoate, hexadecanoate, and octadecanoate), amino acids (serine and pyroglutamate), 2-hydroxybutanoate, glucose, etc. We propose that these alterations may be related to the aggravated inflammatory response and elevated oxidative stress in the acute stage. Moreover, serum 2-hydroxybutanoate and hexadecanoate had good efficacy for the stage division of AS. All the aforementioned findings could provide a basis for the staging treatment and pathogenesis studies of AS.

Supplementary Materials: The following supporting information can be downloaded at: https://www.mdpi.com/article/10.3390/metabo13101058/s1, Figure S1: Analytical characteristics of the serum metabolic profiling method; Figure S2: Validation plots for OSC PLS–DA models.

Author Contributions: Conceptualization, J.Z. and C.W.; methodology, S.W.; software, Y.G. and M.Y.; data curation, J.Z. and D.C.; writing—original draft preparation, Y.G. and S.W.; writing—review and editing, J.Z. and Y.L.; project administration, C.W.; funding acquisition, J.Z. All authors have read and agreed to the published version of the manuscript.

Funding: This research was funded by the Zhejiang Provincial Natural Science Foundation of China (no. LY22H290006) and the Research Project of Zhejiang Chinese Medical University (nos. 2022JKZKTS05 and 2022JKZKTS43).

Institutional Review Board Statement: The study was conducted in accordance with the Declaration of Helsinki and approved by the Ethics Committee of the First Affiliated Hospital of Zhejiang Chinese Medical University (2021–KLJ–010–03, 6 July 2021).

Informed Consent Statement: Informed consent was obtained from all subjects involved in the study.

Data Availability Statement: The data presented in this study are available on request from the corresponding author. The data are not publicly available due to ethical restrictions.

Acknowledgments: We sincerely appreciate the technical support from the Public Platform of Medical Research Center, Academy of Chinese Medical Science, Zhejiang Chinese Medical University.

Conflicts of Interest: The authors declare no conflict of interest.

References

1. Liu, L.; Yuan, Y.; Zhang, S.; Xu, J.; Zou, J. Osteoimmunological insights into the pathogenesis of ankylosing spondylitis. *J. Cell. Physiol.* **2021**, *236*, 6090–6100. [CrossRef]
2. Smith, J.A. Update on ankylosing spondylitis: Current concepts in pathogenesis. *Curr. Allergy Asthma Rep.* **2015**, *15*, 489. [CrossRef]
3. Watad, A.; Bridgewood, C.; Russell, T.; Marzo-Ortega, H.; Cuthbert, R.; McGonagle, D. The Early Phases of Ankylosing Spondylitis: Emerging Insights From Clinical and Basic Science. *Front. Immunol.* **2018**, *9*, 2668. [CrossRef]
4. Costantino, F.; Talpin, A.; Said-Nahal, R.; Goldberg, M.; Henny, J.; Chiocchia, G.; Garchon, H.J.; Zins, M.; Breban, M. Prevalence of spondyloarthritis in reference to HLA-B27 in the French population: Results of the GAZEL cohort. *Ann. Rheum. Dis.* **2015**, *74*, 689–693. [CrossRef]
5. Zochling, J.; Braun, J. Assessments in ankylosing spondylitis. *Best Pract. Res. Clin. Rheumatol.* **2007**, *21*, 699–712. [CrossRef]
6. Braun, J.; Sieper, J. Ankylosing spondylitis. *Lancet* **2007**, *369*, 1379–1390. [CrossRef]
7. Agca, R.; Heslinga, S.C.; Rollefstad, S.; Heslinga, M.; McInnes, I.B.; Peters, M.J.; Kvien, T.K.; Dougados, M.; Radner, H.; Atzeni, F.; et al. EULAR recommendations for cardiovascular disease risk management in patients with rheumatoid arthritis and other forms of inflammatory joint disorders: 2015/2016 update. *Ann. Rheum. Dis.* **2017**, *76*, 17–28. [CrossRef]
8. Tan, S.; Yao, J.; Ward, M.M.; Yao, L.; Summers, R.M. Computer aided evaluation of ankylosing spondylitis using high-resolution CT. *IEEE Trans. Med. Imaging* **2008**, *27*, 1252–1267. [CrossRef]
9. Van der Heijde, D.; Landewe, R. Imaging in spondylitis. *Curr. Opin. Rheumatol.* **2005**, *17*, 413–417. [CrossRef]
10. De Vlam, K. Soluble and tissue biomarkers in ankylosing spondylitis. *Best Pract. Res. Clin. Rheumatol.* **2010**, *24*, 671–682. [CrossRef]
11. Dunn, W.B.; Broadhurst, D.; Begley, P.; Zelena, E.; Francis-McIntyre, S.; Anderson, N.; Brown, M.; Knowles, J.D.; Halsall, A.; Haselden, J.N.; et al. Procedures for large-scale metabolic profiling of serum and plasma using gas chromatography and liquid chromatography coupled to mass spectrometry. *Nat. Protoc.* **2011**, *6*, 1060–1083. [CrossRef]
12. Johnson, C.H.; Ivanisevic, J.; Siuzdak, G. Metabolomics: Beyond biomarkers and towards mechanisms. *Nat. Rev. Mol. Cell Biol.* **2016**, *17*, 451–459. [CrossRef] [PubMed]
13. Rinschen, M.M.; Ivanisevic, J.; Giera, M.; Siuzdak, G. Identification of bioactive metabolites using activity metabolomics. *Nat. Rev. Mol. Cell Biol.* **2019**, *20*, 353–367. [CrossRef] [PubMed]
14. Bujak, R.; Struck-Lewicka, W.; Markuszewski, M.J.; Kaliszan, R. Metabolomics for laboratory diagnostics. *J. Pharm. Biomed. Anal.* **2015**, *113*, 108–120. [CrossRef] [PubMed]
15. Rutowski, J.; Zhong, F.; Xu, M.; Zhu, J. Metabolic shift of Staphylococcus aureus under sublethal dose of methicillin in the presence of glucose. *J. Pharm. Biomed. Anal.* **2019**, *167*, 140–148. [CrossRef] [PubMed]
16. Hu, C.; Zhang, J.; Hong, S.; Li, H.; Lu, L.; Xie, G.; Luo, W.; Du, Y.; Xie, Z.; Han, X.; et al. Oxidative stress-induced aberrant lipid metabolism is an important causal factor for dysfunction of immunocytes from patients with systemic lupus erythematosus. *Free. Radic. Biol. Med.* **2021**, *163*, 210–219. [CrossRef]
17. Arneth, B.; Arneth, R.; Shams, M. Metabolomics of Type 1 and Type 2 Diabetes. *Int. J. Mol. Sci.* **2019**, *20*, 2467. [CrossRef]
18. Ferrarini, A.; Di Poto, C.; He, S.; Tu, C.; Varghese, R.S.; Kara Balla, A.; Jayatilake, M.; Li, Z.; Ghaffari, K.; Fan, Z.; et al. Metabolomic Analysis of Liver Tissues for Characterization of Hepatocellular Carcinoma. *J. Proteome Res.* **2019**, *18*, 3067–3076. [CrossRef]
19. Guma, M.; Dadpey, B.; Coras, R.; Mikuls, T.R.; Hamilton, B.; Quehenberger, O.; Thorisdottir, H.; Bittleman, D.; Lauro, K.; Reilly, S.M.; et al. Xanthine oxidase inhibitor urate-lowering therapy titration to target decreases serum free fatty acids in gout and suppresses lipolysis by adipocytes. *Arthritis Res. Ther.* **2022**, *24*, 175. [CrossRef]
20. Rizzo, C.; Camarda, F.; Donzella, D.; La Barbera, L.; Guggino, G. Metabolomics: An Emerging Approach to Understand Pathogenesis and to Assess Diagnosis and Response to Treatment in Spondyloarthritis. *Cells* **2022**, *11*, 549. [CrossRef]
21. Fischer, R.; Trudgian, D.C.; Wright, C.; Thomas, G.; Bradbury, L.A.; Brown, M.A.; Bowness, P.; Kessler, B.M. Discovery of candidate serum proteomic and metabolomic biomarkers in ankylosing spondylitis. *Mol. Cell Proteomics* **2012**, *11*, M111.013904. [CrossRef]
22. Gao, P.; Lu, C.; Zhang, F.; Sang, P.; Yang, D.; Li, X.; Kong, H.; Yin, P.; Tian, J.; Lu, X.; et al. Integrated GC-MS and LC-MS plasma metabonomics analysis of ankylosing spondylitis. *Analyst* **2008**, *133*, 1214–1220. [CrossRef] [PubMed]
23. Wang, W.; Yang, G.J.; Zhang, J.; Chen, C.; Jia, Z.Y.; Li, J.; Xu, W.D. Plasma, urine and ligament tissue metabolite profiling reveals potential biomarkers of ankylosing spondylitis using NMR-based metabolic profiles. *Arthritis Res. Ther.* **2016**, *18*, 244. [CrossRef]
24. Doğan, H.O.; Şenol, O.; Karadağ, A.; Yıldız, S.N. Metabolomic profiling in ankylosing spondylitis using time-of-flight mass spectrometry. *Clin. Nutr. ESPEN* **2022**, *50*, 124–132. [CrossRef]

25. Berlinberg, A.J.; Regner, E.H.; Stahly, A.; Brar, A.; Reisz, J.A.; Gerich, M.E.; Fennimore, B.P.; Scott, F.I.; Freeman, A.E.; Kuhn, K.A. Multi 'Omics Analysis of Intestinal Tissue in Ankylosing Spondylitis Identifies Alterations in the Tryptophan Metabolism Pathway. *Front. Immunol.* **2021**, *12*, 587119. [CrossRef] [PubMed]
26. Sundstrom, B.; Johansson, G.; Kokkonen, H.; Cederholm, T.; Wallberg-Jonsson, S. Plasma phospholipid fatty acid content is related to disease activity in ankylosing spondylitis. *J. Rheumatol.* **2012**, *39*, 327–333. [CrossRef] [PubMed]
27. Eryavuz Onmaz, D.; Sivrikaya, A.; Isik, K.; Abusoglu, S.; Albayrak Gezer, I.; Humeyra Yerlikaya, F.; Abusoglu, G.; Unlu, A.; Tezcan, D. Altered kynurenine pathway metabolism in patients with ankylosing spondylitis. *Int. Immunopharmacol.* **2021**, *99*, 108018. [CrossRef] [PubMed]
28. Chong, J.; Soufan, O.; Li, C.; Caraus, I.; Li, S.; Bourque, G.; Wishart, D.S.; Xia, J. MetaboAnalyst 4.0: Towards more transparent and integrative metabolomics analysis. *Nucleic Acids Res.* **2018**, *46*, W486–W494. [CrossRef]
29. Mauro, D.; Thomas, R.; Guggino, G.; Lories, R.; Brown, M.A.; Ciccia, F. Ankylosing spondylitis: An autoimmune or autoinflammatory disease? *Nat. Rev. Rheumatol.* **2021**, *17*, 387–404. [CrossRef]
30. Chen, R.; Han, S.; Dong, D.M.; Wang, Y.S.; Liu, Q.P.; Xie, W.; Li, M.; Yao, M. Serum fatty acid profiles and potential biomarkers of ankylosing spondylitis determined by gas chromatography-mass spectrometry and multivariate statistical analysis. *Biomed. Chromatogr.* **2015**, *29*, 604–611. [CrossRef]
31. Wagner, C.; Visvanathan, S.; Braun, J.; van der Heijde, D.; Deodhar, A.; Hsu, B.; Mack, M.; Elashoff, M.; Inman, R.D. Serum markers associated with clinical improvement in patients with ankylosing spondylitis treated with golimumab. *Ann. Rheum. Dis.* **2012**, *71*, 674–680. [CrossRef] [PubMed]
32. Boden, G.; She, P.; Mozzoli, M.; Cheung, P.; Gumireddy, K.; Reddy, P.; Xiang, X.; Luo, Z.; Ruderman, N. Free fatty acids produce insulin resistance and activate the proinflammatory nuclear factor-kappaB pathway in rat liver. *Diabetes* **2005**, *54*, 3458–3465. [CrossRef] [PubMed]
33. Nakamura, M.T.; Yudell, B.E.; Loor, J.J. Regulation of energy metabolism by long-chain fatty acids. *Prog. Lipid Res.* **2014**, *53*, 124–144. [CrossRef] [PubMed]
34. He, G.; Chen, Y.; Wang, Z.; He, H.; Yu, P. Cellular Uptake, Metabolism and Sensing of Long-Chain Fatty Acids. *Front. Biosci. Landmark* **2023**, *28*, 10. [CrossRef] [PubMed]
35. Mulukutla, B.C.; Yongky, A.; Le, T.; Mashek, D.G.; Hu, W.S. Regulation of Glucose Metabolism—A Perspective From Cell Bioprocessing. *Trends Biotechnol.* **2016**, *34*, 638–651. [CrossRef]
36. Possik, E.; Al-Mass, A.; Peyot, M.L.; Ahmad, R.; Al-Mulla, F.; Madiraju, S.R.M.; Prentki, M. New Mammalian Glycerol-3-Phosphate Phosphatase: Role in beta-Cell, Liver and Adipocyte Metabolism. *Front. Endocrinol.* **2021**, *12*, 706607. [CrossRef]
37. Sousa, A.P.; Cunha, D.M.; Franco, C.; Teixeira, C.; Gojon, F.; Baylina, P.; Fernandes, R. Which Role Plays 2-Hydroxybutyric Acid on Insulin Resistance? *Metabolites* **2021**, *11*, 835. [CrossRef]
38. Nie, C.; He, T.; Zhang, W.; Zhang, G.; Ma, X. Branched Chain Amino Acids: Beyond Nutrition Metabolism. *Int. J. Mol. Sci.* **2018**, *19*, 954. [CrossRef]
39. Dimou, A.; Tsimihodimos, V.; Bairaktari, E. The Critical Role of the Branched Chain Amino Acids (BCAAs) Catabolism-Regulating Enzymes, Branched-Chain Aminotransferase (BCAT) and Branched-Chain alpha-Keto Acid Dehydrogenase (BCKD), in Human Pathophysiology. *Int. J. Mol. Sci.* **2022**, *23*, 4022. [CrossRef]
40. Mattick, J.S.A.; Kamisoglu, K.; Ierapetritou, M.G.; Androulakis, I.P.; Berthiaume, F. Branched-chain amino acid supplementation: Impact on signaling and relevance to critical illness. *Wiley Interdiscip. Rev. Syst. Biol. Med.* **2013**, *5*, 449–460. [CrossRef]
41. Mero, A. Leucine supplementation and intensive training. *Sports Med.* **1999**, *27*, 347–358. [CrossRef] [PubMed]
42. Liang, T.; Chen, J.; Xu, G.; Zhang, Z.; Xue, J.; Zeng, H.; Jiang, J.; Chen, T.; Qin, Z.; Li, H.; et al. Platelet-to-Lymphocyte Ratio as an Independent Factor Was Associated With the Severity of Ankylosing Spondylitis. *Front. Immunol.* **2021**, *12*, 760214. [CrossRef] [PubMed]
43. Miyajima, M. Amino acids: Key sources for immunometabolites and immunotransmitters. *Int. Immunol.* **2020**, *32*, 435–446. [CrossRef] [PubMed]
44. Pakula, M.M.; Maier, T.J.; Vorup-Jensen, T. Insight on the impacts of free amino acids and their metabolites on the immune system from a perspective of inborn errors of amino acid metabolism. *Expert. Opin. Ther. Targets* **2017**, *21*, 611–626. [CrossRef] [PubMed]
45. Li, P.; Yin, Y.L.; Li, D.; Kim, S.W.; Wu, G. Amino acids and immune function. *Br. J. Nutr.* **2007**, *98*, 237–252. [CrossRef]
46. Grohmann, U.; Bronte, V. Control of immune response by amino acid metabolism. *Immunol. Rev.* **2010**, *236*, 243–264. [CrossRef]
47. Stone, T.W.; Williams, R.O. Modulation of T cells by tryptophan metabolites in the kynurenine pathway. *Trends Pharmacol. Sci.* **2023**, *44*, 442–456. [CrossRef]
48. Zhang, H.; Chen, Y.; Li, Y.; Zhang, T.; Ying, Z.; Su, W.; Zhang, L.; Wang, T. l-Threonine improves intestinal mucin synthesis and immune function of intrauterine growth-retarded weanling piglets. *Nutrition* **2019**, *59*, 182–187. [CrossRef]
49. Ye, G.; Xie, Z.; Zeng, H.; Wang, P.; Li, J.; Zheng, G.; Wang, S.; Cao, Q.; Li, M.; Liu, W.; et al. Oxidative stress-mediated mitochondrial dysfunction facilitates mesenchymal stem cell senescence in ankylosing spondylitis. *Cell Death Dis.* **2020**, *11*, 775. [CrossRef]
50. Kiranatlioglu-Firat, F.; Demir, H.; Cuce, I.; Altin-Celik, P.; Eciroglu, H.; Bayram, F.; Donmez-Altuntas, H. Increased oxidative and chromosomal DNA damage in patients with ankylosing spondylitis: Its role in pathogenesis. *Clin. Exp. Med.* **2022**, *23*, 1721–1728. [CrossRef]
51. Asantewaa, G.; Harris, I.S. Glutathione and its precursors in cancer. *Curr. Opin. Biotechnol.* **2021**, *68*, 292–299. [CrossRef] [PubMed]

52. Petry, E.R.; Cruzat, V.F.; Heck, T.G.; Leite, J.S.; Homem de Bittencourt, P.I., Jr.; Tirapegui, J. Alanyl-glutamine and glutamine plus alanine supplements improve skeletal redox status in trained rats: Involvement of heat shock protein pathways. *Life Sci.* **2014**, *94*, 130–136. [CrossRef] [PubMed]
53. Pham, C.T. Neutrophil serine proteases: Specific regulators of inflammation. *Nat. Rev. Immunol.* **2006**, *6*, 541–550. [CrossRef] [PubMed]
54. Wiedow, O.; Meyer-Hoffert, U. Neutrophil serine proteases: Potential key regulators of cell signalling during inflammation. *J. Intern. Med.* **2005**, *257*, 319–328. [CrossRef]
55. Emmett, M. Acetaminophen toxicity and 5-oxoproline (pyroglutamic acid): A tale of two cycles, one an ATP-depleting futile cycle and the other a useful cycle. *Clin. J. Am. Soc. Nephrol.* **2014**, *9*, 191–200. [CrossRef]
56. Liu, Y.; Hyde, A.S.; Simpson, M.A.; Barycki, J.J. Emerging regulatory paradigms in glutathione metabolism. *Adv. Cancer Res.* **2014**, *122*, 69–101. [CrossRef]

Disclaimer/Publisher's Note: The statements, opinions and data contained in all publications are solely those of the individual author(s) and contributor(s) and not of MDPI and/or the editor(s). MDPI and/or the editor(s) disclaim responsibility for any injury to people or property resulting from any ideas, methods, instructions or products referred to in the content.

Article

Evaluating the Effects of Omega-3 Polyunsaturated Fatty Acids on Inflammatory Bowel Disease via Circulating Metabolites: A Mediation Mendelian Randomization Study

Xiaojing Jia [1,2,†], Chunyan Hu [1,2,†], Xueyan Wu [1,2,†], Hongyan Qi [1,2], Lin Lin [1,2], Min Xu [1,2], Yu Xu [1,2], Tiange Wang [1,2], Zhiyun Zhao [1,2], Yuhong Chen [1,2], Mian Li [1,2], Ruizhi Zheng [1,2], Hong Lin [1,2], Shuangyuan Wang [1,2], Weiqing Wang [1,2], Yufang Bi [1,2], Jie Zheng [1,2,3,*] and Jieli Lu [1,2,*]

[1] Department of Endocrine and Metabolic Diseases, Shanghai Institute of Endocrine and Metabolic Diseases, Ruijin Hospital, Shanghai Jiao Tong University School of Medicine, Shanghai 200025, China
[2] Shanghai National Clinical Research Center for Metabolic Diseases, Key Laboratory for Endocrine and Metabolic Diseases of the National Health Commission of the PR China, Shanghai Key Laboratory for Endocrine Tumor, State Key Laboratory of Medical Genomics, Ruijin Hospital, Shanghai Jiao Tong University School of Medicine, Shanghai 200025, China
[3] MRC Integrative Epidemiology Unit (IEU), Bristol Medical School, University of Bristol, Oakfield House, Oakfield Grove, Bristol BS8 2BN, UK
* Correspondence: zj12477@rjh.com.cn (J.Z.); ljl11319@rjh.com.cn (J.L.)
† These authors contributed equally to this work.

Abstract: Epidemiological evidence regarding the effect of omega-3 polyunsaturated fatty acid (PUFA) supplementation on inflammatory bowel disease (IBD) is conflicting. Additionally, little evidence exists regarding the effects of specific omega-3 components on IBD risk. We applied two-sample Mendelian randomization (MR) to disentangle the effects of omega-3 PUFAs (including total omega-3, α-linolenic acid, eicosapentaenoic acid (EPA), or docosahexaenoic acid (DHA)) on the risk of IBD, Crohn's disease (CD) and ulcerative colitis (UC). Our findings indicated that genetically predicted increased EPA concentrations were associated with decreased risk of IBD (odds ratio 0.78 (95% CI 0.63–0.98)). This effect was found to be mediated through lower levels of linoleic acid and histidine metabolites. However, we found limited evidence to support the effects of total omega-3, α-linolenic acid, and DHA on the risks of IBD. In the *fatty acid desaturase 2* (*FADS2*) region, robust colocalization evidence was observed, suggesting the primary role of the *FADS2* gene in mediating the effects of omega-3 PUFAs on IBD. Therefore, the present MR study highlights EPA as the predominant active component of omega-3 fatty acids in relation to decreased risk of IBD, potentially via its interaction with linoleic acid and histidine metabolites. Additionally, the *FADS2* gene likely mediates the effects of omega-3 PUFAs on IBD risk.

Keywords: eicosapentaenoic acid; inflammatory bowel disease; Mendelian randomization; mediation; omega-3 polyunsaturated fatty acids

1. Introduction

Inflammatory bowel disease (IBD) is a group of chronic inflammatory disorders affecting the gastrointestinal tract, and its prevalence has increased worldwide, reaching up to 0.5% of the general population in the western world [1,2]. The two primary types of IBD are Crohn's disease (CD) and ulcerative colitis (UC), each with different clinical and histopathological characteristics [3]. The economic burden of IBD is substantial, with over €4.6 billion in annual medical costs in Europe and US$6 billion in the USA, putting a strain on healthcare systems and resources [2]. To alleviate this burden, a comprehensive approach is needed, including the development of preventive care to delay the progression of this disease. Omega-3 polyunsaturated fatty acids (PUFAs) are commonly used nutritional supplements and show beneficial effects on coronary heart disease [4] and asthma [5]. Due to their anti-inflammatory properties, PUFAs have been proposed as potential targets for preventing and treating autoimmune diseases [6]. Omega-3 PUFAs can be quantified based on a shift in the signal induced by the position of the omega-3 double bond. The sum of concentrations of α-linolenic acid, eicosapentaenoic acid (EPA), docosahexaenoic acid (DHA), and other omega-3 PUFAs is expressed as total omega-3 fatty acids. Long-chain omega-3 PUFAs (EPA and DHA) are derived from α-linolenic acid through a series of elongation, desaturation, and β-oxidation events during fatty acid metabolism. The *fatty acid desaturase 2* (*FADS2*) gene encodes delta-6 desaturase and plays a key regulatory role in this metabolism process [7].

In randomized controlled trials (RCTs), EPA and DHA have often been combined as the active components of omega-3 fatty acids and consumed together, despite their distinct molecular functions and clinical impacts [8]. Daily supplementation with EPA and DHA were reported to be effective in reducing the clinical relapse of CD [9]. However, in the large-scaled vitamin D and omega 3 trial (VITAL) with approximately five years of randomized follow-up, fish oil containing EPA and DHA did not significantly reduce the rate of a composite outcome consisting of rheumatoid arthritis, IBD, autoimmune thyroid disease, and all other autoimmune diseases [10]. Moreover, there was a lack of detailed information on IBD in this study. Additionally, observational studies did not provide convincing and consistent evidence of the relationship between dietary intakes of omega-3 PUFAs and the risk of IBD [11–13]. Information on usual diet relied on self-reported dietary questionnaires, which may produce errors or bias in recall. The existing evidence makes it challenging to confirm the causal effect of omega-3 PUFAs on IBD; and identify the key supplement among the omega-3 PUFA component (α-linolenic acid, EPA, and DHA) that may exhibit the protective effect.

Mendelian randomization (MR) is an approach that could estimate causal effect of an exposure on an outcome and overcome issues related to residual confounding or reverse causality [14]. Moreover, this method allows for investigating the effects of each omega-3 PUFA component on IBD, which may be challenging to achieve in an RCT setting. Recently, He et al. reported that total omega-3 fatty acid had a protective effect against increased UC risk instead of CD [15], but the evidence on IBD was not addressed. In addition, their analysis involved only 21 omega-3 instruments after eliminating SNPs associated with potential confounders and outcomes, which might have reduced the power of the analysis. More critically, some instruments in key regulatory genes such as *FADS2* gene were eliminated, which may have important influences on the reliability of the findings. Meanwhile, there remains a knowledge gap in evaluating the separate biological effects of α-linolenic acid, EPA, and DHA, with their metabolic mechanisms being unexplored.

In this study, we aimed to explore the effects of omega-3 PUFAs (i.e., total omega-3, α-linolenic acid, EPA, and DHA) on the risk of IBD and its subtypes, and the potential metabolic pathways linking omega-3 PUFAs with IBD. Given the central role of the *FADS2* gene in omega-3 PUFAs' metabolism, further analyses in this specific region were essential through genetic colocalization. This approach allowed us to assess whether there were shared causal variants within the *FADS2* gene region that could influence both omega-3 PUFAs and IBD risk [16].

2. Materials and Methods

2.1. Study Design

A schematic overview of the study design was detailed in Figure 1. We employed the univariable MR analysis to assess whether total omega-3 fatty acid, α-linolenic acid, EPA, and DHA showed causal effects on IBD and its subtypes (CD and UC), using summary-level data from publicly available genome-wide association studies (GWASs). Colocalization analysis was further conducted in the *FADS2* gene region to test for pleiotropic effect and investigate the underlying mechanisms. A bidirectional MR analysis was applied to estimate the effect of genetic liability to IBD on omega-3 PUFAs. Mediation MR analysis estimated the effect of potential metabolites linking omega-3 PUFAs with the IBD. All datasets were publicly available, and ethical approval was acquired for all original studies.

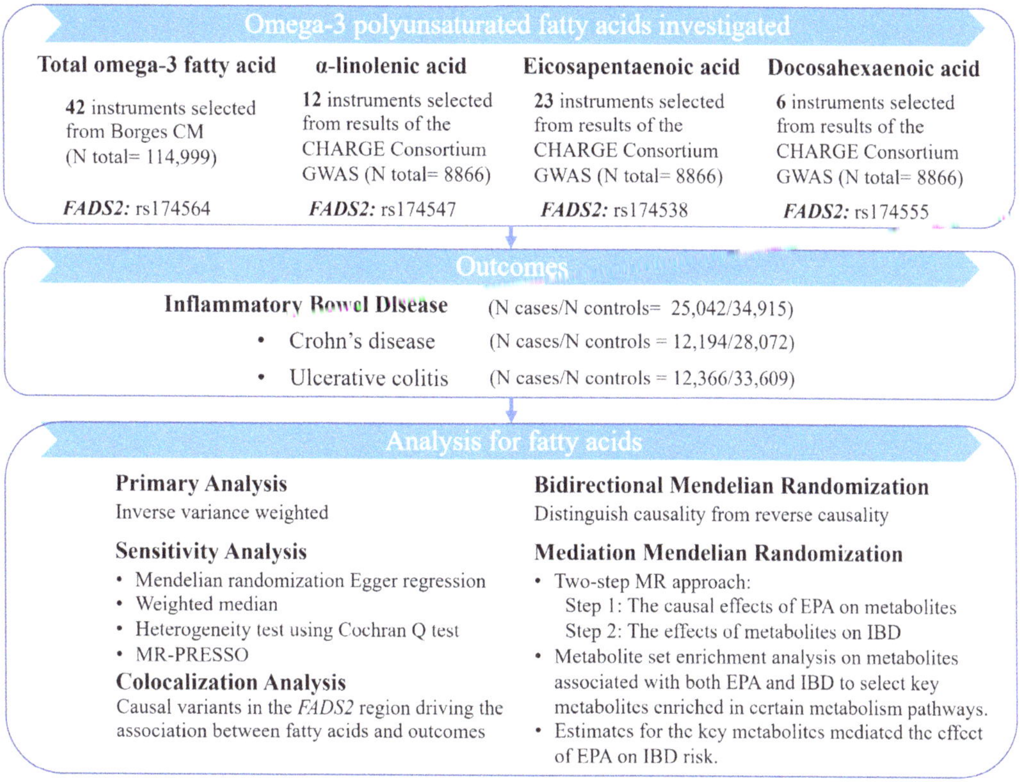

Omega-3 polyunsaturated fatty acids investigated

Total omega-3 fatty acid
42 instruments selected from Borges CM (N total= 114,999)
FADS2: rs174564

α-linolenic acid
12 instruments selected from results of the CHARGE Consortium GWAS (N total= 8866)
FADS2: rs174547

Eicosapentaenoic acid
23 instruments selected from results of the CHARGE Consortium GWAS (N total= 8866)
FADS2: rs174538

Docosahexaenoic acid
6 instruments selected from results of the CHARGE Consortium GWAS (N total= 8866)
FADS2: rs174555

Outcomes

Inflammatory Bowel Disease (N cases/N controls= 25,042/34,915)
- Crohn's disease (N cases/N controls = 12,194/28,072)
- Ulcerative colitis (N cases/N controls = 12,366/33,609)

Analysis for fatty acids

Primary Analysis
Inverse variance weighted

Sensitivity Analysis
- Mendelian randomization Egger regression
- Weighted median
- Heterogeneity test using Cochran Q test
- MR-PRESSO

Colocalization Analysis
Causal variants in the *FADS2* region driving the association between fatty acids and outcomes

Bidirectional Mendelian Randomization
Distinguish causality from reverse causality

Mediation Mendelian Randomization
- Two-step MR approach:
  Step 1: The causal effects of EPA on metabolites
  Step 2: The effects of metabolites on IBD
- Metabolite set enrichment analysis on metabolites associated with both EPA and IBD to select key metabolites enriched in certain metabolism pathways.
- Estimates for the key metabolites mediated the effect of EPA on IBD risk.

Figure 1. Study design of this MR study.

2.2. Data Sources and Genetic Instruments for Omega-3 PUFAs

Single-nucleotide polymorphisms (SNPs) associated with total omega-3 fatty acid were derived from UK Biobank, which collected deep genetic and phenotypic data from approximately 500,000 individuals aged between 40 and 69 [17]. Genetic associations of α-linolenic acid, EPA, and DHA were obtained from a GWAS meta-analysis in 8866 participants of European ancestry from the Cohorts for Heart and Aging Research in Genomic Epidemiology (CHARGE) Consortium [18]. Details of the data sources and sample sizes of the exposures are listed in the Table S1.

In this study, the genetic variants that showed robust association with total omega-3 fatty acid (with genetic association p value $< 5 \times 10^{-8}$) and showed independence (with linkage disequilibrium (LD) $r^2 < 0.01$ in European ancestry) were selected as candidate instruments. Given the limited sample size of the α-linolenic acid, EPA, and DHA GWASs, a slightly more relaxing threshold ($p < 5 \times 10^{-6}$) was used to select instruments for these exposures. After harmonization with outcome data and removing palindromic or mismatching alleles, 42 independent SNPs for total omega-3 fatty acid, 12 independent SNPs for α-linolenic acid, 23 independent SNPs for EPA, and 6 independent SNPs for DHA were selected as instruments (Figure S1). One SNP was selected to represent the effect of each omega-3 PUFA in the *FADS2* region (rs174564 for total omega-3 fatty acid, rs174547 for α-linolenic acid, rs174538 for EPA, and rs174555 for DHA; all these SNPs are in strong LD to each other (LD $r^2 > 0.7$), which represents the same signal in this region).

2.3. Outcome Data Sources

Summary statistics for IBD were obtained from the study by the International Inflammatory Bowel Disease Genetics Consortium (IIBDGC) [19], which contained a total of 59,957 European participants (cases/controls for IBD: 25,042/34,915; UC: 12,366/33,609; CD: 12,194/28,072). All the cases were diagnosed using accepted endoscopic, histopathological and radiological criteria and all the control samples were obtained from the Understanding Society Project.

2.4. Metabolite Data Sources

The full GWAS summary statistics of the 974 circulating metabolites were derived from the IEU OpenGWAS database with GWAS identifier met-a, met-b, met-c, and met-d.

2.5. Statistical Analysis

2.5.1. Two-Sample MR Analysis

The inverse weighted variance (IVW) method was used as a primary analysis to estimate the causal effects of omega-3 PUFAs on IBD and its subtypes, which were calculated by a weighted linear regression of the instrument–outcome association estimates on the instrument–exposure association estimates assuming that all genetic variants were valid instruments [14].

MR analysis had several assumptions. The genetic instruments need to (1) be robustly associated with the exposure ("relevance"), (2) be independent of potential confounders of the instrument–outcome association ("exchangeability"), and (3) only affect the outcome through the exposure being tested and not through alternative pathways (that is, through pleiotropy; "exclusion restriction").

The relevance MR assumption was assessed from the mean F statistics within univariable MR, which was greater than ten for every instrument–exposure association, demonstrating the small possibility of weak instrumental variable bias [20].

We attempted using colocalization to test exchangeability MR assumption whether SNPs associated with two traits are possibly in LD, or a single shared signal (colocalization) [16]. The *FADS2* gene encoded key fatty acid desaturase enzymes, which are pivotal for omega-3 PUFA biosynthesis. Therefore, we performed colocalization analyses in this gene region. A generated colocalization posterior probability greater than 0.70 indicated the same variant causal for both traits and indirectly denied the possibility of "exchangeability".

To check for violation of the "exclusion restriction" assumption of MR and assess pleiotropy, we made different assumptions regarding MR instrument validity using several sensitivity analyses, that included the weighted median method which permitted up to 50% of the information in the MR analysis to come from invalid instruments [21], and the MR-Egger approach which accounted for pleiotropy [22]. The MR pleiotropy residual sum and outlier (MR-PRESSO) method [23], together with MR-Egger intercept could be used to examine the level of horizontal pleiotropy, and reduce the level of horizontal pleiotropy via outlier removal. In addition, heterogeneity of the estimates was detected using Cochran's Q [24]. A leave-one-out analysis was conducted by removing each SNP from the analysis in turn and performing an IVW method on the remaining SNPs to assess the potential influence of a particular variant on the estimates [25].

We also sought to evaluate whether there was a reverse causal effect where liability to IBD consequently altered the levels of omega-3 PUFAs by performing a bidirectional MR analysis [26]. Therefore, we took independent instruments robustly associated with IBD, CD, and UC ($p < 5 \times 10^{-8}$) as exposures to assess their effect on total omega-3 fatty acid, α-linolenic acid, EPA, and DHA, respectively.

2.5.2. Mediation MR Analysis Linking EPA with IBD via Metabolites

We further estimated the mediation effects of circulating metabolites linking EPA with IBD risk. We used a novel analytical pipeline that integrated mediation MR with metabolite set enrichment analyses. First, we used a two-step MR approach to: (1) assess the causal effect of EPA on 974 potential metabolites (step 1) that have publicly available GWAS datasets in the IEU OpenGWAS database, which selected 237 metabolites with FDR < 0.05; and (2) estimate the effect of 237 metabolites on IBD (step 2), which further selected 211 metabolites associated with both EPA and IBD as candidate mediation metabolites. Second, we performed the metabolite set enrichment analysis on the 211 selected candidate metabolites, which aimed to select key metabolites enriched in certain metabolic pathways (Figure S1). For the metabolites that showed evidence of enrichment in the enrichment analysis, we further performed multivariable MR (MVMR) to determine their mediation effects on IBD which was adjusted for the effect of EPA [27]. We used IVW as our main approach to estimate the effect of EPA on the metabolites ($\beta 1$). Additionally, MVMR was applied to estimate: (1) the effect of each metabolite on risk of IBD with adjustment for the genetic effect of EPA ($\beta 2$); and (2) the direct effect of EPA on IBD with adjustment for each mediator individually (βdirect). To calculate the indirect mediation effect of EPA on IBD outcome, we used the difference of coefficients method as our main method, i.e., the casual effect of EPA on outcomes via metabolites (βtotal − βdirect). The total effect was the estimate of EPA on IBD in univariable MR (βtotal). Thus, the proportion of the total effect mediated by each metabolite was separately estimated by dividing the indirect effect by the total effect ((βtotal − βdirect)/βtotal). Standard errors were derived by using the delta method, using effect estimates obtained from 2SMR analysis.

Univariable, bidirectional, and multivariable MR analyses were considered significant with a 2-sided $p \leq 0.05$. Metabolites associated with omega-3 PUFAs or IBD were considered significant with an FDR < 0.05. Enrichment analysis was performed using the online MetaboAnalyst software (version 5.0, Mcgill University, Montreal, QC, Canada; https://www.metaboanalyst.ca, accessed on 17 November 2022) [28]. All analyses were performed using 'TwoSampleMR' and 'MR-PRESSO' package in R Software 3.6.0.

3. Results

We selected 42, 12, 23, and 6 SNPs as instruments to proxy life-long effect of total omega-3 fatty acid, α-linolenic acid, EPA, and DHA, respectively. In bidirectional MR, there were 117, 89, and 62 independent instruments incorporated for IBD, CD, and UC, respectively. Mean F statistics of the exposures ranged from 29.82 to 262.21 indicating that the MR estimates were not likely to be influenced by weak instrument bias (Table S2).

3.1. Genetically Predicted Omega-3 PUFAs on Risk of IBD (Including CD and UC)

Table 1 shows the effects of omega-3 PUFAs on IBD risks. Considering total omega-3 fatty acid as a whole, little evidence indicated its protective effect on IBD risk (odds ratio (OR) of IVW, 0.94; 95% confidence interval (CI), 0.82–1.07). Meanwhile, higher concentrations of α-linolenic acid showed a potential effect on increasing risk of IBD, although the evidence was weaker due to the wide confidence interval (OR of IVW, 1.54; 95% CI, 0.72–3.29). In contrast, genetically increased levels of EPA showed a causal effect on the lower risk of IBD (OR of IVW, 0.78; 95% CI, 0.63–0.98). There was little evidence for the presence of heterogeneity (Cochran's Q-test P_h = 0.10), pleiotropy (MR-Egger intercept $P_{intercept}$ = 0.97), or any outliers (MR-PRESSO P of global test = 0.099). Estimated effect was consistent using the weighted median approach (OR, 0.59; 95% CI, 0.45–0.78). However, there was little evidence to support the effect of DHA on IBD (OR of IVW, 1.05; 95% CI, 0.86–1.28).

The results of the primary MR analyses of CD and UC are presented in Figure 2. Results of sensitivity analyses are listed in Table S3. In consistent with the IBD results, there was little evidence to support the effects of total omega-3 fatty acid, α-linolenic acid, and DHA on the risk of CD and UC (Figure 2A,B,D). Meanwhile, increased levels of genetically proxied EPA still showed a strong effect on a lower risk of CD (OR of IVW, 0.67; 95% CI, 0.50–0.91), but with little effect on UC (OR of IVW, 0.88; 95% CI, 0.68–1.14) (Figure 2C).

Table 1. Two-sample Mendelian randomization estimations showing the effect of omega-3 PUFAs on inflammatory bowel disease.

Exposure	No. of SNPs	Methods	Estimate			Heterogeneity		Pleiotropy	
			OR	95% CI	P	Q	P_h	MR Egger int P	MR-PRESSO P
Total omega-3 fatty acid	42	IVW	0.94	(0.82, 1.07)	0.35	232.9	<0.001	0.06	<0.001
		MR-Egger	0.83	(0.69, 0.99)	0.05				
		Weighted median	0.85	(0.80, 0.92)	<0.001				
		MR-PRESSO outlier test	0.88	(0.81, 0.95)	0.003				
α-linolenic acid	12	IVW	1.54	(0.72, 3.29)	0.26	46.7	<0.001	0.65	<0.001
		MR-Egger	1.40	(0.58, 3.39)	0.48				
		Weighted median	1.42	(0.89, 2.28)	0.14				
		MR-PRESSO outlier test	1.24	(0.79, 1.95)	0.38				
EPA	23	IVW	0.78	(0.63, 0.98)	0.03	30.8	0.099	0.97	0.099
		MR-Egger	0.78	(0.45, 1.34)	0.37				
		Weighted median	0.59	(0.45, 0.78)	<0.001				
		MR-PRESSO outlier test	NA	NA	NA				
DHA	6	IVW	1.05	(0.86, 1.28)	0.65	21.6	<0.001	0.56	0.012
		MR-Egger	1.20	(0.75, 1.93)	0.49				
		Weighted median	1.12	(0.98, 1.28)	0.09				
		MR-PRESSO outlier test	1.11	(0.99, 1.25)	0.43				

Abbreviations: CI, confidence interval; DHA, docosahexaenoic acid; EPA, eicosapentaenoic acid; Egger int, egger intercept; IVW, inverse variance weighted; MR, Mendelian randomization; OR, odds ratio; PUFAs, polyunsaturated fatty acids; P_h, p-value for heterogeneity.

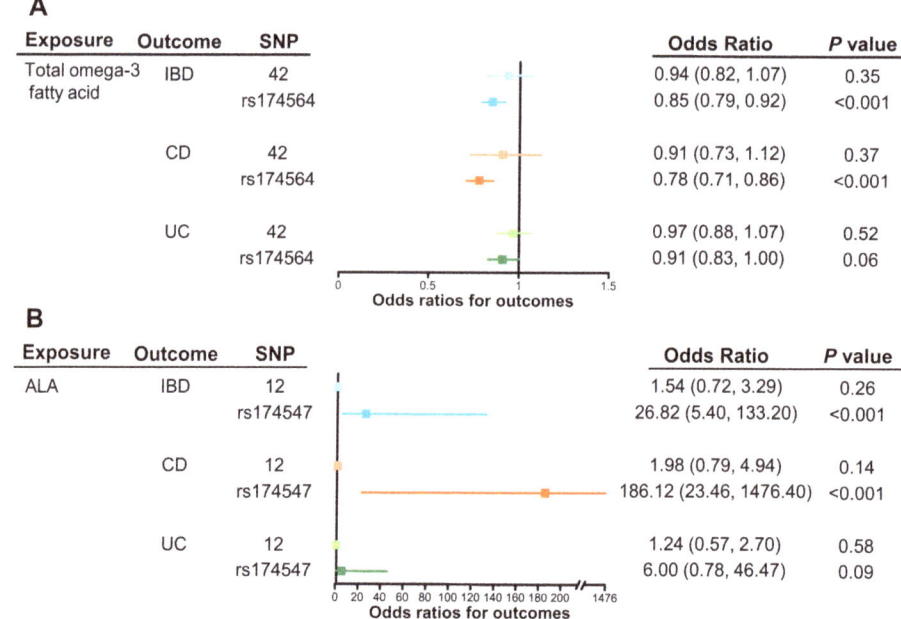

Figure 2. *Cont.*

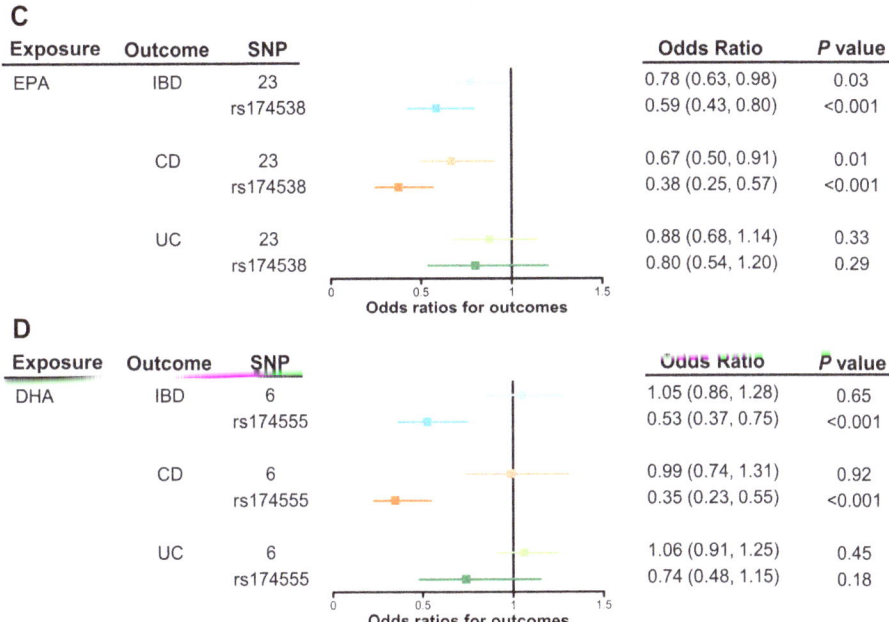

Figure 2. Causal effects of omega-3 polyunsaturated fatty acids on inflammatory bowel disease as a whole, on Crohn's disease, and ulcerative colitis or via the *FADS2* gene cluster. Univariable causal effects of (**A**) total omega-3, (**B**) α-linolenic acid, (**C**) EPA, and (**D**) DHA on investigated outcomes (light shades of blue, orange and green). Causal effects of each fatty acid on investigated outcomes via the *FADS2* gene (blue, orange and green). Abbreviations: ALA, α-linolenic acid; CD, Crohn's disease; DHA, docosahexaenoic acid; EPA, eicosapentaenoic acid; *FADS2, fatty acid desaturase 2*; IBD, inflammatory bowel disease; UC, ulcerative colitis.

3.2. Sensitivity Analysis in FADS2 Gene Region

As shown in the leave-one-out analyses, the MR estimates of omega-3 PUFAs on IBD, CD, and UC were mainly driven by SNP effects in the *FADS2* gene region (rs174564 for total omega-3 fatty acid, rs174547 for α-linolenic acid, rs174538 for EPA, and rs174555 for DHA) (Figure S2). As shown in Figure 2, the MR results of the single *FADS2* SNP showed the causal effects of total omega-3 fatty acid, EPA, and DHA on lower risk of IBD. The ORs (95% CI) were 0.85 (0.79–0.92), 0.59 (0.43–0.80), and 0.53 (0.37–0.75), respectively. On the contrary, α-linolenic acid showed a strong effect on the increasing risk of IBD (OR, 26.82; 95% CI, 5.40–133.20).

As for IBD subtypes, the *FADS2* gene showed a stronger effect on lowering the risk of CD (ORs (95% CI) were 0.78 (0.71–0.86) for total omega-3, 0.38 (0.25–0.57) for EPA, and 0.35 (0.23–0.55) for DHA) but was absent for UC. Meanwhile, a *FADS2* single-SNP in α-linolenic acid had a positive effect on increasing CD risk (OR, 186.12; 95% CI, 23.46–1476.40), but with less effect on UC risk.

Aligning with the MR estimates of a single-SNP in the *FADS2* region, we observed compelling evidence of colocalization for α-linolenic acid with CD (colocalization probability, 98.90%), but with little evidence for UC (colocalization probability, 2.61%; Figure 3A). A similar pattern of colocalization evidence was observed for EPA (colocalization probability of CD, 98.80%; colocalization probability of UC, 2.44%; Figure 3B), as well as DHA (colocalization probability of CD, 94.50%; colocalization probability of UC, 6.56%; Figure 3C). Collectively, colocalization analyses further supported distinct effects of omega-3 PUFAs on CD and UC.

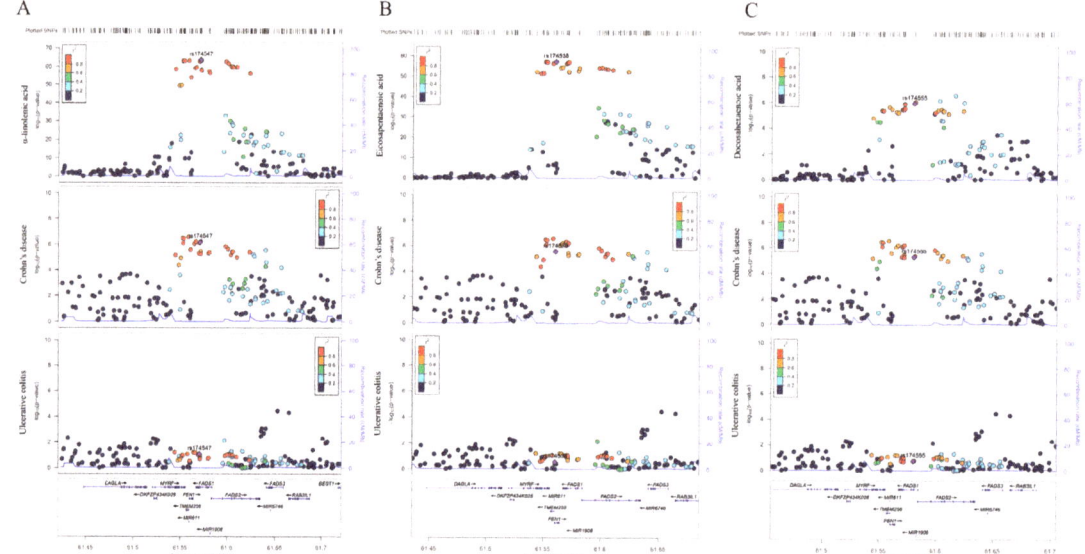

Figure 3. Regional association plots of α-linolenic, eicosapentaenoic, and docosahexaenoic acids with Crohn's disease and ulcerative colitis in the *FADS2* region. (**A**) Regional plots of α-linolenic acid and Crohn's disease and ulcerative colitis in the *FADS2* region without conditional analysis. (**B**) Regional plots of eicosapentaenoic acid and Crohn's disease and ulcerative colitis in the *FADS2* region without conditional analysis. (**C**) Regional plots of docosahexaenoic acid and Crohn's disease and ulcerative colitis in the *FADS2* region without conditional analysis. This figure was obtained from http://locuszoom.org/. Abbreviations: FADS2, fatty acid desaturase 2.

3.3. Effects of Genetic Liability to IBD, CD, and UC on the Levels of Omega-3 PUFAs

We further estimated whether genetic liability to IBD was a causal factor on changing levels of omega-3 PUFAs using bidirectional MR. There was little evidence to suggest the causal effect of genetic liability to IBD and CD on omega-3 PUFAs by using the IVW method (Table 2). However, genetic liability to UC showed an effect on lowering levels of DHA (β −0.05 (95% CI −0.09, −0.002)).

Table 2. Bidirectional Mendelian randomization estimates for causal effects of genetic liability to IBD, CD, and UC on the levels of omega-3 PUFAs.

Exposure	No. of SNPs	Outcome	No. of SNPs	IVW			Heterogeneity		Pleiotropy	
				Beta	95% CI	P	Q	P_h	MR Egger int P	MR-PRESSO P
IBD	117	Total omega-3 fatty acid	105	−0.002	(−0.012, 0.009)	0.76	200.7	<0.001	0.86	<0.001
		α-linolenic acid	39	−0.001	(−0.004, 0.001)	0.36	35.9	0.57	0.94	0.416
		EPA	39	0.001	(−0.019, 0.020)	0.92	57.9	0.02	0.87	0.009
		DHA	39	−0.010	(−0.059, 0.040)	0.71	47.7	0.13	0.69	0.036
CD	89	Total omega-3 fatty acid	83	0.004	(−0.005, 0.013)	0.39	174.5	<0.001	0.28	<0.001
		α-linolenic acid	28	−0.001	(−0.003, 0.001)	0.28	28.7	0.38	0.41	0.463
		EPA	28	0.011	(−0.005, 0.026)	0.17	46.4	0.01	0.85	0.012
		DHA	28	0.029	(−0.011, 0.070)	0.15	39.3	0.06	0.81	0.090
UC	62	Total omega-3 fatty acid	53	−0.005	(−0.018, 0.008)	0.45	131.0	<0.001	0.27	<0.001
		α-linolenic acid	27	0.002	(−0.001, 0.004)	0.23	28.5	0.34	0.34	0.446
		EPA	27	−0.005	(−0.021, 0.010)	0.50	26.5	0.44	0.38	0.088
		DHA	27	−0.045	(−0.089, −0.002)	0.04	26.7	0.42	0.61	0.095

Abbreviations: CI, confidence interval; CD, Crohn's disease; DHA, docosahexaenoic acid; EPA, eicosapentaenoic acid; Egger int, Egger intercept; IBD, inflammatory bowel disease; IVW, inverse variance weighted; MR, Mendelian randomization; PUFAs, polyunsaturated fatty acids; P_h, p-value for heterogeneity; UC, ulcerative colitis.

3.4. Mediation MR of EPA, Metabolites, and IBD Risk

Given that genetically predicted increased EPA had significant benefit on lowering IBD risks, we further estimated whether there were some metabolites or metabolic pathways linking the EPA with IBD risk. For 211 candidate mediation metabolites (selected by the two-step MR described in the Section 2), metabolite set enrichment analysis indicated that α-linolenic acid and linoleic acid metabolism, and methylhistidine metabolism were the top two metabolic pathways that have been significantly enriched (Figure 4). DHA, linoleic acid, and histidine were major metabolites determined in the two pathways, respectively.

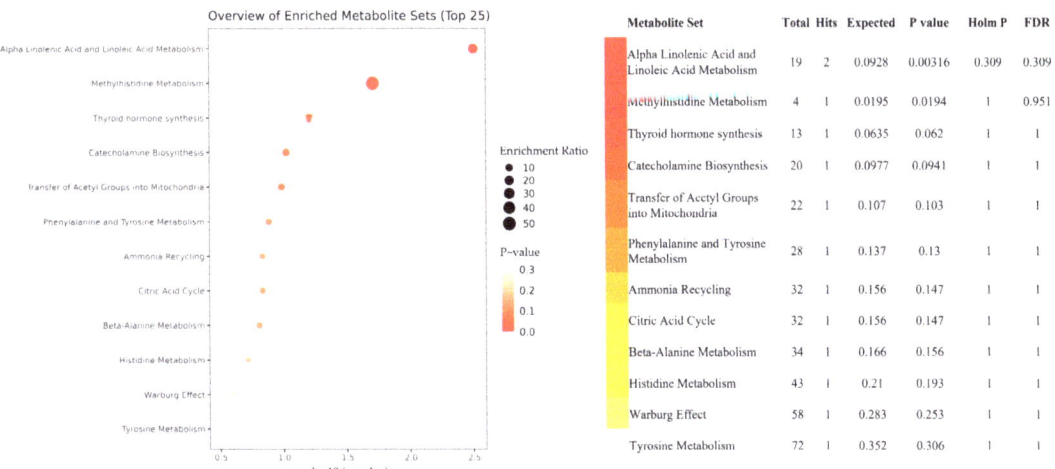

Figure 4. Metabolite set enrichment analysis of 211 selected candidate metabolites associated with both EPA and risk of IBD. The figure shows a graphical representation of the pathway-associated metabolite sets by enrichment analysis in the effect of EPA on IBD. Abbreviations: EPA, eicosapentaenoic acid; IBD, inflammatory bowel disease.

The effect of EPA on each intermediate metabolite (linoleic acid, DHA, and histidine) is shown in Figure 5A, higher levels of EPA were associated with lower linoleic acid (β, −0.51; 95% CI −0.91, −0.11), higher DHA (β, 0.61; 95% CI 0.27, 0.95), and lower histidine (β, −0.10; 95% CI −0.17, −0.03). The effect of each intermediate metabolite on IBD risk was separately adjusted for the EPA effect in the MVMR model, presented as β with 95% CI and was shown in Figure 5B. Linoleic acid and histidine showed effects on increasing risk of IBD, although the result for histidine was with a wide confidence interval. Figure 5C displays the proportion of the mediation effect of EPA on IBD explained by each intermediate metabolite separately. Linoleic acid explained 58.33% (95% CI 32.97%, 83.69%) of the total effect of EPA on IBD, while DHA explained 50.00% (95% CI 25.76%, 74.24%). Histidine explained 66.67% (95% CI 43.34%, 90.00%) of the total effect. Given the large proportion of mediation of these intermediate metabolites, the direct effects of EPA on IBD were massively attenuated after conditioning on each of the intermediate metabolites (Figure 5B).

A. Effect of EPA on each metabolite.		
Metabolites	β1 (95% CI)	P value
LA	−0.51 [−0.91, −0.11]	0.012
DHA	0.61 [0.27, 0.95]	<0.001
Histidine	−0.10 [−0.17, −0.03]	0.005

B. Effect of metabolites on IBD after adjusting for EPA.		
Metabolites	β (95% CI)	P value
EPA (βtotal)	−0.24 [−0.47, −0.02]	0.032
EPA + LA		
EPA (βdirect)	−0.10 [−0.32, 0.13]	0.401
LA (β2)	0.25 [0.04, 0.46]	0.019
EPA + DHA		
EPA (βdirect)	−0.12 [−0.36, 0.12]	0.314
DHA (β2)	−0.20 [−0.43, 0.03]	0.081
EPA + Histidine		
EPA (βdirect)	−0.08 [−0.35, 0.19]	0.554
Histidine (β2)	1.17 [−0.26, 2.59]	0.108

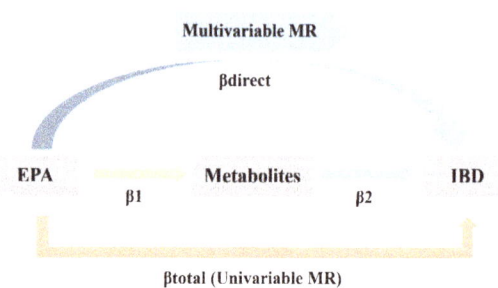

C. Effects of indirect effects of each mediator separately.		
Mediating pathway	Indirect effect (βtotal−βdirect)	Proportion mediated (95% CI)
EPA to IBD via LA	−0.14	58.33 [32.97, 83.69]
EPA to IBD via DHA	−0.12	50.00 [25.76, 74.24]
EPA to IBD via histidine	−0.16	66.67 [43.34, 90.00]

Figure 5. Estimates for the metabolites that mediated the effect of EPA on the risk of IBD. (**A**) MR-estimated effects of EPA on each intermediate metabolite (linoleic acid, DHA, and histidine) separately, presented as β with 95% CI. (**B**) MR-estimated effects of each intermediate metabolite separately on IBD after MVMR adjustment for EPA, presented as β with 95% CI. (**C**) MR-estimated effects of indirect effects of each intermediate metabolite separately, by using the difference of coefficients method with delta method-estimated 95% CIs. MR-estimated proportions mediated (%) are presented with 95% CIs. The sum of proportions mediated (%) were higher than 100%, due to the strong correlation among these intermediate metabolites (linoleic acid, DHA, and histidine). Abbreviations: CI, confidence interval; DHA, docosahexaenoic acid; EPA, eicosapentaenoic acid; IBD, inflammatory bowel disease; linoleic acid, linoleic acid; MR, Mendelian randomization; MVMR, multivariable Mendelian randomization.

4. Discussion

The present study employed a comprehensive analysis using MR to strengthen the inferences regarding the effects of different omega-3 PUFAs (including total omega-3, α-linolenic acid, EPA, and DHA) on IBD risk. We provided evidence supporting that increased levels of EPA are causally associated with a lower risk of IBD and CD, but the effect on UC is relatively weaker. The mediation MR analysis further suggested that EPA may influence IBD via α-linolenic acid, linoleic acid and methylhistidine metabolism pathways. Linoleic acid and histidine were estimated to mediate the effect of EPA on IBD. However, we found limited evidence to support the effects of total omega-3, α-linolenic acid, and DHA on the risk of IBD. Furthermore, leave-one-out, single-locus, and colocalization analyses indicated that the effects of omega-3 PUFAs on IBD were massively driven by SNP effects in the *FADS2* gene region. Therefore, desaturation steps during omega-3 PUFAs' biosynthesis might play a critical role in the relationship between omega-3 PUFAs and IBD. Meanwhile, higher genetic liability to UC might be associated with lower levels of DHA, potentially indicating a weaker absorption or abnormal metabolism of omega-3 PUFAs in UC. Collectively, our results suggest that supplementation with EPA (rather than α-linolenic acid or DHA) might be a more effective strategy to prevent the onset of IBD, especially CD, rather than UC with high probability of weak absorption or abnormal metabolism on omega-3 PUFAs. These findings shed light on the potential differential impacts of specific omega-3 PUFAs on IBD risk and highlight the importance of considering individual PUFA components in designing prevention strategies for this complex disease.

Previous systematic reviews and meta-analysis of RCTs have not yielded firm recommendations regarding the usefulness of omega-3 PUFAs in treating IBD [29,30]. In a study that included 19 RCTs, the results showed no significant benefits of omega-3 PUFA supplementation in maintaining remission of disease [29]. Another study of 9 RCTs, found insufficient data to support the routine use of omega-3 fatty acids for the maintenance of remission in CD and UC [30]. Similarly, a prospective investigation in the Nurses' Health Study cohort reported that the risk of IBD was not influenced by long-term intake of omega-3 PUFAs [31]. Meanwhile, our findings showed weak evidence of protective effects of genetically predicted higher total omega-3 fatty acid against the risk of IBD and its subtypes (both CD and UC) by using MR analysis. In spite of the known anti-inflammatory properties of omega-3 PUFAs, attributed to their ability to reduce the production of cytokines [32,33] and C-reactive protein (CRP) [34], the available data provided less convincing evidence to support the use of omega-3 PUFAs in the prevention or treatment of IBD. One plausible explanation for these findings is that total omega-3 fatty acid comprises various fatty acids with different carbon chain lengths, bond saturation, and diverse biochemical mechanisms [35]. This complexity may lead to an overall effect of total omga-3 fatty acid that is diminished or challenging to decipher in relation to IBD and its subtypes. Hence, the specific roles and effects of individual omega-3 PUFAs, such as EPA and DHA, need to be explored more comprehensively to understand their potential benefits in IBD management.

α-linolenic acid serves as a substrate for other essential omega-3 PUFAs in the body. In our study, genetically predicted α-linolenic acid levels showed a trend toward an increased risk of IBD, although the statistical power of the analysis was relatively low. Observational studies have also provided inconclusive evidence regarding the relationship between α-linolenic acid and IBD. For instance, a case-control study has reported higher dietary α-linolenic acid intakes in newly diagnosed UC patients compared with healthy controls [12]. However, in consistent with our findings, previous studies did not find any association between higher dietary intake of α-linolenic acid and an increased risk of IBD [36,37]. Well powered studies are needed to investigate the effect of α-linolenic acid on IBD and other autoimmune diseases in the future.

EPA and DHA are the main components of long-chain omega-3 fatty acids, which are derived from α-linolenic acid through a series of elongation and desaturation steps and β-oxidation. The beneficial effects of EPA and DHA have been investigated as a combination or as part of omega-3 supplementation in observational studies and experimental trials. However, the distinct effects of EPA and DHA on the risk of IBD have been relatively unexplored. In our study, we conducted separate evaluations and found evidence suggesting that increased levels of EPA were associated with a lower risk of IBD and CD.

Interestingly, our findings indicate that EPA might play a more important role than DHA in relation to IBD risk. Although direct comparative studies on the effects of EPA and DHA on IBD risk are limited, other research has provided insights that align with our results [38]. In twenty-one asthmatic adults, EPA reduced the production of interleukin-1b and tumor necrosis factor from alveolar macrophages to a much greater extent than DHA [39]. Meanwhile, the Cardiovascular Health Study reported that plasma phospholipid EPA, but not DHA, was associated with lower concentrations of CRP [40]. These findings, when integrated with our results, suggest that EPA may be more relevant for prevention of IBD.

We further demonstrated that the protective effect of EPA on risk of IBD was mainly influenced by α-linolenic acid, linoleic acid, and methylhistidine metabolism pathways. These findings are consistent with a previous study that has indicated that krill oil, rich in omega-3 PUFAs, exerts an inhibitory effect on histidine metabolism, leading to attenuated intestinal inflammation [41]. Moreover, significantly increased levels of histidine have been found in IBD patients compared to controls, which implied an association between histidine and an increased risk of IBD [42]. Therefore, EPA might reduce IBD risk through the regulation of histidine levels. Additionally, since there is competition for shared enzymes and metabolic substrates in the synthesis of omega-3 and omega-6 PUFAs, EPA might also influence the levels of linoleic acid. A previous study indicated that higher levels of linoleic acid, which are involved in the production of proinflammatory mediators, were found in IBD patients compared with controls, thereby implicating an increased risk of IBD [43]. Lower levels of linoleic acid might mediate the protective effects of EPA and IBD. In this study, we showed the causal effects of EPA on α-linolenic acid, linoleic acid, and methylhistidine metabolic pathways and three key metabolites (DHA, linoleic acid, and histidine). These results provide valuable insights into the metabolic mechanism through which EPA influences IBD risk.

Collectively, the increased risk of IBD is primarily associated with higher levels of α-linolenic acid or lower levels of EPA, as the differences in desaturation steps driven by the *FADS2* gene will lead to changes in both upstream α-linolenic acid and downstream EPA concentrations [7]. Thus, the role of the *FADS2* gene is crucial and merits further investigation.

Our study also revealed a massive influence of *FADS2* variants on IBD and CD, but not on UC. Furthermore, we found robust colocalization evidence between omega-3 PUFAs and CD in the *FADS2* gene region, but little colocalization evidence for UC. These findings suggest that the key link between omega-3 PUFAs and IBD is driven by effects in the *FADS2* gene cluster. Several lines of evidence support our observations and indicate that the *FADS2* gene is associated with inflammation [44] and CD risk [45,46]. For instance, the *FADS2* gene regulated immune functions and showed colocalization evidence on PUFAs and CD (posterior probability = 0.94) [45]. In addition, integrated data from metabolomics profiling and experiments revealed the role of *FADS2* against chronic inflammation among CD patients [47]. Therefore, *FADS2* is a crucial gene linking omega-3 PUFAs and IBD risk, particularly in the case of CD.

Despite the protective role on CD, our study provided little evidence to support the effect of omega-3 PUFAs on UC risk. Previous epidemiological studies also indicated that an increasing dietary intake of EPA or DHA had no association with a decreased risk or maintenance of remission in UC [37,48]. It is possible that inadequate supplementation or absorption resulted in lower concentrations of fatty acids in UC patients, thereby limiting their ability to trigger protective effects. For example, the inflamed colonic mucosa of patients with UC was linked to a significant decrease in EPA [49]. Similarly, a significant reduction in DHA derivatives was observed in active inflammatory UC [50]. As our bidirectional MR analysis showed, genetic liability to UC had an effect on decreased concentrations of DHA. Therefore, whether it is rational for UC patients to increase supplementation of fish oil or enhance intestinal absorption ability is worth further investigation. In contrast, He et al. recently reported that total omega-3 fatty acid had no causal effect on CD, but decreased UC risk using MR [15]. We believe the discrepant association observed for UC in our study compared with theirs was partly driven by the different instrument selection process. After applying a similar instrument selection as our study, He et al. further eliminated SNPs associated with potential confounders between total omega-3 fatty acid and outcomes. This selection process eliminated over half of the genetic variants from the instrument list for total omega-3 fatty acid. He et al. claimed that this selection was used to satisfy the second assumption of MR (exchangeability). However, this assumption suggested that the instruments are not associated with common causes (confounders) of the instrument–outcome association. MR estimates are generally less susceptible to confounders because human DNA is stable across the life course. Therefore, excluding SNPs associated with confounders between total omega-3 and IBD (e.g., body mass index) will reduce the power of the analysis rather than satisfying the exchangeability assumption of MR. In fact, such an overly stringent selection resulted in the deprivation of genetic variants in the *FADS2* region. As mentioned above, the *FADS* gene cluster plays a central role on PUFAs' metabolism, where genetic effects in the *FADS2* region massively influenced the MR estimates of omega-3 PUFAs on IBD and its subtypes. The effects of total omega-3 fatty acid were found to potentially increase IBD risks after removing the *FADS2* instrument (Figure S2). In summary, the previously reported effect of total omega-3 fatty acid on a lower risk of UC was methodologically arguable and did not align with the evidence from our MR study and other observational studies.

There were several strengths of the present study. First, our study comprehensively explored the causal effects of the different components of omega-3 PUFAs on IBD risk by using a robust MR setting, which reduced bias from residual confounding and excluded reverse causality. Current data contributed to produce informed recommendations based on the relative importance of EPA in preventing IBD. Second, our investigation of the metabolic pathways involving linoleic acid and histidine metabolites provided valuable insights into the mechanisms underlying the effect of EPA on IBD risk, which may have implications for future clinical practice. Third, our findings suggest that supplementation policies should consider the different subtypes of IBD, as EPA demonstrated a significant effect on reducing the risk of CD but not UC, and genetic liability to UC was associated with lower concentrations of DHA. Additionally, we used colocalization methods to thoroughly explore the possibility of a single shared effect signal in the *FADS2* gene region, thus validating the underlying mechanism linking omega-3 PUFAs with CD.

However, there were some limitations that should be considered when interpreting our findings. First, we used different data sources for the exposure variables. The genetic instruments for total omega-3 were obtained from the UK Biobank study, while instruments for α-linolenic acid, EPA, and DHA were derived from the CHARGE Consortium. Although both datasets involved participants with European ancestry, there could still be potential biases introduced by using different sources. Second, we assumed that the relationships between omega-3 fatty acids and IBD risk were linear. Non-linear relationships were not taken into consideration and further investigation is needed to explore potential non-linear effects. Finally, although we used univariable MR analyses to estimate the effect of each fatty acid, we were unable to directly estimate the effect of EPA-to-DHA ratio. The EPA-to-DHA ratio is considered important in the clinical application of fish oil, and its potential impact on IBD risk merits further exploration.

5. Conclusions

In conclusion, our comprehensive MR analyses identified that EPA was the key component among the omega-3 PUFAs that may exhibit a protective effect on IBD and CD, but not on UC. There was little evidence to support the effect of total omega-3, α-linolenic acid, or DHA on IBD risks. We also provided novel insights into the underlying mechanisms of EPA, which may influence IBD via α-linolenic acid, linoleic acid and methylhistidine metabolic pathways. Furthermore, the FADS2 gene is likely to be a core gene that mediates the effects of omega-3 PUFAs on IBD risk. Based on these findings, our study recommended the supplementation or dietary intake of EPA, rather than α-linolenic acid or DHA, might be beneficial for preventing the onset of IBD. The proposed mediators have provided novel insights into the underlying mechanisms of EPA. More well powered epidemiological studies and clinical trials are needed to explore the potential benefits of high EPA concentration or EPA/DHA in IBD and its subtypes. Moreover, further research is needed to investigate the role of histidine metabolites in the context of IBD.

Supplementary Materials: The following supporting information can be downloaded at: https://www.mdpi.com/article/10.3390/metabo13101041/s1, Figure S1: Selection process for the data included in the study; Figure S2: Leave one out plots of the causal effects of omega-3 polyunsaturated fatty acids on inflammatory bowel disease, Crohn's disease, and ulcerative colitis showing inverse variance weighted estimates after omitting each SNP; Table S1: Data sources of genome-wide association studies included in the Mendelian randomization analysis; Table S2: Statistics used to assess instrument strength; Table S3: Sensitivity analyses used to assess causal effects of omega-3 polyunsaturated fatty acids on the risk of Crohn's disease, and ulcerative colitis.

Author Contributions: Conceptualization, J.L., Y.B. and J.Z.; formal analysis, X.J., C.H. and X.W.; writing—original draft preparation, X.J.; writing—review and editing, H.Q., L.L., M.X., Y.X., T.W., Z.Z., Y.C., M.L., R.Z., H.L., S.W., W.W., J.Z. and J.L. All authors have read and agreed to the published version of the manuscript.

Funding: This research was funded by the National Natural Science Foundation of China (grant numbers 81930021, 81970728, 81970691, 82170819, 82370810, and 21904084); Shanghai Outstanding Academic Leaders Plan (grant number 20XD1422800); Shanghai Medical and Health Development Foundation (grant number DMRFP_I_01); Clinical Research Plan of SHDC (grant numbers SHDC2020CR3064B and SHDC2020CR1001A); Science and Technology Committee of Shanghai (grant numbers 20Y11905100 and 19411964200); Clinical Research Project of Shanghai Municipal Health Commission (grant number 20214Y0002); Ministry of Science and Technology of China (grant number 2022YFC2505202); and Innovative research team of high-level local universities in Shanghai. J.Z. was funded by the Academy of Medical Sciences (AMS) Springboard Award; the Wellcome Trust; the Government Department of Business; Energy and Industrial Strategy (BEIS); the British Heart Foundation and Diabetes UK (grant number SBF006\1117); and the Vice-Chancellor Fellowship from the University of Bristol.

Institutional Review Board Statement: Not applicable.

Informed Consent Statement: Not applicable.

Data Availability Statement: The summary statistics of total omega-3 fatty acid were obtained from UK Biobank study at https://doi.org/10.1186/s12916-022-02399-w, accessed on 6 November 2022, and instruments for α-linolenic acid, eicosapentaenoic acid, and docosahexaenoic acid were derived from Cohorts for Heart and Aging Research in Genomic Epidemiology Consortium at https://www.chargeconsortium.com/main/results, accessed on 6 November 2022. Genetic association estimates for inflammatory bowel disease were obtained from the study by the International Inflammatory Bowel Disease Genetics Consortium (IIBDGC) at https://doi.org/10.1038/ng.3760, accessed on 6 November 2022. The full summary statistics of the circulating metabolites were derived from the IEU OpenGWAS database at https://gwas.mrcieu.ac.uk/, accessed on 6 November 2022.

Acknowledgments: The authors thank all investigators for the publicly available summary data.

Conflicts of Interest: The authors declare no conflict of interest.

References

1. Baumgart, D.C.; Carding, S.R. Inflammatory bowel disease: Cause and immunobiology. *Lancet* **2007**, *369*, 1627–1640. [CrossRef] [PubMed]
2. Kaplan, G.G. The global burden of IBD: From 2015 to 2025. *Nat. Rev. Gastroenterol. Hepatol.* **2015**, *12*, 720–727. [CrossRef] [PubMed]
3. Xavier, R.J.; Podolsky, D.K. Unravelling the pathogenesis of inflammatory bowel disease. *Nature* **2007**, *448*, 427–434. [CrossRef] [PubMed]
4. Del Gobbo, L.C.; Imamura, F.; Aslibekyan, S.; Marklund, M.; Virtanen, J.K.; Wennberg, M.; Yakoob, M.Y.; Chiuve, S.E.; dela Cruz, L.; Frazier-Wood, A.C.; et al. ω-3 Polyunsaturated Fatty Acid Biomarkers and Coronary Heart Disease: Pooling Project of 19 Cohort Studies. *JAMA Intern. Med.* **2016**, *176*, 1155–1166. [CrossRef]
5. Talaei, M.; Sdona, E.; Calder, P.C.; Jones, L.R.; Emmett, P.M.; Granell, R.; Bergström, A.; Melén, E.; Shaheen, S.O. Intake of n-3 polyunsaturated fatty acids in childhood, FADS genotype and incident asthma. *Eur. Respir. J.* **2021**, *58*, 2003633. [CrossRef]
6. Calder, P.C. Polyunsaturated fatty acids, inflammatory processes and inflammatory bowel diseases. *Mol. Nutr. Food Res.* **2008**, *52*, 885–897. [CrossRef]
7. Glaser, C.; Heinrich, J.; Koletzko, B. Role of FADS1 and FADS2 polymorphisms in polyunsaturated fatty acid metabolism. *Metabolism* **2010**, *59*, 993–999. [CrossRef]
8. Gorjão, R.; Azevedo-Martins, A.K.; Rodrigues, H.G.; Abdulkader, F.; Arcisio-Miranda, M.; Procopio, J.; Curi, R. Comparative effects of DHA and EPA on cell function. *Pharmacol. Ther.* **2009**, *122*, 56–64. [CrossRef]
9. Belluzzi, A.; Brignola, C.; Campieri, M.; Pera, A.; Boschi, S.; Miglioli, M. Effect of an enteric-coated fish-oil preparation on relapses in Crohn's disease. *N. Engl. J. Med.* **1996**, *334*, 1557–1560. [CrossRef]
10. Hahn, J.; Cook, N.R.; Alexander, E.K.; Friedman, S.; Walter, J.; Bubes, V.; Kotler, G.; Lee, I.-M.; Manson, J.E.; Costenbader, K.H. Vitamin D and marine omega 3 fatty acid supplementation and incident autoimmune disease: VITAL randomized controlled trial. *BMJ* **2022**, *376*, e066452. [CrossRef]

11. Chan, S.S.; Luben, R.; Olsen, A.; Tjonneland, A.; Kaaks, R.; Lindgren, S.; Grip, O.; Bergmann, M.M.; Boeing, H.; Hallmans, G.; et al. Association between high dietary intake of the n-3 polyunsaturated fatty acid docosahexaenoic acid and reduced risk of Crohn's disease. *Aliment Pharmacol. Ther.* **2014**, *39*, 834–842. [CrossRef] [PubMed]
12. Rashvand, S.; Somi, M.H.; Rashidkhani, B.; Hekmatdoost, A. Dietary fatty acid intakes are related to the risk of ulcerative colitis: A case–control study. *Int. J. Color. Dis.* **2015**, *30*, 1255–1260. [CrossRef] [PubMed]
13. Kobayashi, Y.; Ohfuji, S.; Kondo, K.; Fukushima, W.; Sasaki, S.; Kamata, N.; Yamagami, H.; Fujiwara, Y.; Suzuki, Y.; Hirota, Y.; et al. Association of Dietary Fatty Acid Intake with the Development of Ulcerative Colitis: A Multicenter Case-Control Study in Japan. *Inflamm. Bowel Dis.* **2021**, *27*, 617–628. [CrossRef] [PubMed]
14. Lawlor, D.A.; Harbord, R.M.; Sterne, J.A.; Timpson, N.; Davey Smith, G. Faculty Opinions recommendation of Mendelian randomization: Using genes as instruments for making causal inferences in epidemiology. *Stat. Med.* **2008**, *27*, 1133–1163. [CrossRef] [PubMed]
15. He, J.; Luo, X.; Xin, H.; Lai, Q.; Zhou, Y.; Bai, Y. The Effects of Fatty Acids on Inflammatory Bowel Disease: A Two-Sample Mendelian Randomization Study. *Nutrients* **2022**, *14*, 2883. [CrossRef]
16. Zuber, V.; Grinberg, N.F.; Gill, D.; Manipur, I.; Slob, E.A.; Patel, A.; Wallace, C.; Burgess, S. Combining evidence from Mendelian randomization and colocalization: Review and comparison of approaches. *Am. J. Hum. Genet.* **2022**, *109*, 767–782. [CrossRef]
17. Borges, M.C.; Haycock, P.C.; Zheng, J.; Hemani, G.; Holmes, M.V.; Davey Smith, G.; Hingorani, A.D.; Lawlor, D.A. Role of circulating polyunsaturated fatty acids on cardiovascular diseases risk: Analysis using Mendelian randomization and fatty acid genetic association data from over 114,000 UK Biobank participants. *BMC Med.* **2022**, *20*, 210. [CrossRef]
18. Lemaitre, R.N.; Tanaka, T.; Tang, W.; Manichaikul, A.; Foy, M.; Kabagambe, E.K.; Nettleton, J.A.; King, I.B.; Weng, L.-C.; Bhattacharya, S.; et al. Genetic loci associated with plasma phospholipid n-3 fatty acids: A meta-analysis of genome-wide association studies from the CHARGE consortium. *PLoS Genet.* **2011**, *7*, e1002193. [CrossRef]
19. de Lange, K.M.; Moutsianas, L.; Lee, J.C.; Lamb, C.A.; Luo, Y.; Kennedy, N.A.; Jostins, L.; Rice, D.L.; Gutierrez-Achury, J.; Ji, S.-G.; et al. Genome-wide association study implicates immune activation of multiple integrin genes in inflammatory bowel disease. *Nat. Genet.* **2017**, *49*, 256–261. [CrossRef]
20. Burgess, S.; Thompson, S.G. Avoiding bias from weak instruments in Mendelian randomization studies. *Int. J. Epidemiol.* **2011**, *40*, 755–764. [CrossRef]
21. Bowden, J.; Smith, G.D.; Haycock, P.C.; Burgess, S. Consistent Estimation in Mendelian Randomization with Some Invalid Instruments Using a Weighted Median Estimator. *Genet. Epidemiol.* **2016**, *40*, 304–314. [CrossRef]
22. Bowden, J.; Davey Smith, G.; Burgess, S. Mendelian randomization with invalid instruments: Effect estimation and bias detection through Egger regression. *Int. J. Epidemiol.* **2015**, *44*, 512–525. [CrossRef]
23. Verbanck, M.; Chen, C.Y.; Neale, B.; Do, R. Detection of widespread horizontal pleiotropy in causal relationships inferred from Mendelian randomization between complex traits and diseases. *Nat. Genet.* **2018**, *50*, 693–698. [CrossRef]
24. Greco, M.F.; Minelli, C.; Sheehan, N.A.; Thompson, J.R. Detecting pleiotropy in Mendelian randomisation studies with summary data and a continuous outcome. *Stat. Med.* **2015**, *34*, 2926–2940. [CrossRef]
25. Hemani, G.; Zheng, J.; Elsworth, B.; Wade, K.H.; Haberland, V.; Baird, D.; Laurin, C.; Burgess, S.; Bowden, J.; Langdon, R.; et al. The MR-Base platform supports systematic causal inference across the human phenome. *Elife* **2018**, *7*, e34408. [CrossRef]
26. Timpson, N.J.; Nordestgaard, B.G.; Harbord, R.M.; Zacho, J.; Frayling, T.M.; Tybjærg-Hansen, A.; Smith, G.D. C-reactive protein levels and body mass index: Elucidating direction of causation through reciprocal Mendelian randomization. *Int. J. Obes.* **2011**, *35*, 300–308. [CrossRef]
27. Carter, A.R.; Sanderson, E.; Hammerton, G.; Richmond, R.C.; Smith, G.D.; Heron, J.; Taylor, A.E.; Davies, N.M.; Howe, L.D. Mendelian randomisation for mediation analysis: Current methods and challenges for implementation. *Eur. J. Epidemiol.* **2021**, *36*, 465–478. [CrossRef]
28. Pang, Z.; Chong, J.; Zhou, G.; de Lima Morais, D.A.; Chang, L.; Barrette, M.; Gauthier, C.; Jacques, P.-É.; Li, S.; Xia, J. MetaboAnalyst 5.0: Narrowing the gap between raw spectra and functional insights. *Nucleic Acids Res.* **2021**, *49*, W388–W396. [CrossRef]
29. Cabré, E.; Mañosa, M.; Gassull, M.A. Omega-3 fatty acids and inflammatory bowel diseases—A systematic review. *Br. J. Nutr.* **2012**, *107* (Suppl. S2), S240–S252. [CrossRef]
30. Turner, D.; Shah, P.S.; Steinhart, A.H.; Zlotkin, S.; Griffiths, A.M. Maintenance of remission in inflammatory bowel disease using omega-3 fatty acids (fish oil): A systematic review and meta-analyses. *Inflamm. Bowel Dis.* **2011**, *17*, 336–345. [CrossRef]
31. Ananthakrishnan, A.N.; Khalili, H.; Konijeti, G.G.; Higuchi, L.M.; de Silva, P.; Fuchs, C.S.; Willett, W.C.; Richter, J.M.; Chan, A.T. Long-term intake of dietary fat and risk of ulcerative colitis and Crohn's disease. *Gut* **2014**, *63*, 776–784. [CrossRef]
32. Calder, P.C.; Yaqoob, P.; Thies, F.; Wallace, F.A.; Miles, E.A. Fatty acids and lymphocyte functions. *Br. J. Nutr.* **2002**, *87* (Suppl. S1), S31–S48. [CrossRef]
33. Endres, S.; Ghorbani, R.; Kelley, V.E.; Georgilis, K.; Lonnemann, G.; van der Meer, J.W.M.; Cannon, J.G.; Rogers, T.S.; Klempner, M.S.; Weber, P.C.; et al. The effect of dietary supplementation with n-3 polyunsaturated fatty acids on the synthesis of interleukin-1 and tumor necrosis factor by mononuclear cells. *N. Engl. J. Med.* **1989**, *320*, 265–271. [CrossRef]
34. Li, K.; Huang, T.; Zheng, J.; Wu, K.; Li, D. Effect of marine-derived n-3 polyunsaturated fatty acids on C-reactive protein, interleukin 6 and tumor necrosis factor α: A meta-analysis. *PLoS ONE* **2014**, *9*, e88103. [CrossRef]

35. Koletzko, B.; Reischl, E.; Tanjung, C.; Gonzalez-Casanova, I.; Ramakrishnan, U.; Meldrum, S.; Simmer, K.; Heinrich, J.; Demmelmair, H. FADS1 and FADS2 Polymorphisms Modulate Fatty Acid Metabolism and Dietary Impact on Health. *Annu. Rev. Nutr.* **2019**, *39*, 21–44. [CrossRef]
36. Mozaffari, H.; Daneshzad, E.; Larijani, B.; Bellissimo, N.; Azadbakht, L. Dietary intake of fish, n-3 polyunsaturated fatty acids, and risk of inflammatory bowel disease: A systematic review and meta-analysis of observational studies. *Eur. J. Nutr.* **2020**, *59*, 1–17. [CrossRef]
37. Tjonneland, A.; Overvad, K.; Bergmann, M.M.; Nagel, G.; Linseisen, J.; Hallmans, G.; Palmqvist, R.; Sjodin, H.; Hagglund, G.; Berglund, G.; et al. Linoleic acid, a dietary n-6 polyunsaturated fatty acid, and the aetiology of ulcerative colitis: A nested case-control study within a European prospective cohort study. *Gut* **2009**, *58*, 1606–1611.
38. Morin, C.; Blier, P.U.; Fortin, S. Eicosapentaenoic acid and docosapentaenoic acid monoglycerides are more potent than docosahexaenoic acid monoglyceride to resolve inflammation in a rheumatoid arthritis model. *Arthritis Res. Ther.* **2015**, *17*, 142. [CrossRef]
39. Mickleborough, T.D.; Tecklenburg, S.L.; Montgomery, G.S.; Lindley, M.R. Eicosapentaenoic acid is more effective than docosahexaenoic acid in inhibiting proinflammatory mediator production and transcription from LPS-induced human asthmatic alveolar macrophage cells. *Clin. Nutr.* **2009**, *28*, 71–77. [CrossRef]
40. Mozaffarian, D.; Lemaitre, R.N.; King, I.B.; Song, X.; Spiegelman, D.; Sacks, F.M.; Rimm, E.B.; Siscovick, D.S. Circulating long-chain ω-3 fatty acids and incidence of congestive heart failure in older adults: The cardiovascular health study: A cohort study. *Ann. Intern. Med.* **2011**, *155*, 160–170. [CrossRef]
41. Liu, F.; Smith, A.D.; Solano-Aguilar, G.; Wang, T.T.Y.; Pham, Q.; Beshah, E.; Tang, Q.; Urban, J.F.; Xue, C.; Li, R.W. Mechanistic insights into the attenuation of intestinal inflammation and modulation of the gut microbiome by krill oil using in vitro and in vivo models. *Microbiome* **2020**, *8*, 83. [CrossRef] [PubMed]
42. Bosch, S.; Struys, E.A.; van Gaal, N.; Bakkali, A.; Jansen, E.W.; Diederen, K.; Benninga, M.A.; Mulder, C.J.; de Boer, N.K.; de Meij, T.G. Fecal Amino Acid Analysis Can Discriminate de Novo Treatment-Naïve Pediatric Inflammatory Bowel Disease from Controls. *J. Pediatr. Gastroenterol. Nutr.* **2018**, *66*, 773–778. [CrossRef] [PubMed]
43. Ueda, Y.; Kawakami, Y.; Kunii, D.; Okada, H.; Azuma, M.; Le, D.S.N.; Yamamoto, S. Elevated concentrations of linoleic acid in erythrocyte membrane phospholipids in patients with inflammatory bowel disease. *Nutr. Res.* **2008**, *28*, 239–244. [CrossRef]
44. O'Neill, C.M.; Minihane, A.-M. The impact of fatty acid desaturase genotype on fatty acid status and cardiovascular health in adults. *Proc. Nutr. Soc.* **2017**, *76*, 64–75. [CrossRef]
45. Chu, X.; Jaeger, M.; Beumer, J.; Bakker, O.B.; Aguirre-Gamboa, R.; Oosting, M.; Smeekens, S.P.; Moorlag, S.; Mourits, V.P.; Koeken, V.A.C.M.; et al. Integration of metabolomics, genomics, and immune phenotypes reveals the causal roles of metabolites in disease. *Genome Biol.* **2021**, *22*, 198. [CrossRef] [PubMed]
46. Costea, I.; Mack, D.R.; Lemaitre, R.N.; Israel, D.; Marcil, V.; Ahmad, A.; Amre, D.K. Interactions between the dietary polyunsaturated fatty acid ratio and genetic factors determine susceptibility to pediatric Crohn's disease. *Gastroenterology* **2014**, *146*, 929–931. [CrossRef]
47. Liu, R.; Qiao, S.; Shen, W.; Liu, Y.; Lu, Y.; Liangyu, H.; Guo, Z.; Gong, J.; Shui, G.; Li, Y.; et al. Disturbance of Fatty Acid Desaturation Mediated by FADS2 in Mesenteric Adipocytes Contributes to Chronic Inflammation of Crohn's Disease. *J. Crohn's Colitis* **2020**, *14*, 1581–1599. [CrossRef]
48. Turner, D.; Steinhart, A.H.; Griffiths, A.M. Omega 3 fatty acids (fish oil) for maintenance of remission in ulcerative colitis. *Cochrane Database Syst. Rev.* **2007**, *3*, Cd006443.
49. Pearl, D.S.; Masoodi, M.; Eiden, M.; Brümmer, J.; Gullick, D.; McKeever, T.M.; Whittaker, M.A.; Nitch-Smith, H.; Brown, J.F.; Shute, J.K.; et al. Altered colonic mucosal availability of n-3 and n-6 polyunsaturated fatty acids in ulcerative colitis and the relationship to disease activity. *J. Crohns Colitis* **2014**, *8*, 70–79. [CrossRef]
50. Ungaro, F.; Tacconi, C.; Massimino, L.; Corsetto, P.A.; Correale, C.; Fonteyne, P.; Piontini, A.; Garzarelli, V.; Calcaterra, F.; Della Bella, S.; et al. MFSD2A Promotes Endothelial Generation of Inflammation-Resolving Lipid Mediators and Reduces Colitis in Mice. *Gastroenterology* **2017**, *153*, 1363–1377.e6. [CrossRef]

Disclaimer/Publisher's Note: The statements, opinions and data contained in all publications are solely those of the individual author(s) and contributor(s) and not of MDPI and/or the editor(s). MDPI and/or the editor(s) disclaim responsibility for any injury to people or property resulting from any ideas, methods, instructions or products referred to in the content.

Article

Metabolomic Signatures Associated with Radiation-Induced Lung Injury by Correlating Lung Tissue to Plasma in a Rat Model

Liming Gu [†], Wenli Wang [†], Yifeng Gu [†], Jianping Cao * and Chang Wang *

State Key Laboratory of Radiation Medicine and Protection, School of Radiation Medicine and Protection, Medical College of Soochow University, School for Radiological and Interdisciplinary Sciences (RAD-X), Jiangsu Provincial Key Laboratory of Radiation Medicine and Protection, Suzhou Industrial Park Renai Road 199, Suzhou 215123, China; 20211102001@stu.suda.edu.cn (L.G.); 20214220035@stu.suda.edu.cn (W.W.); guyifeng19950830@126.com (Y.G.)
* Correspondence: jpcao@suda.edu.cn (J.C.); wangchang@suda.edu.cn (C.W.); Tel.: +86-512-65880067 (C.W.)
[†] These authors contributed equally to this work.

Abstract: The lung has raised significant concerns because of its radiosensitivity. Radiation-induced lung injury (RILI) has a serious impact on the quality of patients' lives and limits the effect of radiotherapy on chest tumors. In clinical practice, effective drug intervention for RILI remains to be fully elucidated. Therefore, an in-depth understanding of the biological characteristics is essential to reveal the mechanisms underlying the complex biological processes and discover novel therapeutic targets in RILI. In this study, Wistar rats received 0, 10, 20 or 35 Gy whole-thorax irradiation (WTI). Lung and plasma samples were collected within 5 days post-irradiation. Then, these samples were processed using liquid chromatography–mass spectrometry (LC-MS). A panel of potential plasma metabolic markers was selected by correlation analysis between the lung tissue and plasma metabolic features, followed by the evaluation of radiation injury levels within 5 days following whole-thorax irradiation (WTI). In addition, the multiple metabolic dysregulations primarily involved amino acids, bile acids and lipid and fatty acid β-oxidation-related metabolites, implying disturbances in the urea cycle, intestinal flora metabolism and mitochondrial dysfunction. In particular, the accumulation of long-chain acylcarnitines (ACs) was observed as early as 2 d post-WTI by dynamic plasma metabolic data analysis. Our findings indicate that plasma metabolic markers have the potential for RILI assessment. These results reveal metabolic characteristics following WTI and provide new insights into therapeutic interventions for RILI.

Keywords: metabolomics; whole-thorax irradiation (WTI); radiation-induced lung injury (RILI); metabolic marker; dynamic

Citation: Gu, L.; Wang, W.; Gu, Y.; Cao, J.; Wang, C. Metabolomic Signatures Associated with Radiation-Induced Lung Injury by Correlating Lung Tissue to Plasma in a Rat Model. *Metabolites* **2023**, *13*, 1020. https://doi.org/10.3390/metabo13091020

Academic Editor: Federico Tesio Torta

Received: 15 August 2023
Revised: 7 September 2023
Accepted: 14 September 2023
Published: 17 September 2023

Copyright: © 2023 by the authors. Licensee MDPI, Basel, Switzerland. This article is an open access article distributed under the terms and conditions of the Creative Commons Attribution (CC BY) license (https://creativecommons.org/licenses/by/4.0/).

1. Introduction

With the development of nuclear power and the widespread application of nuclear technology, the potential risk for radiation damage to people has greatly increased. A rapid, sensitive and accurate assay to assess the severity of the critical organ systems, as well as radiation dose estimation of possible exposed individuals, is one of the key links to emergency medical assistance after radiation damage. The lung has raised significant concerns because of its radiosensitivity. Radiation-induced lung injury (RILI) has a serious impact on the quality of patients' lives and limits the effect of radiotherapy (RT) for chest tumors, with 5–20% of patients experiencing this adverse effect [1]. The Clinical symptoms of radiation pneumonitis include a persistent dry cough, shortness of breath, mild fever or, occasionally, a high fever that may be secondary to radiation-induced pulmonary fibrosis and may even be the direct cause of death [2]. In the presence of extensive pulmonary fibrosis, antibiotic and corticosteroid therapeutics are limited, and there is no effective clinical treatment, which severely affects the patient's quality of life and even their survival [3]. Macrophages, fibroblasts and T lymphocytes, as well as other inflammatory and immune cells, have key roles in the development of RILI. In addition, TGF-β, IL-4, IL-13 and IFN-γ have also been implicated in this process [4,5]. However, specific biological mechanisms and effective drug interventions for RILI remain to be fully elucidated. Current diagnosis methods for RILI, such as clinical biochemical indicators, lung function and medical imaging, have the drawbacks of sensitivity, specificity and lag effects [6]. Thus, identifying biomarkers for early diagnosis and revealing the molecular mechanisms of RILI is crucial for preventing disease progression, reducing patient mortality and taking effective measures as early as possible.

As an important component of systems biology, metabolomics is a comprehensive analysis of small-molecule metabolites and may reflect pathophysiologic states [7]. Metabolomics technologies have been developed over the past two decades to enable reliable identification, detection and quantification of novel metabolites in food, plant, environmental, animal and human studies [8] and have been widely adopted as a new approach for biomarker discovery and comprehensive understanding of the underlying pathogenesis [9]. To meet the demands of rapid radiation damage assessment in large-scale nuclear accidents, metabolomics has been attempted to identify biomarkers of radiation injury in various biological samples. Although the majority of studies have focused on a variety of biofluids derived from animal models (like mice, rats and non-human primates), as well as humans, the overlap in biomarkers of radiation injury across species has highlighted the metabolic pathways that are most perturbed, including β-oxidation of fatty acids (acylcarnitines), energy metabolism (TCA cycle intermediates), purines and pyrimidines metabolism, pro-inflammatory pathways (the omega-6 constituents, polyunsaturated fatty acids) and amino acids metabolism [10–14]. In the case of RILI, metabolomics has been utilized to reveal the metabolic characteristics of RILI in different genotypes of mice [15] and to explore the metabolic changes in serum and lung tissues exposed to irradiation [16,17]. These studies, however, primarily focus on the metabolic changes at a single time point post irradiation, which reflect the metabolic characteristics at a certain stage of RILI development and ignore the influence of time. In contrast, dynamic metabolomics could capture the variation generated by time and truly reveal dynamic metabolic changes during the development of RILI. Therefore, the combination of static and dynamic analyses is necessary to obtain the key metabolic characteristics related to RILI and discover the pathology of RILI.

Biofluid samples, such as plasma, serum, urine and saliva, are common sample types due to their convenient and minimally invasive collection. On the other hand, most biomarkers in biofluid samples only reflect the overall metabolic changes and cannot reflect the pathophysiologic change in injured tissues. In metabolomic studies of biofluid samples, unwanted confounding factors unrelated to diseases may lead to the discovery of false positive biomarkers [18]. For example, the metabolic characteristics of blood and urine are heavily influenced by gender, lifestyle, diet and other factors [19,20], which are difficult to unify. Therefore, metabolites found in biofluid samples are sometimes unable to accurately reflect the pathological status of disease. Nevertheless, tissue metabolomics can provide more abundant physiological or pathological information, which is important for diagnosis and treatment. Therefore, it is of importance to conjointly analyze differential metabolites in plasma and lung tissues.

In this study, metabolomics signatures of lung tissues and serial plasma specimens within 5 days of exposure to WTI in a rat model were performed. Furthermore, a panel of potential minimally invasive plasma metabolic markers for RILI was selected, followed by the assessment of radiation injury. Our findings will throw light on the molecular mechanism and serve as a strategy to aid in discovering minimally invasive diagnosis markers for RILI.

2. Materials and Methods

2.1. Chemicals and Reagents

Mass-grade methanol and acetonitrile were from Fisher Chemical (Thermo Fisher Scientific, Boston, MA, USA). Ammonium bicarbonate and formic acid were purchased from Fluka (Dresden, Germany). The ultrapure water was prepared with a Milli Q purification system (Millipore, Burlington, MA, USA). The chemical standards for compound identification were obtained from Sigma-Aldrich (St. Louis, MO, USA), Adamas (Hong Kong, China) or JK Chemical Ltd. (Shanghai, China). The deuterium-labeled internal standards (ISs), including cholic acid-d4, chenodeoxycholic acid-d4, succinic acid-d4, L-leucine-d10, L-phenylalanine-d5, L-tryptophan-d5, L-citrulline-d4, acylcarnitine C10:0-d3 and acylcarnitine C10:0-d3 were from Cambridge Isotope Laboratories, and the natural lipid analogs, including palmitic acid-d3 and 1stearic acid-d3, were supplied by Avanti Polar Lipids.

2.2. Animals, Irradiation and Sample Collection

Female Wistar rats (170–190 g) were obtained from the Shanghai SLAC Laboratory Animal Ltd. (Shanghai, China), which were randomized into control ($n = 13$) and irradiated cohorts ($n = 33$). Prior to treatment, these animals were allowed to acclimate to the facility for one week. Then, these animals were anesthetized with 100 mg/kg ketamine and 10 mg/kg xylazine. To develop a radiation-induced lung injury rat model, we used a small animal radiotherapy treatment plan (X-RAD SmART) system. Images acquired through cone beam computed tomography (CT) were used to reconstruct and delineate targets. Multi-beam and CT-guided Monte Carlo-based plans were performed to optimize doses to targets. The terminal dose of WTI that the rats received was equivalent to either 10 Gy ($n = 10$), 20 Gy ($n = 11$) or 35 Gy ($n = 12$) at a dose rate of 2.7 Gy/min.

Plasma was obtained through periorbital bleeding at time points of 1, 2, 3 and 5 days post-irradiation, while lung tissue was collected on the fifth day post-irradiation. As some lung tissue samples have been exhausted for other analysis, a total of 39 samples (11 for controls, 10 for 10 Gy, 9 for 20 Gy, 9 for 35 Gy) were available from the cohort that was dedicated to the metabolomics study. All of the plasma and lung tissues were stored at $-80\ °C$ before LC-MS analysis.

The study was approved by the Ethics Committee of Soochow University.

2.3. Histology

Lung tissues from each group were immersed in 10% neutral buffered formalin and allowed to fix for a minimum of 24 h. The fixed lung specimens were embedded in paraffin, sliced into 4 μm thick sections and stained with hematoxylin and eosin (H&E) for analysis of tissue morphology changes following WTI.

2.4. LC-MS Pseudotargeted Metabolomics Analysis

To obtain more comprehensive metabolic characteristics, pseudotargeted metabolomics analyses based on LC-MS were used to determine lung or plasma metabolites. Meanwhile, considering that only known compounds can be biologically explained, the identified metabolites in the samples were kept for the metabolic analysis.

The composition and concentration of internal standards (ISs) for plasma and lung tissue are listed in Tables S1 and S2.

Plasma preparation: 200 μL ISs was added into 50 μL of sample for the protein precipitation. After vortexing, the sample was centrifuged at 13,000 rpm/min for 10 min (4 °C); the supernatant was taken and divided into two parts and dried by vacuum. Before LC-MS analysis, two dried supernatants were redissolved with 50 μL ACN: H_2O (1:3, v/v).

Lung tissue preparation: About 20 mg of lung tissue samples were homogenized with ceramic beads in 1.5 mL ISs solution two times using a Tissue Lyser homogenizer (Gene Ready Ultracool, Life Real, Hangzhou, China). The homogenization took 5 min with 15 s intervals each time (45 HZ). Then, the homogenized tissue sample was centrifuged at 13,000 rpm/min for 15 min (4 °C), the supernatant was taken and divided into two parts, and dried by vacuum. Before LC-MS analysis, two dried supernatants were redissolved with 50 µL ACN: H$_2$O (1:3, v/v). Before LC-MS analysis, two reconstituted samples were used for positive ion mode and negative ion mode, respectively.

Pseudotargeted analysis of plasma and lung tissue metabolites was performed on TSQ Vantage HPLC-MS/MS (Thermo Fisher, USA) with ESI, which was developed according to the proposed strategy described by Zheng et al. [21]. In the positive ion mode, a BEH C8 100 × 2.1 mm column (1.7 µm particle size, Waters) was employed for the separation. Mobile phase A was 0.1% aqueous formic acid in water. Mobile phase B consisted of 0.1% formic acid in acetonitrile. The linear gradient elution was set: 0–0.5 min, 5% B; 0.5–24 min, 5–100% B; 24–28 min, 100% B; 28–28.5 min, 100% B back to 5% B; 28.5–31.5 min, 5% B. An HSS T3 100 × 2.1 mm column (1.8 µm particle size, Waters) was utilized in the negative ion mode for the separation. The mobile phases were composed of 6.5 mM ammonium bicarbonate in water (C) and 6.5 mM ammonium bicarbonate in 95% methanol/water (D). The linear gradient elution followed: 0–1 min, 2% B; 1–20 min, 2–100% B; 20–24 min, 100% B; 24–24.5 min, 100% B back to 2% B; 24.5–27.5 min, 2% B. The flow rate was 0.25 mL/min in both positive and negative ion modes. The column temperature was kept at 50 °C, and the sample injection volume was 5 µL. The mass parameters with electrospray ionization were set as follows: 350 °C capillary temperature, 300 °C vaporize temperature, 35 arbitrary unit sheath gas flow rate, 10 arbitrary unit auxiliary gas flow rate, 3.0 kV capillary voltage for ESI+ mode and −2.5 kV for ESI mode.

2.5. Urea Detection

Urea contents were detected by Urea (BUN) Colorimetric Assay Kit (Urease Method). Samples and working reagent were added to the 96-well plate, and then absorbance at 580 nm was measured by microplate reader (BioTek, Winooski, VT, USA).

2.6. Quantitative Real-Time Polymerase Chain Reaction (q-RT-PCR)

Total RNA of the lung samples of SD rats were homogenized and isolated using RNA-Quick Purification Kit (ES Science, Shanghai, China), and cDNA synthesis was performed by the Reverse Transcription Reagent Kit (ABM, Vancouver, BC, Canada) according to the specification. Vii7 PCR system and SYBR® Green PCR kit (QIAGEN, Hilden, Germany) were used for quantitative Real-Time Chain Reaction (q-RT-PCR). Data were normalized to the expression of α-tubulin in each sample. The forward and reverse primers used for qPCR were as follows:

CPT1A (Forward: 5′-CCTACCACGGCTGGATGTTT-3′, Reverse: 5′-TACAACATGGG CTTCCGACC-3′); CPT1B (Forward: 5′-ACAGGCATAAGGGGTGGCAT-3′, Reverse: 5′-CACTCCAATCCCACCTCGACC-3′).

2.7. Data Processing and Statistical Analysis

The integration of the peaks from pseudotargeted analysis was conducted by Xcalibur (LC-MS/MS). The metabolites with less than 20% missing values and relative standard deviation (RSD) below 30% in QC samples remained. Then, the peak area of the metabolite was normalized to ISs (for plasma) or ISs and tissue weight (for lung tissue), which was utilized for following data processing. A paired analysis with nonparametric test (two-tailed Wilcoxon signed-rank test) was performed to discover the differential features ($p < 0.05$) using the SPSS 16.0 software. False-discovery rate (FDR < 0.2) was used to reduce false-discovery rate. Heat map employing MeV 4.9.0 was used to visualize the metabolic regulations of the differential metabolites associated with ionizing radiation exposure.

The panel of potential biomarkers for radiation exposure was further refined by variable importance in projection (VIP) of partial least squares–discriminant analysis (PLS-DA). Multivariate statistical models, including principal component analysis (PCA), partial least-squares discriminant analysis (PLS-DA) and nonlinear kernel partial least squares (KPLS) combined with a preprocessing technique of orthogonal signal correction (OSC) [22], were carried out using SIMCA-P 11.5 demo version (Umetrics AB, Umeå, Sweden). Response permutation test with 200 times was conducted to assess whether the model was overfitting.

2.8. Metabolic Correlation Network Analysis

Metabolic correlation network was performed using Cytoscape software (version 2.8.3). In the correlation network map, the nodes represent the metabolites. The solid black and red edge lines show positive and negative relationships, respectively. Metabolites with Pearson correlation coefficients above a threshold ($r \geq 0.7$, $p < 0.05$) were connected by lines.

3. Results

3.1. Histological Destruction of Rat Lung Tissues in Response to WTI

The hematoxylin–eosin (H&E) staining results showed that the structure of the lung tissue had severe destruction after exposure to radiation. As shown in Figure 1, the control group had a regular alveoli structure of lung tissue with slender alveoli and blood vessel walls. Compared with the control group, there was more diffuse hyperemia in the lung tissue at 5 d after 10 Gy irradiation. In the 20 Gy group, more inflammatory cells in the alveolar wall, thickening of the blood vessel wall and increased exudate in the alveoli could be observed. When it comes to 35 Gy, the lung tissue structure became more disorganized, with significant aggregation of lymphocytes, which reflected dose-dependent damage in lung tissues.

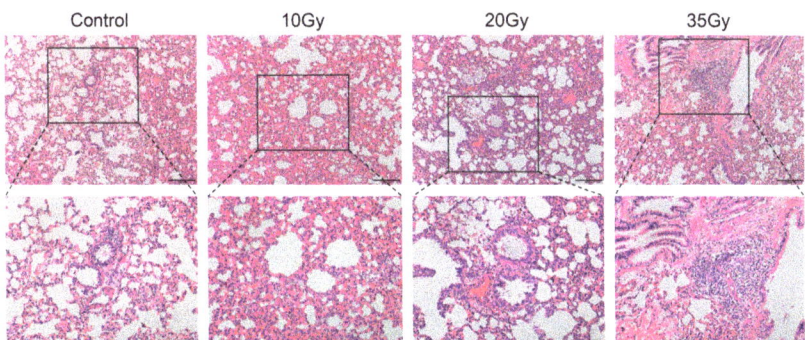

Figure 1. Representative images of HE staining in the control group and irradiated groups (first row: 40×, second row: 100×).

3.2. Ratlung Metabolic Signatures Exposed to WTI

To evaluate the stability of the analytical systems, quality control (QC) samples were evenly inserted into the analytical queue during the run of samples in LC-MS metabolomics analysis. QC samples were prepared similarly to the other samples. As shown in Figure S1A, the RSDs of 82.76, 94.58 and 99.01% of metabolites detected in QC samples were less than 10, 20 and 30%, respectively. In addition, QC samples were all within two times the standard deviation (SD) (Figure S1B). All of these results confirmed the reproducibility and stability of the metabolic profiling in LC-MS.

To highlight the separation of the study groups, a PLS-DA model was used to perform a multivariate pattern recognition analysis, and two principal components (PCs) were calculated based on this PLS-DA model. As shown in Figure 2A, except for a lesser overlap between the 10 and 20 Gy irradiated groups, there was a clear clustering trend and dose-dependent distance in different groups, which indicated that metabolic disorders in lung tissues were positively related to radiation doses.

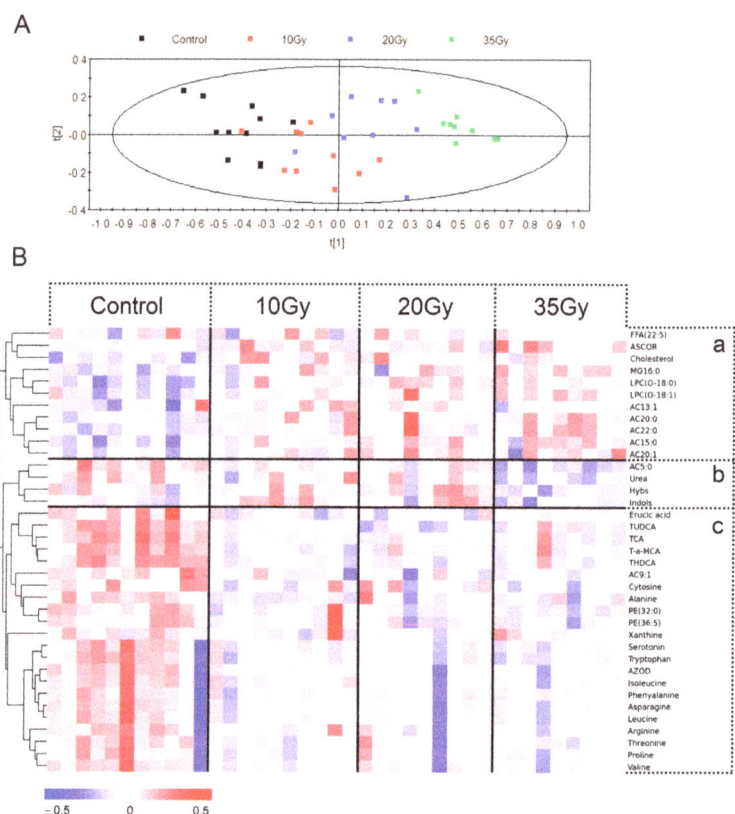

Figure 2. Statistical analysis for the data set of lung tissues at 5 d after WTI. (A) PLS-DA score plot comparing control and individual WTI doses. (B) Heatmap of the 37 differential metabolites in lung tissues, with the degree of changes compared with control group marked with colors. AS-COR: ascorbic acid; MG16:0: palmitoylglycerol; AC13:1: acylcarnitine C13:1; AC20:0: acylcarnitine C20:0; AC22:0: acylcarnitine C22:0; AC15:0: acylcarnitine C15:0; AC20:1: acylcarnitine C20:1; AC5:0: acylcarnitine C5:0; Hybs: 4−Hydroxybenzenesulfonic acid; Indols: indoxyl sulfate; TUDCA: tauroursodeoxycholic acid; TCA: taurocholic acid; T−α−MCA: tauro−α−Muricholic acid; AZOD: 3−Amino−2−oxazolidinone. The red and blue colors represent significant increases and decreases in response to WTI. According to the clustering, the metabolic alternations induced by WTI could be divided into three zones (a, b, c).

Based on the analysis of variance p-value (ANOVA, $p < 0.05$) and false-discovery rate (FDR < 0.2), 37 differential metabolites associated with RILI were selected; the effects of irradiation on the differential metabolites are listed in Table S3 and visualized in Figure 2B. The metabolic alternations induced by WTI could be divided into three zones (a, b, c) according to the clustering. Metabolites in zone A were significantly up-regulated following WTI irradiation, mainly including lipid metabolites and fatty acid β-oxidation-related metabolites, such as cholesterol, lysophosphatidylcholine and long-chain AC. Meanwhile, most metabolites in panel C (bile acid and lipid metabolites) showed a significant decrease following irradiation. Additionally, metabolites in region B, containing acylcarnitine C5:0 (AC5:0), urea, 4-Hydroxybenzenesulfonic acid (Hybs) and indoxyl sulfate (Indols) were only down-regulated in the 35 Gy group.

3.3. Plasma Metabolic Signatures Exposed to WTI

Similarly, we assess the stability of the total metabolic profiling analytical systems in plasma. The RSDs of 43.60%, 83.72% and 95.93% of metabolites were less than 10%, 20% and 30%, respectively (Figure S2A), and QC samples were also all within two times the SD (Figure S2B). These results confirmed the reliability of metabolic profiling in LC-MS.

The plasma metabolomic profiles from 1 d to 5 d after were subsequently depicted on the basis of the PLS-DA model (Figure 3). There were two types of metabolic derangements, including radiation-induced changes and time-associated changes. Individual doses shown in Figure 3A could not be distinguished at 1 d after WTI, reflecting the lack of sensitivity of differential metabolites to identify the WTI doses. Differently from 1 d, the 2 d PLS-DA plots could distinguish the control and irradiated groups, although different WTI doses could not be clearly divided (Figure 3B). At 3 d after WTI, individual doses could be distinguished clearly, whereas the control group overlapped with the 10 Gy group (Figure 3C). Compared with 3 d, PLS-DA plots at 5 d after WTI showed that the control group clustered closely apart from the irradiated groups, and the high-WTI dose (35 Gy) group dispersed from the moderate-WTI dose (10 and 20 Gy) groups (Figure 3D). It is clear that the distance between the control area and the irradiated group became farther with the extension of time and increase in radiation exposure doses. These results suggested that irradiation could lead to metabolic disorders, and the degree of disorders in plasma was also positively related to radiation doses. Finally, there were 40, 84, 109 and 128 differential metabolites reaching significance in plasma selected at 1 d, 2 d, 3 d and 5 d after irradiation, respectively (Tables S4–S7).

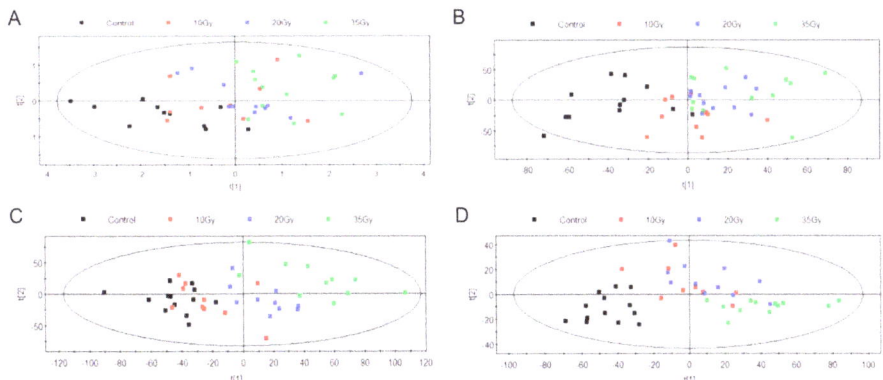

Figure 3. Statistical analysis for the data set in plasma at different times after WTI. PLS–DA score plot comparing control and individual WTI doses ((**A**), 1 d; (**B**), 2 d; (**C**), 3 d; (**D**), 5 d).

3.4. Potential Plasma Metabolite Markers of Radiation-Induced Lung Injury

To select a panel of biomarkers for clinical RILI diagnosis and prognosis, metabolic data in plasma and lung tissues were analyzed conjointly. We initially selected 37 and 123 differential metabolites in lung tissue and plasma, respectively. Among these metabolites, 23 metabolites exhibited significant changes in both comparison groups simultaneously (Figure S3A).

In order to reduce the risk of false positives, checking the metabolites before identifying the biomarker candidates in this discovery phase is necessary. The first step was to construct the PLS model based on the intersection of 23 metabolites and analyze the variable importance (VIP) of each metabolite. We used two principal components (PCs) in plasma and lung tissues to screen differential metabolites. As shown in Figure S3B, 10 metabolites were obtained to preserve the statistical importance of the classification in two PCs based on VIP (VIP > 1). Secondly, a correlation analysis was performed on the above 10 metabolites to select metabolites with high correlation in plasma and lung tissue, and all these metabolites were obtained based on the correlation (Table S8). Finally, the VIP values of the 10 elected metabolites were analyzed again to assess the contribution to classification. Then, the top seven metabolites (taurocholic acid (TCA), acylcarnitine C5:0 (AC5:0), Leucine, tauroursodeoxycholic acid (THDCA), tauro-α-Muricholic acid (T-α-MCA), acylcarnitine C9:1 (AC9:1) and urea) with the highest VIP scores were selected as the most potential panel of plasma metabolic makers for RILI (Figure S4).

Subsequently, the potential panel of metabolic markers was assessed by the OSC-KPLS model to discriminate different dose groups at different stages of radiation exposure. Figure S5 shows the clustering graph of the control and irradiated groups at 1 d, 2 d, 3 d and 5 d after WTI. Each data point represents a real sample, with the vertical coordinate representing the actual radiation dose received and the horizontal coordinate representing the injury classification. The comparison between the predicted radiation doses and the observed values based on the panel is displayed in Table S9. Table 1 shows the classification results at different time points after irradiation. Compared with the early stage (1 d, 2 d after WTI), nearly all predicted values were close to the observed values at a later stage (3 d, 5 d after WTI), with the accuracies of classification all more than 80%. The result indicated the potential of the panel for estimating the approximate radiation dose and being a biomarker of RILI.

Table 1. Classification of radiation injury at different time points after WTI based on OSC-KPLS model and the panel of potential biomarkers.

Triage	Control	Mild	Moderate	Severe
Accuracy of classification at 1 d	72.7%	90.0%	58.3%	91.7%
Accuracy of classification at 2 d	92.3%	50.0%	72.7%	66.7%
Accuracy of classification at 3 d	100.0%	90.0%	100.0%	100.0%
Accuracy of classification at 5 d	100.0%	80.0%	81.8%	100.0%

In order to directly trace the changes of the potential metabolic markers in the early stages of RILI, we analyzed dynamic plasma metabolic data within 5 days post WTI. As shown in Figure S6, the majority of screened metabolites began to show a significant difference at 2 d after WTI, except urea. Urea showed a down-regulated trend from 1 d to 5 d. AC5:0 and AC9:1 began to decrease at 2 d and 3 d, respectively. Aminoacids (leucine) began to decrease at a later period (3 d after WTI). Cholic acid levels, including TCA, THDCA and T-a-MCA, began to show significant decreases until 5 d.

To further explore the temporal trajectory of these metabolites, the levels for each irradiated rat were divided by the average controls to rule out metabolic derangement due to time-related changes (Figure S7). Interestingly, despite the complexity of radiation regulations, most of them displayed a monotonic response in 20 Gy and 35 Gy-irradiated cohorts from 2 d to 5 d post-irradiation. These features are considered the key metabolites with consistent changing tendencies when comparing high doses versus low doses and low doses versus Pre, reflecting the temporal variations with RILI progression and indicating their potential for assessment of radiation injury for RILI.

3.5. Metabolic Correlation Network Analysis

Due to complex physicochemical reactions, not only did metabolite levels show significant changes, but linkages between metabolites could also be altered [23]. The correlation network could provide an overview of a given status of the complex biological system and reveal dysregulated biochemical mechanisms associated with the stimulus [24]. A positive correlation between two metabolites indicates the adjacent relationship in a metabolic pathway, whereas a negative correlation indicates that one of two metabolites is used to generate the other one directly or indirectly [25].

Then, the correlation network analysis between differential metabolites was performed to reveal the metabolic regulations following WTI using Cytoscape software. The linkage line of metabolites was based on the Pearson correlation coefficient by calculating the relative levels of the metabolites. As shown in Figure 4, most metabolites connected with each other and more connections between different metabolites could be observed after irradiation, which inferred the complex metabolic regulation in RILI. The metabolites with more connections to others may play a more important role in the metabolic regulations of radiation exposure.

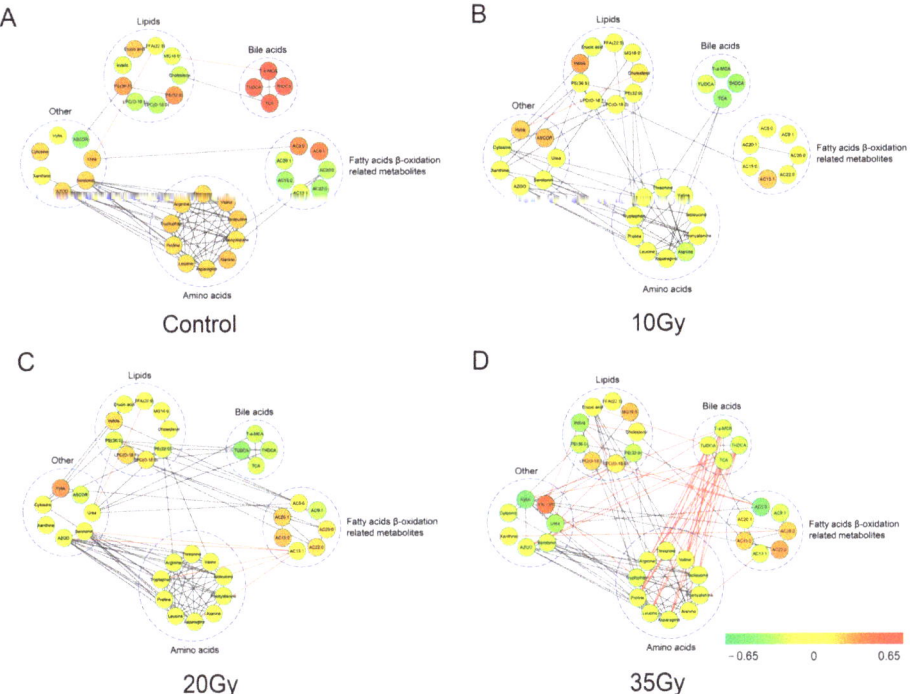

Figure 4. Metabolic correlation network analysis based on differential metabolites associated with radiation injury in the control and irradiated groups at 5 d after WTI ((**A**), Control; (**B**), 10 Gy; (**C**), 20 Gy; (**D**), 35 Gy). Nodes represent the metabolites, and the lines between nodes represent their relationship associated with biochemical reactions. ASCOR: ascorbic acid; MG16:0: palmitoylglycerol; AC13:1: acylcarnitine C13:1; AC20:0: acylcarnitine C20:0; AC22:0: acylcarnitine C22:0; AC15:0: acylcarnitine C15:0; AC20:1: acylcarnitine C20:1; AC5:0: acylcarnitine C5:0; Hybs: 4-Hydroxybenzenesulfonic acid; Indols: indoxyl sulfate; TUDCA: tauroursodeoxycholic acid; TCA: taurocholic acid; T-α-MCA: tauro-α-Muricholic acid; AZOD: 3-Amino-2-oxazolidinone. The metabolites (nodes) in red and green colors represent significant increases and decreases in response to WTI. The black and red lines between metabolites represent positive and negative relationships, respectively.

Lipids were found to correlate more positively with amino acids with increased radiation exposure doses. In contrast, bile acids displayed a more negative correlation with amino acids, such as arginine, proline, leucine and asparagine, in the 35 Gy-irradiated group. In the 20 Gy irradiated group, fatty acid β-oxidation-related metabolites began to show strong correlations with lipids. β-oxidation is a process of generating energy by the formation of ketone bodies from fatty acids, which could explain the correlation between lipids and β-oxidation and indicate the significant contribution in the process of RILI. These results indicated the important roles of bile acids, lipids and fatty acid β-oxidation-related metabolites in the metabolic regulation of RILI.

3.6. CPT1 Gene Expression Level and Enzyme Activity in the Lung Samples of Rats Exposed to WTI

We further found that the levels of Acylcarnitine C20:1 and Acylcarnitine C20:1 increased at 1 d after exposure and maintained the change up to 5 d, indicating the radiosensitivity of fatty acid β-oxidation (Figure S8A). Thus, we evaluated the carnitine acyltransferases (CPT1, presenting in the mitochondrial outer membrane; CPT2, situating at the matrix side of the inner membrane) involved in acylcarnitine metabolism, which regulates this transport system [26]. The activities of two enzymes can be estimated by ratios, such as the CPT1 ratio (carnitine/(C16:1 + C18:0)) and the CPT2 ratio (C16:0 + C18:1/C2). An elevation of the CPT1 ratio indicates CPT1 deficiency or impaired functions [27], reflecting the increased mitochondrial entrance of long-chain FA. Meanwhile, the increase in the CPT2 ratio points to a significant reduction in long-chain fatty acid oxidation or impaired CPT2 functions, which means long-chain acylcarnitine cannot be converted to their corresponding acyl-CoA esters [26]. As can be seen in Table 2, the CPT1 ratio significantly decreased in response to WTI, while the CPT2 ratio markedly increased, indicating the accumulation of long-chain acylcarnitine in mitochondria.

Table 2. Related ratios in control, 10 Gy, 20 Gy and 35 Gy groups.

Enzymes	Control	10 Gy	20 Gy	35 Gy
CPT1	155.4 (139.5–197.4)	129.4 * (109.6–140.2)	112.4 ** (100.2–128.7)	104.1 ** (94.9–112.3)
CPT2	0.004 (0.003–0.004)	0.005 * (0.004–0.005)	0.005 ** (0.005–0.006)	0.006 ***,# (0.005–0.006)

Note: Values are expressed as medians (25th, 75th percentiles). The p-values were calculated based on nonparametric Kruskal–Wallis test. Compared with control, * $p < 0.05$, ** $p < 0.01$, *** $p < 0.001$; compared with 10 Gy, # $p < 0.05$.

In addition, to preliminarily discover the extra accumulation of long-chain acylcarnitine after WTI, the mRNA levels of CPT1A and CPT1B in lung tissues from model rats were analyzed at 5 d after WTI. As shown in Figure S8B, the mRNA levels of CPT1A and CPT1B showed significant dose-dependent elevation compared with the control group. This further substantiates that radiation plays a crucial role in the regulation of CPT1 activity, which may lead to the accumulation of long-chain acylcarnitine.

4. Discussion

In the current study, not only the metabolite levels but also the metabolic correlation networks were significantly altered following WTI. These metabolic abnormalities are mainly involved in amino acids, bile acids, lipids and fatty acid β-oxidation-related metabolites, which are discussed in the following sections.

4.1. Amino Acids

Amino acids are a kind of vital metabolite in the organism, the basic components of proteins and have biological functions such as synthesizing hormones, transmitting cell signals and regulating gene expression [28]. After WTI, the levels of amino acids, including arginine, phenylalanine, tryptophan, valine, leucine, isoleucine, threonine, proline and alanine were significantly reduced in lung tissues (Table S3). These amino acids are involved in multiple metabolic pathways, such as arginine and proline metabolism; valine, leucine and isoleucine degradation;the urea cycle; and aspartate metabolism.

As essential amino acids, branched-chain amino acids (BCAAs, leucine, isoleucine and valine) are widely studied due to their crucial role in the regulation of protein synthesis, primarily through the activation of the mTOR signaling pathway and their growing recognition as players in the regulation of a variety of physiological and metabolic processes [29,30]. Elevated blood BCAA levels in both animal models and humans following total body irradiation (TBI) have been implicated in radiation-induced activated protein breakdown [13,22,31,32]. In contrast to the above findings, BCAAs showed obvious reduced levels in both lung tissues and plasma in response to WTI (Figure S9). BCAAs are ketogenic and glycogenic amino acids, which can be converted to branched-chain keto acids (BCKAs) through BCAA transaminase (BCAT) in body metabolism [33]. Decreased blood BCAA levels have been reported in patients with chronic obstructive pulmonary disease (COPD), while dietary supplementation with BCAAs ameliorates COPD-related weight loss and respiratory muscle weakness [34–38]. It has been found that the inhibition of BCAT can inhibit airway inflammation and remodeling [39], suggesting that BCAT activity is related to pneumonia response. Furthermore, mTOR signaling is closely associated with the dysregulation of autophagy, inflammation, as well as cell growth and survival, resulting in the development of pulmonary fibrosis [40]. Studies suggest that mTOR inhibitors are promising modulators of radiation-induced pulmonary fibrosis (RIPF) [41]. Given that BCAAs have been recognized as having anabolic effects in protein metabolism, which involve the activation of the mTOR pathway, lower BCAAs levels may cause the dysregulation of the mTOR pathway, leading to RILI.Therefore, identifying the BCAA metabolic pathway may be a potential attractive treatment for therapeutic targets in RILI.

Decreases in urea and arginine in lung tissues indicate a urea cycle disorder following WTI. Arginine engages in the urea cycle in the body, promoting the formation of urea, thus transforming ammonia in the human body into non-toxic urea and reducing blood ammonia concentrations [42]. Ionizing radiation could cause the urea-to-ammonia ratio to drop precipitously and thus give rise to hyperammonemia [43,44], suggesting the disturbance of the urea cycle and agreeing with our findings with declined levels of pulmonary arginine and urea. Arginine is also involved in the nitric oxide (NO) pathway and is a substrate for the synthesis of endogenous NO catalyzed by the enzyme NO synthase (NOS) [45]. Elevated exhaled NO following thoracic radiation has been reported to be predictive of RILI. Recent studies have shown that arginine has an important protective effect on pulmonary inflammation and fibrosis [46–48]. Moreover, the supplementation of arginine can significantly downregulate procollagen mRNA transcription and hydroxyproline content in lung tissues [49]. Consequently, we conclude that the decline of pulmonary arginine may be related to the injury repair of the body in response to WTI, which further results in the depletion of proline.

Phenylalanine and tryptophan belong to aromatic amino acids, which can synthesize acute proteins in response to inflammation [50]. Such proteins can play an anti-inflammatory effect through immune regulation [51]. Therefore, the disturbance of aromatic amino acid metabolism in WTI rats may be related to the immune response of RILI. In addition, tryptophan is the only precursor of serotonin, which is a key monoamine neurotransmitter that participates in the modulation of central neurotransmission and physiological function in the enteric system [52]. In addition, tryptophan can be metabolized to kynurenine, tryptamine and indole, modulating neuroendocrine and gut immune responses [52]. In our study, phenylalanine, tryptophan and serotonin were significantly decreased in lung tissue in response to WTI, while indoxyl sulfate (Indols) was increased at low doses (10 Gy, 20 Gy) and decreased at the high dose (35 Gy) (Figure S10). The changed levels of all these metabolites in lung tissues implied a role for gut microflora in the lung tissue exposed to WTI. Chen et al. reported that fecal microbiota transplantation (FMT) attenuated radiation pneumonia, scavenged oxidative stress and ameliorated lung function in mouse models following local chest irradiation [53]. This research further indicates the relationship between radiation pneumonia and intestinal flora metabolism, which is consistent with the results of this study. Furthermore, the decreased levels of plasma metabolites associated with tryptophan, including kynurenic acid, serotonin and indole at 3–5 d post-irradiation, reinforced the importance of gut microflora in RILI.

4.2. Bile Acids

Bile acids (BAs) are synthesized in the liver and secreted into the digestive tract, where they facilitate the digestion and absorption of lipids. They are associated with chronic inflammation and remodeling of the lung microbiota. Furthermore, BAs can regulate the composition of the microbiota in indirect or direct ways and protect the gut barrier [54]. In recent years, increasing evidence has demonstrated that there is a close connection between gut microbes and the lung by modulating the transmission route of the gut–lung axis [55]. Many lung diseases often present with dysbiosis of the gut flora, which may refer to the development of disease [54]. Additionally, BAs participate in the interactions between the intestinal microbiota and the host's immunity [56]. It has been shown that BAs can act as signaling molecules via the activation of dedicated receptors, such as nuclear receptor Farnesoid X Receptor (FXR) and membrane-bound receptor Takeda-G protein receptor 5 (TGR5). In addition, the FXR for BAs has been shown to be expressed in human airway epithelial cells [57], and the agonists have been proven to have beneficial effects in a wide range of pulmonary diseases, such as chronic obstructive pulmonary disease (COPD) and idiopathic pulmonary fibrosis [58]. Recent reports verified the therapeutic effects of the natural agonists of FXR (DCA and LCA) on inflammatory bowel disease by restoring intestinal barrier function and alleviating inflammatory reactions [59]. In our research, taurocholic acid (TCA), taurohyodeoxycholic acid (THDCA), Taurohyodeoxycholic acid (TUDCA) and tauro-α-Muricholic acid (T-α-MCA) were significantly decreased in the lung tissues of rats in response to WTI, which revealed that RILI induced the disturbance of bile acid metabolism and gut barrier dysfunction. These results are in agreement with a recent study by Li et al., who reported that the bile acid pool had a marked reduction after whole chest irradiation and was recovered bycryptotanshinone (CPT) treatment in large part [55].

4.3. Lipids and Fatty Acid β-Oxidation

It is well known that lipids are the major constituents of cell membrane bilayers, playing a major role in cell signaling, membrane anchorage and substrate transport.Radiation exposure causes dysfunction of the cell membrane and disrupted lipid metabolism, together with changes in lipid concentration and increased lipid peroxidation [60,61]. Increasing studies have indicated that irradiation resulted in lipid accumulation, evidenced by elevated triacylglycerol and cholesterol levels in plasma, liver or lung tissues [12,62–64]. In accordance with the lipid accumulation revealed in the above reports, pulmonary palmitoylglycerol, cholesterol, unsaturated free fatty acid 22:5 (FFA 22:5), LPC (LPC(O-18:0) and LPC(O-18:1)) were significantly increased in response to WTI (Figure S11). Although the exact mechanism of radiation alters lipid mechanism is unclear, increased glucose catabolism by providing increased levels of glycerophosphate as a lipid precursor and up-regulated the lipoprotein lipase and fatty acid binding protein expression have been considered important contributors to this lipid accumulation [65,66]. Moreover, pulmonary lipid metabolites could induce chronic inflammation in tissues primarily by promoting the infiltration and activation of macrophages [4]. Among the affected lipids, the alterations of two PEs, including PE 32:0 and PE 36:5, deserve attention (Figure S11). As the second most abundant membrane phospholipid in mammals, PE plays an essential role in mammalian development and cellular processes, including metabolism and signaling [67]. It has been demonstrated that PE strongly contributes to surfactant-induced inhibition of collagen expression in human lung fibroblasts via a Ca^{2+} signal, and early administration of PE-enriched Beractant decreases lung fibrosis in mice [68]. Thus, decreased levels of pulmonary PEs in the irradiated groups may be indicative of injury repair in RILI at the expense of consumption. Overall, these findings support the idea that an alteration in lipid metabolism is important in RILI pathology.

The β-oxidation of fatty acids is one of the main methods of energy metabolism in organisms [69], including the activation, transfer and oxidation of fatty acids, culminating in the production of acetyl-CoA and direct involvement in the tricarboxylic acid cycle (TCA) or generation of metabolites, such as ketone bodies for energy metabolism. Acylcarnitines (ACs) are formed when fatty acid enters the mitochondria for β-oxidation through the carnitine shuttle. ACs can be divided into short (C3–C5), medium (C6–C12) and long-chain (>C12) ACs depending on the length of the acyl groups. Due to the large number and special structure, ACs play an important role in the physiological activities of cells and become a key substance for cellular metabolism [70]. Levels of ACs can vary depending on the metabolic conditions but may accumulate when rates of β-oxidation exceed those of tricarboxylic acid cycle (TCA). ACs play a major role in the β-oxidation of long-chain fatty acids (LCFAs) and serve as a carrier to transport activated long-chain acyl-CoAs into the mitochondria for subsequent β-oxidation to provide energy for cellular activities [71].

Prior studies have implicated carnitine metabolites as potential biomarkers of radiation injury in biofluids derived from animals and humans [10,11,72]. Meanwhile, enhancement of AC levels has been reported in the small intestine of abdominal-irradiated rats [73]. In the current study, the decreased CPT1 ratio and increased CPT2 ratio in response to WTI (Table 2) reflected increased mitochondrial entrance of long-chain ACs and incomplete fatty acid β-oxidation, accounting for the accumulation of long-chain ACs in both lung tissues and plasma (Tables S3–S7).

5. Conclusions

In the present study, early time-point plasma and lung metabolic signatures following WTI were revealed. To identify minimally invasive markers for RILI, the metabolic features of the lung tissue in response to WTI were cross-correlated with plasma metabolic features. In the combined multivariate PLS model, the panel of potential plasma metabolic markers was selected and used to assess the radiation injury levels within 5 days following WTI. Our data implied that plasma metabolites can potentially be used to estimate radiation doses associated with RILI. Moreover, the significant difference in metabolite levels and metabolic correlation network in the lung tissue revealed that multiple metabolic dysregulation primarily involved amino acids, bile acids, lipids and fatty acid β-oxidation-related metabolites. In particular, the accumulation of long-chain ACs deserves attention by jointly analyzing dynamic plasma metabolic characteristics. These findings provide insight into the metabolic characteristics associated with RILI. Further extensive studies, including the validation of the panel of potential metabolic markers for RILI by another cohort of animals and thoracic radiotherapy patients and exploration of the effect of fatty acid β-oxidation metabolism in RILI, are required.

Supplementary Materials: The following supporting information can be downloaded at: https://www.mdpi.com/article/10.3390/metabo13091020/s1, Figure S1: Reproducibility and stability of metabolic profiling in lung tissues were assessed. (A) RSD distribution of the QC samples. (B) PCA score plots of the QC sample distributions in LC-MS; Figure S2: Reproducibility and stability of metabolic profiling in plasma. (A) RSD distribution of the QC samples. (B) PCA score plots of the QC sample distributions in LC-MS; Figure S3: Screen of biomarkers in plasma. (A) Venn diagram showing the shared and unique metabolites between plasma and lung tissues. (B–E) Column plots of VIP value in plasma and lung tissues, metabolites with VIP [1] and VIP [2] > 1 were selected; Figure S4: Column plots of VIP value in plasma; metabolites with more importance were selected as biomarkers; Figure S5: (A,C,E,G) Comparison between the predicted radiation doses and observed values in plasma at 1 d, 2 d, 3 d and 5 d after WTI. (B,D,F,H) Validation plots obtained from 200 permutation tests based on the panel of these 7 biomarkers, and the model parameters showed the predictive ability of model; Figure S6: Dynamic changes of these 7 potential biomarkers in plasma.Compared with Control, * $p < 0.05$, ** $p < 0.01$; Figure S7: Dynamic changes of these 7 potential biomarkers in plasma (normalized by control group); Figure S8: (A,B) Dynamic changes of Acylcarnitine C20:0 and Acylcarnitine C20:1 in plasma (normalized by control group). (C,D) The mRNA level of CPT1A and CPT1B in lung tissues at 5 d after different radiation dose exposures.Compared with Control, * $p < 0.05$, ** $p < 0.01$, *** $p < 0.001$; Figure S9: (A,B) Dynamic changes of valine and leucine in plasma. (C–E) Changes of valine, leucine and isoleucine in lung tissues. Compared with Control, * $p < 0.05$, ** $p < 0.01$; Figure S10: (A,B) Changes in aromatic amino acids (phenylalanine and tryptophan) in lung tissues. (C,D) Changes intryptophan-related metabolites (serotonin and indoxyl sulfate) in lung tissues.Compared with Control, * $p < 0.05$, ** $p < 0.01$; compared with 10 Gy, ## $p < 0.01$; compared with 20 Gy, && $p < 0.01$; Figure S11: Changes in lipids (LPC (0–18:0, LPC (O-18:1), PE (32:0) and PE (36:5)) in lung tissues.Compared with control, * $p < 0.05$, ** $p < 0.01$; Table S1: Concentration of internal target used for pretreatment of plasma samples; Table S2: Concentration of internal target used for pretreatment of lung tissue samples; Table S3: Statistical information of differential metabolites in lung tissues at 5 d after WTI (X ± SD); Table S4: Statistical information of differential metabolites in plasma at 1 d after WTI (X ± SD); Table S5: Statistical information of differential metabolites in plasma at 2 d after WTI (X ± SD); Table S6: Statistical information of differential metabolites in plasma at 3 d after WTI (X ± SD); Table S7: Statistical information of differential metabolites in plasma at 5 d after WTI (X ± SD); Table S8: Correlation analysis of differential metabolites in plasma and lung tissues at 5 d after WTI; Table S9: Stimulation of RILI classification after WTI.

Author Contributions: C.W. designed the project, collected and interpreted the data; C.W. and L.G. wrote the first version of the manuscript; L.G. and W.W. performed data analysis; L.G., W.W. and Y.G. collected the samples, conducted experiments and collected data. J.C. revised the final version of the manuscript. All authors have read and agreed to the published version of the manuscript.

Funding: The study has been supported by the foundation (No. 81773361, No. U1967220) from the National Natural Science Foundation of China, Collaborative Innovation Center of Radiological Medicine of Jiangsu Higher Education Institutions and A Project Funded by the Priority Academic Program Development of Jiangsu Higher Education Institutions (PAPD).

Institutional Review Board Statement: The study was conducted in accordance with the Declaration of Helsinki, and approved by the Ethics Committee of Soochow University (protocol code: 201804A557).

Informed Consent Statement: Not applicable.

Data Availability Statement: All data that support the findings of the study are within the manuscript or in Supplemental Materials.

Conflicts of Interest: The authors declare no conflict of interest.

References

1. Hanania, A.N.; Mainwaring, W.; Ghebre, Y.T.; Hanania, N.A.; Ludwig, M. Radiation-Induced Lung Injury: Assessment and Management. *Chest* **2019**, *156*, 150–162. [CrossRef] [PubMed]
2. Inoue, A.; Kunitoh, H.; Sekine, I.; Sumi, M.; Tokuuye, K.; Saijo, N. Radiation pneumonitis in lung cancer patients: A retrospective study of risk factors and the long-term prognosis. *Int. J. Radiat. Oncol. Biol. Phys.* **2001**, *49*, 649–655. [CrossRef]
3. Liu, X.; Shao, C.; Fu, J. Promising Biomarkers of Radiation-Induced Lung Injury: A Review. *Biomedicines* **2021**, *9*, 1181. [CrossRef]
4. Sunil Gowda, N.S.; Raviraj, R.; Nagarajan, D.; Zhao, W. Radiation-induced lung injury: Impact on macrophage dysregulation and lipid alteration—A review. *Immunopharmacol. Immunotoxicol.* **2019**, *41*, 370–379. [CrossRef]
5. Yan, Y.; Fu, J.; Kowalchuk, R.O.; Wright, C.M.; Zhang, R.; Li, X.; Xu, Y. Exploration of radiation-induced lung injury, from mechanism to treatment: A narrative review. *Transl. Lung Cancer Res.* **2022**, *11*, 307–322. [CrossRef]
6. Zhang, T.; Zhou, Z.; Wen, L.; Shan, C.; Lai, M.; Liao, J.; Zeng, X.; Yan, G.; Cai, L.; Zhou, M.; et al. Gene Signatures for Latent Radiation-Induced Lung Injury Post X-ray Exposure in Mouse. *Dose Response* **2023**, *21*, 15593258231178146. [CrossRef] [PubMed]
7. Liu, R.; Bao, Z.X.; Zhao, P.J.; Li, G.H. Advances in the Study of Metabolomics and Metabolites in Some Species Interactions. *Molecules* **2021**, *26*, 3311. [CrossRef]
8. Di Minno, A.; Gelzo, M.; Caterino, M.; Costanzo, M.; Ruoppolo, M.; Castaldo, G. Challenges in Metabolomics-Based Tests, Biomarkers Revealed by Metabolomic Analysis, and the Promise of the Application of Metabolomics in Precision Medicine. *Int. J. Mol. Sci.* **2022**, *23*, 5213. [CrossRef] [PubMed]
9. Trivedi, D.K.; Hollywood, K.A.; Goodacre, R. Metabolomics for the masses: The future of metabolomics in a personalized world. *NewHoriz. Transl. Med.* **2017**, *3*, 294–305. [CrossRef] [PubMed]
10. Laiakis, E.C.; Mak, T.D.; Anizan, S.; Amundson, S.A.; Barker, C.A.; Wolden, S.L.; Brenner, D.J.; Fornace, A.J., Jr. Development of a metabolomic radiation signature in urine from patients undergoing total body irradiation. *Radiat. Res.* **2014**, *181*, 350–361. [CrossRef]
11. Pannkuk, E.L.; Laiakis, E.C.; Authier, S.; Wong, K.; Fornace, A.J., Jr. Targeted Metabolomics of Nonhuman Primate Serum after Exposure to Ionizing Radiation: Potential Tools for High-throughput Biodosimetry. *RSC Adv.* **2016**, *6*, 51192–51202. [CrossRef]
12. Feurgard, C.; Bayle, D.; Guezingar, F.; Serougne, C.; Mazur, A.; Lutton, C.; Aigueperse, J.; Gourmelon, P.; Mathe, D. Effects of ionizing radiation (neutrons/gamma rays) on plasma lipids and lipoproteins in rats. *Radiat. Res.* **1998**, *150*, 43–51. [CrossRef]
13. Hong, X.; Tian, L.; Wu, Q.; Gu, L.; Wang, W.; Wu, H.; Zhao, M.; Wu, X.; Wang, C. Plasma metabolomic signatures from patients following high-dose total body irradiation. *Mol. Omics* **2023**, *19*, 492–503. [CrossRef]
14. Pannkuk, E.L.; Fornace, A.J., Jr.; Laiakis, E.C. Metabolomic applications in radiation biodosimetry: Exploring radiation effects through small molecules. *Int. J. Radiat. Biol.* **2017**, *93*, 1151–1176. [CrossRef] [PubMed]

15. Stirling, E.R.; Cook, K.L.; Roberts, D.D.; Soto-Pantoja, D.R. Metabolomic Analysis Reveals Unique Biochemical Signatures Associated with Protection from Radiation Induced Lung Injury by Lack of cd47 Receptor Gene Expression. *Metabolites* **2019**, *9*, 218. [CrossRef] [PubMed]
16. Gao, Y.; Li, X.; Gao, J.; Zhang, Z.; Feng, Y.; Nie, J.; Zhu, W.; Zhang, S.; Cao, J. Metabolomic Analysis of Radiation-Induced Lung Injury in Rats: The Potential Radioprotective Role of Taurine. *Dose Response* **2019**, *17*, 1559325819883479. [CrossRef] [PubMed]
17. Feng, Y.; Gao, Y.; Tu, W.; Feng, Y.; Cao, J.; Zhang, S. Serum Metabolomic Analysis of Radiation-Induced Lung Injury in Rats. *Dose Response* **2022**, *20*, 15593258211067060. [CrossRef]
18. Li, Y.; Li, M.; Jia, W.; Ni, Y.; Chen, T. MCEE: A data preprocessing approach for metabolic confounding effect elimination. *Anal. Bioanal. Chem.* **2018**, *410*, 2689–2699. [CrossRef]
19. Slupsky, C.M.; Rankin, K.N.; Wagner, J.; Fu, H.; Chang, D.; Weljie, A.M.; Saude, E.J.; Lix, B.; Adamko, D.J.; Shah, S.; et al. Investigations of the effects of gender, diurnal variation, and age in human urinary metabolomic profiles. *Anal. Chem.* **2007**, *79*, 6995–7004. [CrossRef]
20. Cross, A.J.; Moore, S.C.; Boca, S.; Huang, W.Y.; Xiong, X.; Stolzenberg-Solomon, R.; Sinha, R.; Sampson, J.N. A prospective study of serum metabolites and colorectal cancer risk. *Cancer* **2014**, *120*, 3049–3057. [CrossRef] [PubMed]
21. Zheng, F.; Zhao, X.; Zeng, Z.; Wang, L.; Lv, W.; Wang, Q.; Xu, G. Development of a plasma pseudotargeted metabolomics method based on ultra-high-performance liquid chromatography-mass spectrometry. *Nat. Protoc.* **2020**, *15*, 2519–2537. [CrossRef]
22. Tang, X.X.; Zheng, M.C.; Zhang, Y.Y.; Fan, S.J.; Wang, C. Estimation value of plasma amino acid target analysis to the acute radiation injury early triage in the rat model. *Metabolomics* **2013**, *9*, 853–863. [CrossRef]
23. Yao, X.T.; Xu, C.; Cao, Y.R.; Lin, L.; Wu, H.X.; Wang, C. Early metabolic characterization of brain tissues after whole body radiation based on gas chromatography-mass spectrometry in a rat model. *Biomed. Chromatogr.* **2019**, *33*, e4448. [CrossRef]
24. Bonelli, R.; Woods, S.M.; Ansell, B.R.E.; Heeren, T.F.C.; Egan, C.A.; Khan, K.N.; Guymer, R.; Trombley, J.; Friedlander, M.; Bahlo, M.; et al. Systemic lipid dysregulation is a risk factor for macular neurodegenerative disease. *Sci. Rep-UK* **2020**, *10*, 12165. [CrossRef]
25. Wang, L.; Hou, E.T.; Wang, L.J.; Wang, Y.J.; Yang, L.J.; Zheng, X.H.; Xie, G.Q.; Sun, Q.; Liang, M.Y.; Tian, Z.M. Reconstruction and analysis of correlation networks based on GC-MS metabolomics data for young hypertensive men. *Anal. Chim. Acta* **2015**, *854*, 95–105. [CrossRef] [PubMed]
26. Yang, Y.; Sadri, H.; Prehn, C.; Adamski, J.; Rehage, J.; Danicke, S.; Saremi, B.; Sauerwein, H. Acylcarnitine profiles in serum and muscle of dairy cows receiving conjugated linoleic acids or a control fat supplement during early lactation. *J. Dairy. Sci.* **2019**, *102*, 754–767. [CrossRef] [PubMed]
27. Fingerhut, R.; Roschinger, W.; Muntau, A.C.; Dame, T.; Kreischer, J.; Arnecke, R.; Superti-Furga, A.; Troxler, H.; Liebl, B.; Olgemoller, B.; et al. Hepatic carnitine palmitoyltransferase I deficiency: Acylcarnitine profiles in blood spots are highly specific. *Clin. Chem.* **2001**, *47*, 1763–1768. [CrossRef] [PubMed]
28. Wu, G. Amino acids: Metabolism, functions, and nutrition. *Amino Acids* **2009**, *37*, 1–17. [CrossRef] [PubMed]
29. Laplante, M.; Sabatini, D.M. mTOR Signaling in Growth Control and Disease. *Cell* **2012**, *149*, 274–293. [CrossRef]
30. Blomstrand, E.; Eliasson, J.; Karlsson, H.K.R.; Kohnke, R. Branched-chain amino acids activate key enzymes in protein synthesis after physical exercise. *J. Nutr.* **2006**, *136*, 269s–273s. [CrossRef]
31. Holecek, M.; Skopec, F.; Sprongl, L.; Mraz, J.; Skalska, H.; Pecka, M. Effect of alanyl-glutamine on leucine and protein metabolism in irradiated rats. *Amino Acids* **2002**, *22*, 95–108. [CrossRef] [PubMed]
32. Khan, A.R.; Rana, P.; Devi, M.M.; Chaturvedi, S.; Javed, S.; Tripathi, R.P.; Khushu, S. Nuclear magnetic resonance spectroscopy-based metabonomic investigation of biochemical effects in serum of gamma-irradiated mice. *Int. J. Radiat. Biol.* **2011**, *87*, 91–97. [CrossRef]
33. Holecek, M. Why Are Branched-Chain Amino Acids Increased in Starvation and Diabetes? *Nutrients* **2020**, *12*, 3087. [CrossRef]
34. Crowley, G.; Kwon, S.; Haider, S.H.; Caraher, E.J.; Lam, R.; St-Jules, D.E.; Liu, M.; Prezant, D.J.; Nolan, A. Metabolomics of World Trade Center-Lung Injury: A machine learning approach. *BMJ Open Respir. Res.* **2018**, *5*, e000274. [CrossRef] [PubMed]
35. Engelen, M.P.K.J.; Rutten, E.P.A.; De Castro, C.L.N.; Wouters, E.F.M.; Schols, A.M.W.J.; Deutz, N.E.P. Supplementation of soy protein with branched-chain amino acids alters protein metabolism in healthy elderly and even more in patients with chronic obstructive pulmonary disease. *Am. J. Clin. Nutr.* **2007**, *85*, 431–439. [CrossRef]
36. Rogers, R.M.; Donahoe, M.; Costantino, J. Physiological-Effects of Oral Supplemental Feeding in Malnourished Patients with Chronic Obstructive Pulmonary-Disease—A Randomized Control Study. *Am. Rev. Respir. Dis.* **1992**, *146*, 1511–1517. [CrossRef] [PubMed]
37. Kutsuzawa, T.; Shioya, S.; Kurita, D.; Haida, M. Plasma branched-chain amino acid levels and muscle energy metabolism in patients with chronic obstructive pulmonary disease. *Clin. Nutr.* **2009**, *28*, 203–208. [CrossRef]
38. Wang, L.L.; Tang, Y.F.; Liu, S.; Mao, S.T.; Ling, Y.; Liu, D.; He, X.Y.; Wang, X.G. Metabonomic Profiling of Serum and Urine by H-1 NMR-Based Spectroscopy Discriminates Patients with Chronic Obstructive Pulmonary Disease and Healthy Individuals. *PLoS ONE* **2013**, *8*, e65675. [CrossRef]
39. Li, J.; Chen, M.; Lu, L.; Wang, J.; Tan, L. Branched-chain amino acid transaminase 1 inhibition attenuates childhood asthma in mice by effecting airway remodeling and autophagy. *Respir. Physiol. Neurobiol.* **2022**, *306*, 103961. [CrossRef]
40. Lawrence, J.; Nho, R. The Role of the Mammalian Target of Rapamycin (mTOR) in Pulmonary Fibrosis. *Int. J. Mol. Sci.* **2018**, *19*, 778. [CrossRef] [PubMed]

41. Chung, E.J.; Sowers, A.; Thetford, A.; McKay-Corkum, G.; Chung, S.I.; Mitchell, J.B.; Citrin, D.E. Mammalian Target of Rapamycin Inhibition with Rapamycin Mitigates Radiation-Induced Pulmonary Fibrosis in a Murine Model. *Int. J. Radiat. Oncol.* **2016**, *96*, 857–866. [CrossRef] [PubMed]
42. Scibior, D.; Czeczot, H. Arginine--metabolism and functions in the human organism. *Postepy. Hig. Med. Dosw.* **2004**, *58*, 321–332.
43. van Rijn, J.; van den Berg, J.; Teerlink, T.; Kruyt, F.A.; Schor, D.S.; Renardel de Lavalette, A.C.; van den Berg, T.K.; Jakobs, C.; Slotman, B.J. Changes in the ornithine cycle following ionising radiation cause a cytotoxic conditioning of the culture medium of H35 hepatoma cells. *Br. J. Cancer* **2003**, *88*, 447–454. [CrossRef]
44. Moroz, B.B.; Vasil'ev, P.S.; Fedorovskii, L.L.; Grozdov, S.P.; Morozova, N.V. Effect of local x-ray irradiation of the abdominal area on the amino acid content of the blood plasma and their urinary excretion in dogs and rats. *Radiobiologiia* **1987**, *27*, 332–338. [PubMed]
45. Stuehr, D.J. Enzymes of the L-arginine to nitric oxide pathway. *J. Nutr.* **2004**, *134*, 2748S–2751S, discussion 2765S–2767S. [CrossRef] [PubMed]
46. Ma, X.; Zhang, Y.; Jiang, D.; Yang, Y.; Wu, G.; Wu, Z. Protective Effects of Functional Amino Acids on Apoptosis, Inflammatory Response, and Pulmonary Fibrosis in Lipopolysaccharide-Challenged Mice. *J. Agric. Food Chem.* **2019**, *67*, 4915–4922. [CrossRef] [PubMed]
47. Chen, J.; Jin, Y.; Yang, Y.; Wu, Z.; Wu, G. Epithelial Dysfunction in Lung Diseases: Effects of Amino Acids and Potential Mechanisms. *Adv. Exp. Med. Biol.* **2020**, *1265*, 57–70. [CrossRef]
48. Gao, L.; Zhang, J.H.; Chen, X.X.; Ren, H.L.; Feng, X.L.; Wang, J.L.; Xiao, J.H. Combination of L-Arginine and L-Norvaline protects against pulmonary fibrosis progression induced by bleomycin in mice. *Biomed. Pharmacother.* **2019**, *113*, 108768. [CrossRef]
49. Song, L.; Wang, D.; Cui, X.; Hu, W. The protective action of taurine and L-arginine in radiation pulmonary fibrosis. *J. Environ. Pathol. Toxicol. Oncol.* **1998**, *17*, 151–157.
50. Reeds, P.J.; Fjeld, C.R.; Jahoor, F. Do the differences between the amino acid compositions of acute-phase and muscle proteins have a bearing on nitrogen loss in traumatic states? *J. Nutr.* **1994**, *124*, 906–910. [CrossRef]
51. Speelman, T.; Dale, L.; Louw, A.; Verhoog, N.J.D. The Association of Acute Phase Proteins in Stress and Inflammation-Induced T2D. *Cells* **2022**, *11*, 2163. [CrossRef] [PubMed]
52. Gao, K.; Mu, C.L.; Farzi, A.; Zhu, W.Y. Tryptophan Metabolism: A Link Between the Gut Microbiota and Brain. *Adv. Nutr.* **2020**, *11*, 709–723. [CrossRef]
53. Chen, Z.Y.; Xiao, H.W.; Dong, J.L.; Li, Y.; Wang, B.; Fan, S.J.; Cui, M. Gut Microbiota-Derived PGF2alpha Fights against Radiation-Induced Lung Toxicity through the MAPK/NF-kappaB Pathway. *Antioxidants* **2021**, *11*, 65. [CrossRef]
54. Budden, K.F.; Gellatly, S.L.; Wood, D.L.; Cooper, M.A.; Morrison, M.; Hugenholtz, P.; Hansbro, P.M. Emerging pathogenic links between microbiota and the gut-lung axis. *Nat. Rev. Microbiol.* **2017**, *15*, 55–63. [CrossRef]
55. Li, Z.; Shen, Y.; Xin, J.; Xu, X.; Ding, Q.; Chen, W.; Wang, J.; Lv, Y.; Wei, X.; Wei, Y.; et al. Cryptotanshinone alleviates radiation-induced lung fibrosis via modulation of gut microbiota and bile acid metabolism. *Phytother. Res.* **2023**. [CrossRef]
56. Hang, S.; Paik, D.; Yao, L.; Kim, E.; Trinath, J.; Lu, J.; Ha, S.; Nelson, B.N.; Kelly, S.P.; Wu, L.; et al. Bile acid metabolites control T(H)17 and T(reg) cell differentiation. *Nature* **2019**, *576*, 143–148. [CrossRef] [PubMed]
57. Hendrick, S.M.; Mroz, M.S.; Greene, C.M.; Keely, S.J.; Harvey, B.J. Bile acids stimulate chloride secretion through CFTR and calcium-activated Cl- channels in Calu-3 airway epithelial cells. *Am. J. Physiol. Lung Cell Mol. Physiol.* **2014**, *307*, L407–L418. [CrossRef]
58. Wu, J.N.; Chen, J.R.; Chen, J.L. Role of Farnesoid X Receptor in the Pathogenesis of Respiratory Diseases. *Can. Respir. J.* **2020**, *2020*, 9137251. [CrossRef]
59. van der Lugt, B.; Vos, M.C.P.; Grootte Bromhaar, M.; Ijssennagger, N.; Vrieling, F.; Meijerink, J.; Steegenga, W.T. The effects of sulfated secondary bile acids on intestinal barrier function and immune response in an inflammatory in vitro human intestinal model. *Heliyon* **2022**, *8*, e08883. [CrossRef]
60. Benderitter, M.; Vincent-Genod, L.; Pouget, J.P.; Voisin, P. The cell membrane as a biosensor of oxidative stress induced by radiation exposure: A multiparameter investigation. *Radiat. Res.* **2003**, *159*, 471–483. [CrossRef] [PubMed]
61. Laiakis, E.C.; Wang, Y.W.; Young, E.F.; Harken, A.D.; Xu, Y.; Smilenov, L.; Garty, G.Y.; Brenner, D.J.; Fornace, A.J., Jr. Metabolic Dysregulation after Neutron Exposures Expected from an Improvised Nuclear Device. *Radiat. Res.* **2017**, *188*, 21–34. [CrossRef]
62. Gould, R.G.; Bell, V.L.; Lilly, E.H. Stimulation of cholesterol biosynthesis from acetate in rat liver and adrenals by whole body x-irradiation. *Am. J. Physiol.* **1959**, *196*, 1231–1237. [CrossRef] [PubMed]
63. Feliste, R.; Dousset, N.; Carton, M.; Douste-Blazy, L. Changes in plasma apolipoproteins following whole-body irradiation in rabbit. *Radiat. Res.* **1981**, *87*, 602–612. [CrossRef] [PubMed]
64. Beheshti, A.; Chakravarty, K.; Fogle, H.; Fazelinia, H.; Silveira, W.A.D.; Boyko, V.; Polo, S.L.; Saravia-Butler, A.M.; Hardiman, G.; Taylor, D.; et al. Multi-omics analysis of multiple missions to space reveal a theme of lipid dysregulation in mouse liver. *Sci. Rep.* **2019**, *9*, 19195. [CrossRef]
65. Gao, F.; Liu, C.; Guo, J.; Sun, W.; Xian, L.; Bai, D.; Liu, H.; Cheng, Y.; Li, B.; Cui, J.; et al. Radiation-driven lipid accumulation and dendritic cell dysfunction in cancer. *Sci. Rep.* **2015**, *5*, 9613. [CrossRef]
66. Levis, G.M.; Efstratiadis, A.A.; Mantzos, J.D.; Miras, C.J. The effect of ionizing radiation on lipid metabolism in bone marrow cells. *Radiat. Res.* **1975**, *61*, 342–349. [CrossRef] [PubMed]

67. Calzada, E.; Onguka, O.; Claypool, S.M. Phosphatidylethanolamine Metabolism in Health and Disease. *Int. Rev. Cell Mol. Biol.* **2016**, *321*, 29–88. [CrossRef]
68. Vazquez-de-Lara, L.G.; Tlatelpa-Romero, B.; Romero, Y.; Fernandez-Tamayo, N.; Vazquez-de-Lara, F.; Justo-Janeiro, J.-M.; Garcia-Carrasco, M.; de-la-Rosa Paredes, R.; Cisneros-Lira, J.G.; Mendoza-Milla, C.; et al. Phosphatidylethanolamine Induces an Antifibrotic Phenotype in Normal Human Lung Fibroblasts and Ameliorates Bleomycin-Induced Lung Fibrosis in Mice. *Int. J. Mol. Sci.* **2018**, *19*, 2758. [CrossRef]
69. Vishwanath, V.A. Fatty Acid Beta-Oxidation Disorders: A Brief Review. *Ann. Neurosci.* **2016**, *23*, 51–55. [CrossRef]
70. Batchuluun, B.; Al Rijjal, D.; Prentice, K.J.; Eversley, J.A.; Burdett, E.; Mohan, H.; Bhattacharjee, A.; Gunderson, E.P.; Liu, Y.; Wheeler, M.B. Elevated Medium-Chain Acylcarnitines Are Associated with Gestational Diabetes Mellitus and Early Progression to Type 2 Diabetes and Induce Pancreatic beta-Cell Dysfunction. *Diabetes* **2018**, *67*, 885–897. [CrossRef]
71. Tarasenko, T.N.; Cusmano-Ozog, K.; McGuire, P.J. Tissue acylcarnitine status in a mouse model of mitochondrial beta-oxidation deficiency during metabolic decompensation due to influenza virus infection. *Mol. Genet. Metab.* **2018**, *125*, 144–152. [CrossRef] [PubMed]
72. Goudarzi, M.; Weber, W.M.; Mak, T.D.; Chung, J.; Doyle-Eisele, M.; Melo, D.R.; Brenner, D.J.; Guilmette, R.A.; Fornace, A.J., Jr. Metabolomic and lipidomic analysis of serum from mice exposed to an internal emitter, cesium-137, using a shotgun LC-MS(E) approach. *J. Proteome Res.* **2015**, *14*, 374–384. [CrossRef] [PubMed]
73. Liu, H.X.; Lu, X.; Zhao, H.; Li, S.; Gao, L.; Tian, M.; Liu, Q.J. Enhancement of Acylcarnitine Levels in Small Intestine of Abdominal Irradiation Rats Might Relate to Fatty Acid beta-Oxidation Pathway Disequilibration. *Dose Response* **2022**, *20*, 15593258221075118. [CrossRef]

Disclaimer/Publisher's Note: The statements, opinions and data contained in all publications are solely those of the individual author(s) and contributor(s) and not of MDPI and/or the editor(s). MDPI and/or the editor(s) disclaim responsibility for any injury to people or property resulting from any ideas, methods, instructions or products referred to in the content.

Article

Rapid Detection of Volatile Organic Metabolites in Urine by High-Pressure Photoionization Mass Spectrometry for Breast Cancer Screening: A Pilot Study

Ming Yang [1,2,3], Jichun Jiang [1,3], Lei Hua [1,3], Dandan Jiang [1,3], Yadong Wang [4], Depeng Li [2], Ruoyu Wang [4], Xiaohui Zhang [2,*] and Haiyang Li [1,3,*]

1. Key Laboratory of Separation Science for Analytical Chemistry, Dalian Institute of Chemical Physics, Chinese Academy of Sciences, Dalian 116023, China; yangm@dicp.ac.cn (M.Y.); jjc@dicp.ac.cn (J.J.); lhua@dicp.ac.cn (L.H.); jiangdandan@dicp.ac.cn (D.J.)
2. College of Environment and Chemical Engineering, Dalian University, Dalian 116000, China; lidepeng@dlu.edu.cn
3. Center for Advanced Mass Spectrometry, Dalian Institute of Chemical Physics, Chinese Academy of Sciences, Dalian 116023, China
4. Department of Oncology Medicine, Affiliated Zhongshan Hospital of Dalian University, Dalian 116023, China; wangruoyu1963@163.com (R.W.); wangyadong@dlu.edu.cn (Y.W.)
* Correspondence: zhangxiaohui@dlu.edu.cn (X.Z.); hli@dicp.ac.cn (H.L.)

Abstract: Despite surpassing lung cancer as the most frequently diagnosed cancer, female breast cancer (BC) still lacks rapid detection methods for screening that can be implemented on a large scale in practical clinical settings. However, urine is a readily available biofluid obtained non-invasively and contains numerous volatile organic metabolites (VOMs) that offer valuable metabolic information concerning the onset and progression of diseases. In this work, a rapid method for analysis of VOMs in urine by using high-pressure photon ionization time-of-flight mass spectrometry (HPPI-TOFMS) coupled with dynamic purge injection. A simple pretreatment process of urine samples by adding acid and salt was employed for efficient VOM sampling, and the numbers of metabolites increased and the detection sensitivity was improved after the acid (HCl) and salt (NaCl) addition. The established mass spectrometry detection method was applied to analyze a set of training samples collected from a local hospital, including 24 breast cancer patients and 27 healthy controls. Statistical analysis techniques such as principal component analysis, partial least squares discriminant analysis, and the Mann–Whitney U test were used, and nine VOMs were identified as differential metabolites. Finally, acrolein, 2-pentanone, and methyl allyl sulfide were selected to build a metabolite combination model for distinguishing breast cancer patients from the healthy group, and the achieved sensitivity and specificity were 92.6% and 91.7%, respectively, according to the receiver operating characteristic curve analysis. The results demonstrate that this technology has potential to become a rapid screening tool for breast cancer, with significant room for further development.

Keywords: high-pressure photoionization mass spectrometry; urine; volatile organic metabolites; breast cancer; rapid detection

1. Introduction

The global prevalence of female breast cancer (BC) has surged to 11.7%, accounting for approximately 2.3 million cases, thus surpassing lung cancer as the most frequently diagnosed malignancy. Additionally, it stands as the fifth major contributor to cancer-related fatalities worldwide, claiming the lives of 685,000 individuals [1,2]. The incidence and mortality rates of breast cancer exhibit an upward trend. Prior research indicates that the fatality rate associated with breast cancer could be significantly reduced through timely detection and comprehensive treatment [1–3]. Presently, mammography serves as the conventional modality for breast screening; however, it exhibits diminished sensitivity towards detecting small tumors, is constrained by patient age limitations, and cannot yield definitive disease outcomes [4,5]. Furthermore, ultrasound and magnetic resonance imaging (MRI) are commonly employed in conjunction with mammography to identify minute lesions that may evade detection through mammography alone. However, these methods exhibit relatively lower specificity, and their costly nature can potentially contribute to instances of overdiagnosis [6–8]. Therefore, there exists a pressing demand for novel operational strategies that can be readily implemented on a wide scale within practical clinical settings for breast cancer screening.

Over the past few decades, a multitude of platforms leveraging omics technology have been developed and extensively employed in the realm of disease diagnosis and screening, encompassing not only cancer but also its distinct subtypes. Numerous molecular constituents, such as genes, proteins, and metabolites, have been proposed as potential biomarkers for breast cancer [9–12]. Metabolomics represents a robust and auspicious avenue for examining the intricate interplay between metabolites and physiopathological alterations through comprehensive qualitative and quantitative analysis of all organismic metabolites [10,12–14]. This approach harbors immense potential to discern and identify heterogeneous tumor diseases during their nascent stages [9]. Urine, serving as an optimal biofluid for metabolomic investigations, boasts several advantages, including non-invasive sampling, easy accessibility, and lower protein content, thereby reducing complexity. In addition, compounds produced by the body's metabolism need to be concentrated by the kidneys before being excreted, making urine a rich source of metabolites [15]. Numerous volatile organic metabolites present in urine offer abundant insights into the onset and progression of diseases. Previous research has demonstrated that tissues generate distinct VOMs or exhibit altered concentrations of VOMs in pathological states, encompassing infections, neoplasms, and metabolic disorders [15–17].

To detect VOMs in urine, certain analytical techniques based on gas chromatography-mass spectrometry (GC-MS) have been utilized by integrating static/dynamic head-space-solid phase microextraction or stir bar extraction methodologies [18]. Some potential biomarkers for cancers, such as lung cancer [13], prostate cancer [19], breast cancer [4], and gastric cancer [20] have been successfully identified. Nevertheless, the GC-MS methods necessitate intricate pretreatment procedures and prolonged analysis durations, rendering them unsuitable for high-throughput and large-scale disease screening. Direct mass spectrometry based on soft ionization techniques, such as proton transfer reaction mass spectrometry (PTR-MS), selected ion flow tube mass spectrometry (SIFT-MS), and photoionization mass spectrometry (PI-MS) has been successfully used for rapid detection of trace volatile organic compounds in a complex matrix. Huang et al. used SIFT-MS to analyze urine headspace of gastro esophageal cancer patients and found seven statistically different VOMs [14]. PTR-MS was used in gastric cancer patients for VOM analysis in breath gas by Yoon et al. [21]. A high-pressure photoionization time-of-flight mass spectrometry (HPPI-TOFMS) has recently been developed with the advantages of high sensitivity, fast response, and good moisture resistance, which is especially suitable for rapid detection of trace volatiles and has been widely used in other fields [22–25]. It has shown excellent performance in the detection of exhaled breath, with the limits of detection (LODs) as low as 0.015 ppb for aliphatic and aromatic hydrocarbons [23], and has been successfully applied in the early screening of lung and gastroesophageal cancers [26,27]. HPPI-TOFMS has also been successfully used in the detection of VOMs in human urine with the LOD for trimethylamine as low as 100 ng L^{-1} under alkaline conditions, and a new biomarker 2,5-dimethylpyrrole was exclusively found in the smoker's urine sample in addition to toluene [24].

In this study, the integration of HPPI-TOFMS with the dynamic purge-injection method was employed for the rapid and highly sensitive detection of volatile compounds in urine. A straightforward pretreatment approach involving the addition of acid and salt was implemented and investigated for VOM sampling from urine samples. After optimizing the experimental conditions, the method was applied to analyze urine samples obtained from 24 breast cancer patients and 27 healthy controls. The resulting MS data were subjected to statistical analysis to identify distinctive VOMs in urine samples between breast cancer patients and the healthy control group. Subsequently, the model's classification efficacy was assessed by constructing a receiver operating characteristic (ROC) curve.

2. Materials and Methods

2.1. Instrumentation

The home-built HPPI-TOFMS was composed of a HPPI ion source, an ion transmission system, and an orthogonal acceleration reflectron mass analyzer (see Supporting Information, S1). As shown in Figure 1, the HPPI ion source consisted of a vacuum ultraviolet krypton (VUV-Kr) lamp (Heraeus Noblelight Ltd., Shenyang, China) and a high-pressure photoionization region, which was constructed by five annular stainless steel electrodes: a repelling electrode (6 mm i.d., 5 mm thick), two identical transmission electrodes (14 mm i.d., 5 mm thick), a focusing electrode (14 mm i.d., 5 mm thick), and a Skimmer-1 electrode (1 mm i.d., 4 mm thick). Three 1 mm thick polyether-ether-ketone (PEEK) insulation annular washers (16 mm i.d.) were employed to separate the electrodes, except for the space between the last focusing electrode and Skimmer-1 electrode for an excess neutral exhaust. All the electrodes were electrically connected by using a 1 MΩ resistor string, and additionally, the Skimmer-1 electrode was further connected by another 1 MΩ resistor to the ground. The voltages applied on the repelling electrode and Skimmer-1 electrode were 18 V and 12 V, respectively, while a voltage of 16 V was applied on the focusing electrode to form a nonuniform electric field in the ionization region, which was utilized for ion focusing and higher ion transmission efficiency. A mass resolving power of 5000 (full width at half-maximum, FWHM) was achieved with a 0.5 m field-free drift tube. All the mass spectra were accumulated for 10 s at a repetition rate of 25 kHz, and all data were obtained by averaging results from six parallel measurements.

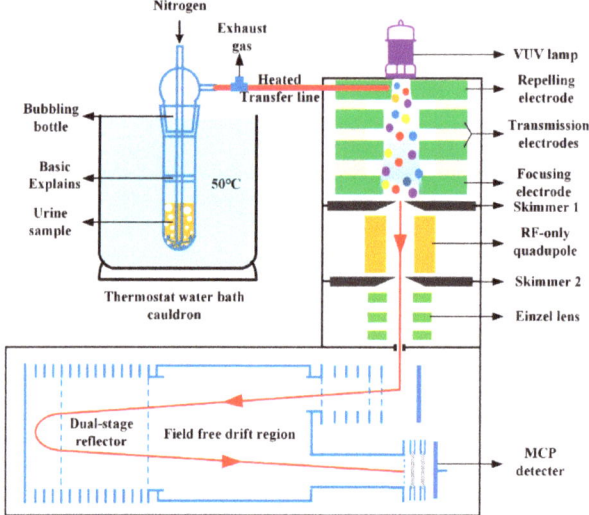

Figure 1. Schematic diagram of the HPPI-TOFMS system combined with dynamic purge-injection apparatus.

A dynamic purge-injection apparatus, composed of a thermostat water bath cauldron and a bubbling bottle with 20 mL inner volume, was employed for VOM sampling from urine samples into gaseous phase, as shown in Figure 1. The structure of the bubbling bottle was basically the same as that in our previous work [24], except for the addition of a porous glass cushion in the middle of the bottle, which was used to prevent the foam generated by bubbling from entering the sampling tube. A heated transfer line, containing a stainless steel capillary, 250 μm i.d. and 50 cm length, was used as the sampling tube to directly introduce gaseous VOMs from the outlet of the bubbling bottle into the ion source.

2.2. Chemicals and Reagents

Concentrated hydrochloric acid (AR, 36~38%) was purchased from Xilong Scientific Co., Ltd. (Guangdong, China). Sodium chloride (GR, 99.8%) was purchased from Shanghai Aladdin Bio-Chem Technology Co., Ltd. (Shanghai, China). Purified water was purchased from Hangzhou Wahaha Group Co., Ltd. (Hangzhou, China). Hydrochloric acid solution (4 mol·L^{-1}) was prepared by diluting concentrated hydrochloric acid with purified water. High-purity nitrogen gas (99.999%) was provided by Dalian Institute of Chemical Physics, Chinese Academy of Sciences (Liaoning, China) and used as the gas source for the dynamic purge system.

2.3. Urine Sample Collection, Preparation, and Detection

The middle stream of morning urine samples was collected from 24 breast cancer patients (BC, age 42–76 years, mean 52) and 27 healthy controls (CTL, age 18–61 years, mean 44) at Affiliated Zhongshan Hospital of Dalian University. All the urine samples were frozen at −80 °C immediately after sampling and thawed at 4 °C before detection. The study protocol was approved by the local ethics committee of Affiliated Zhongshan Hospital of Dalian University, and the method was carried out according to the approved guideline (2022021). Informed consent was obtained from all participants.

The urine samples were analyzed in four different conditions: (1) pure urine; (2) salted condition with addition of 1.0 g NaCl in 4 mL of pure urine; (3) acid condition with addition of 100 μL HCl (4 mol·L^{-1}) in 4 mL of pure urine to adjust pH at 1; and (4) acid–salted condition with addition of 100 μL of HCl (4 mol·L^{-1}) and 1.0 g NaCl in 4 mL of pure urine to adjust pH at 1. These samples were well mixed under ice and water bath conditions, stored at 4 °C and tested within 24 h. A urine pool noted as quality control (QC) was prepared by mixing the urine specimens (each with a volume of 400 μL) of all the participants in this study. The QC sample was processed in the same conditions and detected on every ten samples.

For VOM analysis, 4 mL of each urine sample was loaded into the clean bubbling bottle, which was sealed in 50 °C water bath. Subsequently, a high-purity nitrogen stream with 100 mL·min^{-1} was purged into the urine sample and produced a large number of small bubbles. Large quantities of VOMs were released into the gaseous phase by bubbles bursting and taken into the HPPI source through the stainless steel capillary for MS analysis. As the sampling flow rate of the inlet capillary was 50 mL·min^{-1}, the extra gas was exhausted by a stainless steel tee connected before the capillary. The heated transfer line and ionization region were maintained at 100 °C throughout the whole analysis process to prevent condensation of the VOMs. Data acquisition of each mass spectrum was started from the introduction of purge gas and accumulated for 2 min. The entire experimental process, from the start of sample preparation to the end of data acquisition, took only about 4 min.

2.4. Statistical Analysis

The data were divided into two groups, i.e., BC group and CTL group. All the data points with signal intensity values below 20 counts were set to 0 to avoid interference from the background noise. Variables with non-zero values of intensity in at least 90% of each group were included in the data set; otherwise, the variables were removed. Afterwards, data filtering and normalization were performed to obtain a two-dimensional matrix containing metabolite information (the data can be found in the Excel file named "DATA" provided in the supporting materials). Multivariable analyses were carried out using SIMCA-P software (version 14.0, Umetrics, Umea, Sweden) with unit variance scaling (UV scaling). The principal component analysis (PCA) and partial least squares discriminant analysis (PLS-DA) models were built among different groups. The Mann-Whitney U test was used for the nonparametric test and implemented by Multi Experiment Viewer (MeV, version 4.9.0, TIGR, Boston, MA, USA). Mass peaks with variable importance of the projection (VIP) > 1 and p-value < 0.05 were selected and used to determine the statistically significant VOMs. Binary logistic regression analysis and ROC analysis of combinational VOMs were figured out by using PASW Statistics 25 software (SPSS, Chicago, IL, USA). Ten-fold cross validation was performed by an online metabolomics data analysis website MetaboAnalyst 5.0 to test the discrimination power of the combination of statistically significant VOMs.

3. Results

3.1. Influence of Acid and Salt Addition

Acidification and alkalization of urine are prevalent pretreatment methodologies utilized for the extraction of VOMs during urine sampling. In our previous work, the VOMs identified in alkalized urine predominantly consisted of nitrogen-containing alkaline compounds, including dimethylamine, trimethylamine, piperidine, and dimethyl pyrazine [24], which were absent in the potential biomarker list from previous works by others [28,29]. Therefore, the pretreatment method for acidification (HCl) of urine was employed and investigated in this work. Adding acid can lower the pH of urine, which enhances the volatilization of acidic compounds, such as carboxylic acids, aldehydes, ketones, alcohols, etc., from the urine into the headspace, thus improving the detection sensitivity of these compounds [4,30]. In addition, NaCl was added in the urine sample to promote the volatilization of VOMs in urine, as the solubility of VOMs would decrease when the concentration of salt increased in the solution, known as the "salting-out effect" [31]. The addition of salt modifies the matrix of the sample by increasing ion activity. A significant quantity of the water molecules will exist as hydration associated with the ions in the solution under a high concentration of salt. VOMs do not dissolve well in the solution, which is bonded to the ions. Therefore, the solubility of VOMs in the liquid phase will decrease, and more VOMs move into the gas phase [31]. A mixed urine sample from four healthy volunteers (each with a volume of 20 mL) was used to evaluate the influence of HCl and NaCl addition. The signal intensities of over 33 mass peaks increased by more than 2-fold, and the signal enhancement of mass peaks with m/z 48, 59, 65, 77, and 94 even reached 11- to 21-fold after acidification of the mixed urine, as shown in Figure 2. Furthermore, 19 new peaks appeared in the acidified urine. After adding salt into the acidified urine, the signal intensity of mass peaks was further enhanced up to 62-fold (m/z = 94), compared with the pure mixed urine. Finally, based on putative annotation (level 2) [32], the measured masses of the characteristic ions were compared with their theoretical masses with a mass error of less than 30 ppm, resulting in the identification of several compounds as shown in Table 1.

Table 1. A list of 24 metabolites that appeared in the spectra of QC sample under acid-salted condition.

Measured Mass (Th)	Theoretical Mass (Th)	Mass Error (ppm)	Characteristic Peaks	Chemicals
47.0495	47.0496	−2	$C_2H_6O·H^+$	ethanol
45.0328	45.0340	−27	$C_2H_4O·H^+$	acetaldehyde
48.0031	48.0033	−4	CH_4S^+	methanethiol
49.0106	49.0112	−12	$CH_4S·H^+$	
57.0338	57.0341	−4	$C_3H_4O·H^+$	acrolein
59.0498	59.0496	3	$C_3H_6O·H^+$	acetone
77.0603	77.0602	1	$C_3H_6O·H_3O^+$	
117.0915	117.0915	0	$(C_3H_6O)_2·H^+$	

Table 1. Cont.

Measured Mass (Th)	Theoretical Mass (Th)	Mass Error (ppm)	Characteristic Peaks	Chemicals
61.0280	61.0289	−15	$C_2H_4O_2 \cdot H^+$	acetic acid
79.0398	79.0395	4	$C_2H_4O_2 \cdot H_3O^+$	
73.0652	73.0653	−1	$C_4H_8O \cdot H^+$	2-butanone
91.0754	91.0759	−5	$C_4H_8O \cdot H_3O^+$	
144.1131	144.1150	−13	$C_8H_{16}O_2^+$	octanoic acid
145.1225	145.1228	−2	$C_8H_{16}O_2 \cdot H^+$	
83.0715	83.0735	−24	$C_5H_9N^+$	pentanenitrile
87.0808	87.0810	−2	$C_5H_{10}O \cdot H^+$	
105.0915	105.0915	0	$C_5H_{10}O \cdot H_3O^+$	2-pentanone
173.1519	173.1542	−13	$(C_5H_{10}O)_2 \cdot H^+$	
00.0346	88.0346	0	$C_4H_8S^+$	methyl allyl sulfide
92.0629	92.0626	3	$C_7H_8^+$	toluene
93.0581	93.0578	2	$C_6H_7N^+$	3-methylpyridine
93.9908	93.9910	−2	$C_2H_6S_2^+$	disulfide, dimethyl
96.0576	96.0575	1	$C_6H_8O^+$	2,5-dimethylfuran
97.0507	97.0527	−21	$C_5H_7NO^+$	2,5-dimethyloxazole
101.0599	101.0602	−3	$C_5H_9O_2^+$	2,3-pentanedione
101.0955	101.0966	−11	$C_6H_{12}O \cdot H^+$	2-hexanone
107.0713	107.0735	−21	$C_7H_9N^+$	2,6-lutidine
110.0725	110.0731	−5	$C_7H_{10}O^+$	2-propylfuran
114.0135	114.0139	−4	$C_5H_6OS^+$	2-methoxythiophene
115.1112	115.1123	−10	$C_7H_{14}O \cdot H^+$	2-heptanal
136.1240	136.1252	−9	$C_{10}H_{16}^+$	limonene
139.1120	139.1123	−2	$C_9H_{14}O \cdot H^+$	2-pentylfuran

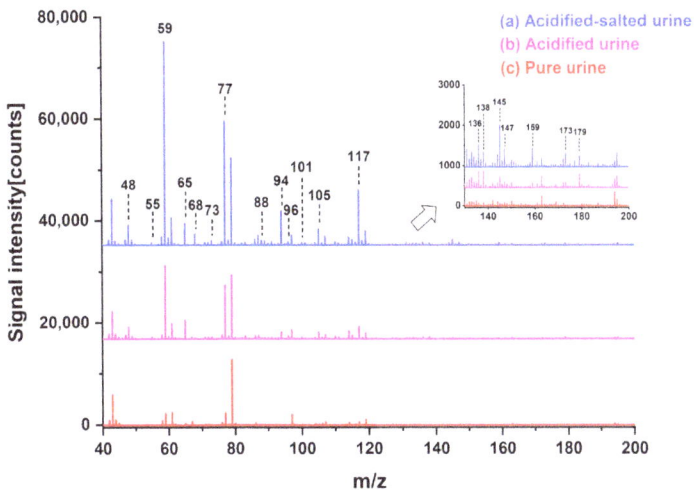

Figure 2. The comparison of mass spectra for different treatment methods of urine.

3.2. Multivariate Statistical Analysis

The processed MS data of BC and CTL groups were imported into SIMCA-P for PCA and PLS-DA analysis. During the urine sample analysis of BC and CTL, a QC detection was inserted for every ten samples. Five QC mass spectra were obtained, and clustered tightly together on the score plot of the PCA (see the Supporting Information, Figure S2a). Furthermore, the relative standard deviations (RSDs) of about 94% of the mass peaks were less than 30% for the QC sample (see the Supporting Information, Figure S2b), which exhibited the satisfactory repeatability and reliability of the method. PLS-DA maximizes the differences between samples by utilizing the biological measurements or category information in the Y-matrix, which could effectively solve the classification problem of metabolic phenotypes. As shown in Figure 3a, the BC group could be well separated from the CTL group from the score plot of PLS-DA, which indicated that the metabolite profiles could be well distinguished between the two groups. The cross validation with 200 iterations was performed, and the result shown in Figure 3b indicated that the PLS-DA model was not overfitted as the R2- and Q2-intercept values were 0.394 and −0.383, respectively.

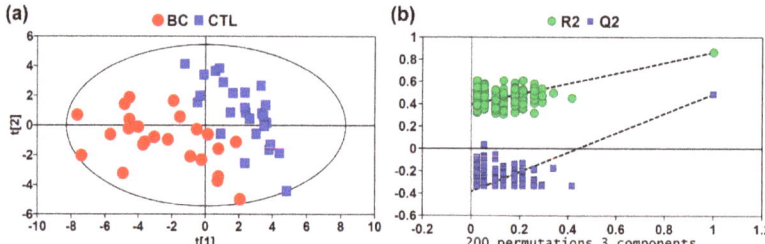

Figure 3. Multivariable analysis: (**a**) PLS-DA score plot (R2Y = 0.864, Q2 = 0.487). (**b**) Cross-validation plot of PLS-DA analysis, with a permutation test repeated 200 times and intercepts: R2 = (0.0, 0.394) and Q2 = (0.0, −0.383).

3.3. Differential Metabolites in Urine of BC Patients

Univariate analysis was performed on the Multi Experiment Viewer, and the Mann-Whitney U test was used here to assess the significance of the selected candidate metabolites. Generally, a p-value < 0.05 was considered significant for the selected metabolite with a statistical significance. Furthermore, the variable importance for the projection (VIP) was plotted to summarize the importance of MS peaks, and only VIP > 1 can be reserved in the end. To further narrow down the range of significant candidate metabolites, the false discovery rate (FDR), based on the Benjamini–Hochberg correction, was introduced as another criterion. Metabolites that ultimately met a VIP > 1 and a p-value < 0.05 were selected as the differential metabolites. Finally, nine VOMs were identified as differential metabolites in the urine samples between BC patients and the CTL group, which could be classified as unsaturated aldehydes, ketones, aromatic hydrocarbons, volatile sulfur compounds, and heterocyclic compounds, as shown in Table 2.

Furthermore, hierarchical cluster analysis (HCA) was performed to better demonstrate the differences at metabolic levels between BC patients and the CTL group. The alteration of these VOMs in the urine of BC patients and the CTL group can be clearly observed in the heatmap as shown in Figure 4. The urine of BC patients had increased amounts of 2-butanone, 3-methylpyridine, and acrolein, but reduced concentrations of 2-pentyfuran, methyl allyl sulfide, 2-pentanone, 2-hexanone, octanoic acid, and 2-methoxythiophene.

Table 2. Identification of differential metabolites in the urine samples between BC patients and healthy controls (CTL).

VOMs	Chemical Formula	Characteristic Peaks	Ratio	p-Value	VIP	Ref.
acrolein	C_3H_4O	$C_3H_4O \cdot H^+$	2.00	2.57×10^{-5}	1.63	[33]
2-butanone	C_4H_8O	$C_4H_8O \cdot H^+$ $C_4H_8O \cdot H_3O^+$	1.92 1.58	8.66×10^{-5} 0.0016	1.54 1.32	[34,35]
2-pentanone	$C_5H_{10}O$	$C_5H_{10}O \cdot H^+$ $(C_5H_{10}O)_2 \cdot H^+$	0.53 0.38	0.0062 2.51×10^{-4}	1.17 1.47	[34,35]
methyl allyl sulfide	C_4H_8S	$C_4H_8S^+$	0.16	0.0012	1.37	[30,36]
3-methylpyridine	C_6H_7N	$C_6H_7N^+$	2.16	0.0043	1.14	[20]
2-hexanone	$C_6H_{12}O$	$C_6H_{12}O \cdot H^+$	0.56	7.30×10^{-4}	1.12	[3,37]
2-methoxythiophene	C_5H_6OS	$C_5H_6OS^+$	0.48	0.0097	1.11	[4,30]
2-pentylfuran	$C_9H_{14}O$	$C_9H_{14}O \cdot H^+$	0.46	2.18×10^{-5}	1.38	[3,37]
octanoic acid	$C_8H_{17}O_2$	$C_8H_{16}O_2 \cdot H^+$	0.45	0.0108	1.21	[4]

Note: The value of "Ratio" is obtained by dividing the average concentration of BC by the average concentration of CTL.

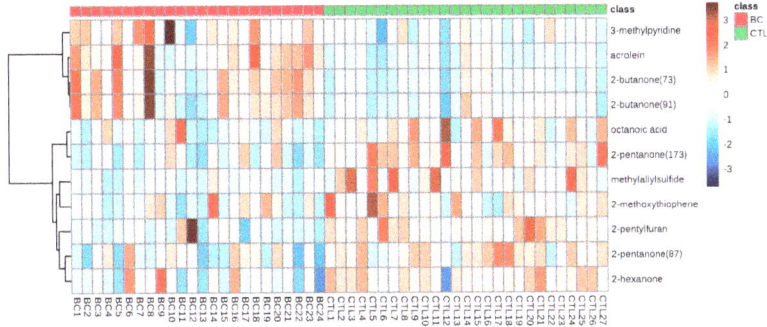

Figure 4. Heat map of the Pearson correlation coefficients between the differential metabolite contents.

3.4. Receiver Operating Characteristic Curve Analysis

The receiver operating characteristic curve is often used to evaluate the classification effectiveness of the model. However, the specificity and sensitivity of models containing a single differential metabolite for distinguishing BC patients from healthy controls were not definitive (see the Supporting Information, Table S1). A feasible solution for this problem is to combine more differential metabolites into a group for higher specificity and sensitivity. Therefore, the binary logistic regression analysis was employed to screen the differential metabolites to obtain an optimal metabolite combination. Eventually, three statistically significant metabolites, including acrolein, 2-pentanone, and methyl allyl sulfide were selected to build a metabolite combination model. This combination of metabolites has not been reported previously. The area under the ROC curve (AUC) of the statistically significant metabolic combination in the discovery set was 0.97, and the sensitivity and specificity were 92.6% and 91.7%, respectively, as shown in Figure 5a. The result indicated that this model has a good ability to identify BC patients. Subsequently, 10-fold cross-validation was performed to evaluate the model, as shown in Figure 5b, with the AUC = 0.88, sensitivity = 85.2%, and specificity = 83.3%, respectively. The results demonstrated the robustness of the model, which has the potential to be a useful tool for early screening of breast cancer.

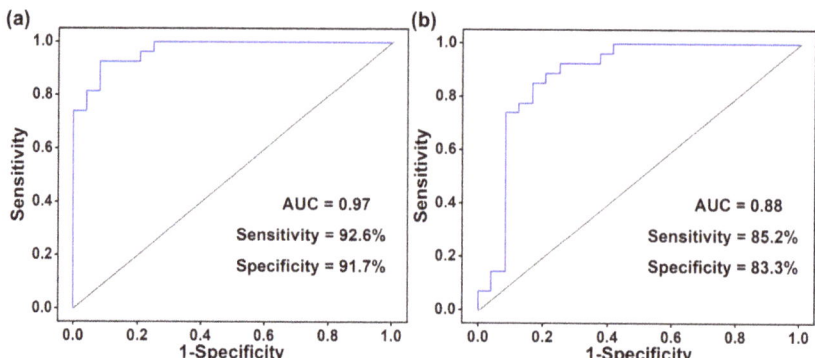

Figure 5. Receiver operating characteristic curve: (**a**) ROC curve of potential metabolic marker combination; (**b**) ROC curve of potential metabolic marker combination based on ten-fold cross validation.

4. Discussion

4.1. Potential Metabolic Pathway Analysis

The metabolic pathways of VOMs are pretty complex. As shown in Figure 4, the concentration of these VOMs were different between the BC and CTL groups, which is probably related to the increased oxidative stress and decreased apoptosis of cancer patients [14]. The relationship between the VOMs and cancer metabolism was not fully understood until now. The potential metabolic pathway of the five classes of the identified differential metabolites in Table 2 were summarized here according to previous studies.

Ketones are very abundant in urine. As shown in Table 2, there are three ketone compounds identified between the BC and CTL groups in this study: 2-butanone, 2-pentanone, and 2-hexanone. Different studies have shown that the ketogenic pathway may be directly related to tumor growth, and some ketones have been assigned as designated biomarkers for different cancers. Two potential pathways could be involved in their production: (i) oxidation of secondary alcohols catalyzed by ADHs (or cytochrome p450 (CYP2E1), and (ii) β-oxidation of fatty acids [20]. Therefore, 2-butanone, 2-pentanone, and 2-hexanone may be derived from 2-butanol, 2-pentanol, and 2-hexanol, respectively. But the source of these secondary alcohols remains unclear. They might stem from the oxidation of n-alkanes catalyzed by cytochrome p450 enzymes, microbial metabolism, or diet. Among them, 2-butanone and 2-pentanone have been detected as potential biomarkers in the breath gas of patients with gastric and ovarian cancers [34,35].

Although, only methyl allyl sulfide was identified as a differential sulfide compound, as listed in Table 2, sulfide compounds are generated by the incomplete metabolism of methionine and cysteine through the transamination pathway with high expression in urine [38]. On the one hand, during the transamination cascade, the methyl mercaptan produced by the conversion of methionine and cysteine is easily oxidized to produce a variety of volatile sulfides [38,39]. On the other hand, gram-negative bacteria can also produce these sulfur metabolites [40].

Additionally, there are volatile aldehydes in Table 2, which are common products of lipid peroxidation [30]. Acrolein is produced from the oxidation of arachidonic, linolenic, and linoleic acids in the presence of hydrogen peroxide and Fe^{2+} [35]. In addition to oxidative stress on unsaturated lipids, spermine and spermidine are potential carbon sources for acrolein. These compounds are oxidized by amine oxidase to corresponding amino aldehydes and spontaneously form acrolein in situ [33].

2-Pentylfuran was identified as the differential furan compound between the BC and CTL groups. Furans can be found in different exogenous sources, such as various foods. Furans are considered to be potential carcinogens, and high concentrations of furans can increase the probability of bile duct tumors in rats [41]. Additionally, furans have also been reported to be involved in anti-cancer defense mechanisms [42]. 2-Pentylfuran was found in the breath of patients with aspergillus fumigatus infections and human skin emanation [43]. Its production by natural dehydration of monosaccharides and oxidation of some fatty acids catalyzed by lipoxygenases could take place in adipocytes in the context of lipid peroxidation [43].

The last two differential metabolites in Table 2 are heterocyclic compounds, 2-methoxythiophene and 3-methylpyridine were detected in several reports and can even be considered as metabolic markers [4,20,30]. In Silva's report, the concentration of 2-pentylfuran in BC patients is significantly higher than that in normal people, and it is considered as a biomarker of BC [4].

4.2. Methods Comparison and Limitations

GC-MS has become a core technology in metabolomic analysis due to its satisfactory performance in sensitivity and specificity [44]. Many researchers have utilized this technique to discover biomarkers for breast cancer in urine, achieving promising results [3,4,37,45–47]. Nevertheless, sample preparation is complex and time-consuming, involving multiple steps that restrict its application in high-throughput analysis and rapid screening. PTR-MS, as a highly sensitive direct MS technique, has also been applied to the detection of VOMs in urine [48,49]. However, the vast amount of water vapor from urine samples makes the ionization process more complicated and increases the difficulty of data.

Compared to other methods, HPPI-TOFMS is more suitable for high-throughput urine sample analysis. Firstly, HPPI-TOFMS offers fast analysis speed and requires simple sample treatment steps such as acidification and salting. There is no enrichment or desorption process, and samples are directly detected after gasification. Secondly, a HPPI ionization source is less affected by humidity, enabling effective ionization of different compound types. As a soft ionization source, it avoids excessive fragmentation ions, making spectrum interpretation simpler. Thirdly, the instrument is easy to operate and has low maintenance costs. However, one drawback of HPPI-TOFMS is its reliance on high-resolution TOFMS for accurate qualitative analysis. Additionally, due to the lack of GC, it is unable to differentiate structural isomers.

Achieving positive results in a pilot study is encouraging; however, there are also some limitations of this study that need to be further addressed. The small sample size and lack of external validation in this study may limit the generalizability of the findings. Increasing the sample size would enhance statistical power and confidence in the results. External validation should be included to improve the reliability of the findings.

Additionally, confounding factors such as diet, medication, lifestyle, and clinical variables may influence metabolomic characteristics and introduce bias. Future research should employ appropriate methods to control for these factors and improve the reliability of the conclusions. Further research is needed to confirm the metabolic pathways and mechanisms underlying the associations between specific VOMs and breast cancer risk. In vitro and in vivo experiments are necessary to establish causal relationships and understand the biological significance of these findings.

5. Conclusions

This pilot study showcases a robust method for high-throughput analysis of VOMs in urine using the integration of high-pressure photoionization time-of-flight mass spectrometry with dynamic purge-injection. Its preliminary application in rapid breast cancer screening is demonstrated. VOMs present in urine samples are effectively volatilized and introduced into the HPPI-TOFMS system through dynamic purge-injection following the simple addition of acid and salt to the samples. The obtained mass spectrometry data were analyzed using partial least squares discriminant analysis and the Mann-Whitney U test, resulting in the identification of nine differential metabolites in the urine samples of 24 breast cancer patients and 27 healthy controls. Furthermore, a metabolite combination model was constructed using acrolein, 2-pentylfuran, and methyl allyl sulfide, which exhibited a satisfactory discriminatory performance (sensitivity = 92.6%, specificity = 91.2%) in distinguishing between breast cancer patients and healthy controls. Currently, the combination of HPPI-TOFMS with dynamic purge-injection has shown potential as a tool for breast cancer screening. In the future, efforts will be focused on expanding the sample size for external validation and employing appropriate methods to control the influence of clinical factors, further enhancing the reliability of this method.

Supplementary Materials: The following supporting information can be downloaded at: https://www.mdpi.com/article/10.3390/metabo13070870/s1, Figure S1: Schematic diagram of the HPPI-TOFMS.; Figure S2: (a) Principal component analysis (PCA) score plot; (b) RSD distribution for ion features in QC samples. Table S1: The result of ROC analysis for individual metabolites.

Author Contributions: Conceptualization, H.L. and D.L.; software, J.J. and Y.W.; formal analysis, X.Z. and R.W.; investigation, L.H. and D.J.; data curation, M.Y.; writing—original draft preparation, M.Y.; writing—review and editing, H.L. All authors have read and agreed to the published version of the manuscript.

Funding: This research was funded by the National Natural Science Foundation of China (22174142), the Scientific Instrument Developing Project of the Chinese Academy of Sciences (ZDKYYQ20210005), the United Foundation for Dalian Institute of Chemical Physics, the Chinese Academy of Sciences and the Second Hospital of Dalian Medical University (DMU-2 and DICP UN202305), and the Innovation Research fund of Dalian Institute of Chemical Physics (DICP I202123).

Institutional Review Board Statement: The institutional review board of the Affiliated Zhongshan Hospital of Dalian University approved the study protocol (approval number: 2022021). The study was performed by following the ethical guidelines expressed in the Declaration of Helsinki and the International Conference on Harmonization Guidelines for Good Clinical Practice.

Informed Consent Statement: Informed consent was obtained from all subjects involved in the study.

Data Availability Statement: All data are contained in the article and Supplementary Materials.

Acknowledgments: Thanks to the Affiliated Zhongshan Hospital of Dalian University for providing the urine samples.

Conflicts of Interest: The authors declare that there are no competing interests associated with the manuscript.

References

1. Sung, H.; Ferlay, J.; Siegel, R.L.; Laversanne, M.; Soerjomataram, I.; Jemal, A.; Bray, F. Global Cancer Statistics 2020: GLOBOCAN Estimates of Incidence and Mortality Worldwide for 36 Cancers in 185 Countries. *CA Cancer J. Clin.* **2021**, *71*, 209–249. [CrossRef] [PubMed]
2. McCormack, V.; McKenzie, F.; Foerster, M.; Zietsman, A.; Galukande, M.; Adisa, C.; Anele, A.; Parham, G.; Pinder, L.F.; Cubasch, H.; et al. Breast cancer survival and survival gap apportionment in sub-Saharan Africa (ABC-DO): A prospective cohort study. *Lancet Glob. Health* **2020**, *8*, e1203–e1212. [CrossRef] [PubMed]
3. Silva, C.L.; Perestrelo, R.; Silva, P.; Tomas, H.; Camara, J.S. Implementing a central composite design for the optimization of solid phase microextraction to establish the urinary volatomic expression: A first approach for breast cancer. *Metabolomics* **2019**, *15*, 64. [CrossRef]
4. Silva, C.L.; Passos, M.; Camara, J.S. Solid phase microextraction, mass spectrometry and metabolomic approaches for detection of potential urinary cancer biomarkers—A powerful strategy for breast cancer diagnosis. *Talanta* **2012**, *89*, 360–368. [CrossRef]
5. Hellquist, B.N.; Czene, K.; Hjälm, A.; Nyström, L.; Jonsson, H. Effectiveness of population-based service screening with mammography for women ages 40 to 49 years with a high or low risk of breast cancer: Socioeconomic status, parity, and age at birth of first child. *Cancer* **2015**, *121*, 251–258. [CrossRef] [PubMed]
6. Onega, T.; Goldman, L.E.; Walker, R.L.; Miglioretti, D.L.; Buist, D.S.; Taplin, S.; Geller, B.M.; Hill, D.A.; Smith-Bindman, R. Facility Mammography Volume in Relation to Breast Cancer Screening Outcomes. *J. Med. Screen.* **2016**, *23*, 31–37. [CrossRef]
7. Ozmen, N.; Dapp, R.; Zapf, M.; Gemmeke, H.; Ruiter, N.V.; van Dongen, K.W. Comparing different ultrasound imaging methods for breast cancer detection. *IEEE Trans. Ultrason. Ferroelectr. Freq. Control* **2015**, *62*, 637–646. [CrossRef]
8. Roganovic, D.; Djilas, D.; Vujnovic, S.; Pavic, D.; Stojanov, D. Breast MRI, digital mammography and breast tomosynthesis: Comparison of three methods for early detection of breast cancer. *Bosn. J. Basic. Med. Sci.* **2015**, *15*, 64–68. [CrossRef]
9. Hassan, A.M.; El-Shenawee, M. Review of electromagnetic techniques for breast cancer detection. *IEEE Rev. Biomed. Eng.* **2011**, *4*, 103–118. [CrossRef]

10. Broza, Y.Y.; Mochalski, P.; Ruzsanyi, V.; Amann, A.; Haick, H. Hybrid volatolomics and disease detection. *Angew. Chem. Int. Ed. Engl.* **2015**, *54*, 11036–11048. [CrossRef]
11. Hu, J.; Liu, F.; Chen, Y.; Shangguan, G.; Ju, H. Mass Spectrometric Biosensing: A Powerful Approach for Multiplexed Analysis of Clinical Biomolecules. *ACS Sens.* **2021**, *6*, 3517–3535. [CrossRef]
12. Hasin, Y.; Seldin, M.; Lusis, A. Multi-omics approaches to disease. *Genome Biol.* **2017**, *18*, 83. [CrossRef] [PubMed]
13. Hanai, Y.; Shimono, K.; Matsumura, K.; Vachani, A.; Albelda, S.; Yamazaki, K.; Beauchamp, G.K.; Oka, H. Urinary volatile compounds as biomarkers for lung cancer. *Biosci. Biotechnol. Biochem.* **2012**, *76*, 679–684. [CrossRef]
14. da Costa, B.R.B.; De Martinis, B.S. Analysis of urinary VOCs using mass spectrometric methods to diagnose cancer: A review. *Clin. Mass. Spectrom.* **2020**, *18*, 27–37. [CrossRef]
15. Taunk, K.; Taware, R.; More, T.H.; Porto-Figueira, P.; Pereira, J.A.M.; Mohapatra, R.; Soneji, D.; Câmara, J.S.; Nagarajaram, H.A.; Rapole, S. A non-invasive approach to explore the discriminatory potential of the urinary volatilome of invasive ductal carcinoma of the breast. *RSC Adv.* **2018**, *8*, 25040–25050. [CrossRef]
16. Filipiak, W.; Filipiak, A.; Sponring, A.; Schmid, T.; Zelger, B.; Ager, C.; Klodzinska, E.; Denz, H.; Pizzini, A.; Lucciarini, P.; et al. Comparative analyses of volatile organic compounds (VOCs) from patients, tumors and transformed cell lines for the validation of lung cancer-derived breath markers. *J. Breath Res.* **2014**, *8*, 027111. [CrossRef] [PubMed]
17. Abaffy, T.; Moller, M.G.; Riemer, D.D.; Milikowski, C.; DeFazio, R.A. Comparative analysis of volatile metabolomics signals from melanoma and benign skin: A pilot study. *Metabolomics* **2013**, *9*, 998–1008. [CrossRef]
18. Di Lena, M.; Porcelli, F.; Altomare, D.F. Volatile organic compounds as new biomarkers for colorectal cancer: A review. *Color. Dis.* **2016**, *18*, 654–663. [CrossRef]
19. Jimenez-Pacheco, A.; Salinero-Bachiller, M.; Iribar, M.C.; Lopez-Luque, A.; Mijan-Ortiz, J.L.; Peinado, J.M. Furan and p-xylene as candidate biomarkers for prostate cancer. *Urol. Oncol.* **2018**, *36*, 243.e21–243.e7. [CrossRef] [PubMed]
20. Mochalski, P.; Leja, M.; Gasenko, E.; Skapars, R.; Santare, D.; Sivins, A.; Aronsson, D.E.; Ager, C.; Jaeschke, C.; Shani, G.; et al. Ex vivo emission of volatile organic compounds from gastric cancer and non-cancerous tissue. *J. Breath Res.* **2018**, *12*, 046005. [CrossRef]
21. Portillo-Estrada, M.; Van Moorleghem, C.; Janssenswillen, S.; Cooper, R.J.; Birkemeyer, C.; Roelants, K.; Van Damme, R.; Durand, P. Proton-transfer-reaction time-of-flight mass spectrometry (PTR-TOF-MS) as a tool for studying animal volatile organic compound (VOC) emissions. *Methods Ecol. Evol.* **2021**, *12*, 748–766. [CrossRef]
22. Wu, C.; Wen, Y.; Hua, L.; Jiang, J.; Xie, Y.; Cao, Y.; Chai, S.; Hou, K.; Li, H. Rapid and highly sensitive measurement of trimethylamine in seawater using dynamic purge-release and dopant-assisted atmospheric pressure photoionization mass spectrometry. *Anal. Chim. Acta* **2020**, *1137*, 56–63. [CrossRef]
23. Wang, Y.; Hua, L.; Li, Q.; Jiang, J.; Hou, K.; Wu, C.; Li, H. Direct Detection of Small n-Alkanes at Sub-ppbv Level by Photoelectron-Induced O_2^+ Cation Chemical Ionization Mass Spectrometry at kPa Pressure. *Anal. Chem.* **2018**, *90*, 5398–5404. [CrossRef] [PubMed]
24. Wang, Y.; Hua, L.; Jiang, J.; Xie, Y.; Hou, K.; Li, Q.; Wu, C.; Li, H. High-pressure photon ionization time-of-flight mass spectrometry combined with dynamic purge-injection for rapid analysis of volatile metabolites in urine. *Anal. Chim. Acta* **2018**, *1008*, 74–81. [CrossRef]
25. Wang, Y.; Jiang, J.; Hua, L.; Hou, K.; Xie, Y.; Chen, P.; Liu, W.; Li, Q.; Wang, S.; Li, H. High-Pressure Photon Ionization Source for TOFMS and Its Application for Online Breath Analysis. *Anal. Chem.* **2016**, *88*, 9047–9055. [CrossRef] [PubMed]
26. Meng, S.; Li, Q.; Zhou, Z.; Li, H.; Liu, X.; Pan, S.; Li, M.; Wang, L.; Guo, Y.; Qiu, M.; et al. Assessment of an Exhaled Breath Test Using High-Pressure Photon Ionization Time-of-Flight Mass Spectrometry to Detect Lung Cancer. *JAMA Netw. Open.* **2021**, *4*, e213486. [CrossRef]
27. Huang, Q.; Wang, S.; Li, Q.; Wang, P.; Li, J.; Meng, S.; Li, H.; Wu, H.; Qi, Y.; Li, X.; et al. Assessment of Breathomics Testing Using High-Pressure Photon Ionization Time-of-Flight Mass Spectrometry to Detect Esophageal Cancer. *JAMA Netw. Open.* **2021**, *4*, e2127042. [CrossRef]
28. Gao, Y.; Zhang, J.; Chen, H.; Wang, Z.; Hou, J.; Wang, L. Dimethylamine enhances platelet hyperactivity in chronic kidney disease model. *J. Bioenerg. Biomembr.* **2021**, *53*, 585–595. [CrossRef] [PubMed]
29. Jankowski, J.; Westhof, T.; Vaziri, N.D.; Ingrosso, D.; Perna, A.F. Gases as uremic toxins: Is there something in the air? *Semin. Nephrol.* **2014**, *34*, 135–150. [CrossRef]
30. Smith, S.; Burden, H.; Persad, R.; Whittington, K.; de Lacy Costello, B.; Ratcliffe, N.M.; Probert, C.S. A comparative study of the analysis of human urine headspace using gas chromatography-mass spectrometry. *J. Breath Res.* **2008**, *2*, 037022. [CrossRef]
31. Aggio, R.B.; Mayor, A.; Coyle, S.; Reade, S.; Khalid, T.; Ratcliffe, N.M.; Probert, C.S. Freeze-drying: An alternative method for the analysis of volatile organic compounds in the headspace of urine samples using solid phase micro-extraction coupled to gas chromatography—Mass spectrometry. *Chem. Cent. J.* **2016**, *10*, 9. [CrossRef] [PubMed]
32. Sumner, L.W.; Amberg, A.; Barrett, D.; Beale, M.H.; Beger, R.; Daykin, C.A.; Fan, T.W.; Fiehn, O.; Goodacre, R.; Griffin, J.L.; et al. Proposed minimum reporting standards for chemical analysis Chemical Analysis Working Group (CAWG) Metabolomics Standards Initiative (MSI). *Metabolomics* **2007**, *3*, 211–221. [CrossRef]
33. Kato, S.; Post, G.C.; Bierbaum, V.M.; Koch, T.H. Chemical ionization mass spectrometric determination of acrolein in human breast cancer cells. *Anal. Biochem.* **2002**, *305*, 251–259. [CrossRef] [PubMed]

34. Amal, H.; Leja, M.; Funka, K.; Skapars, R.; Sivins, A.; Ancans, G.; Liepniece-Karele, I.; Kikuste, I.; Lasina, I.; Haick, H. Detection of precancerous gastric lesions and gastric cancer through exhaled breath. *Gut* **2016**, *65*, 400–407. [CrossRef]
35. Alonso, M.; Castellanos, M.; Besalu, E.; Sanchez, J.M. A headspace needle-trap method for the analysis of volatile organic compounds in whole blood. *J. Chromatogr. A* **2012**, *1252*, 23–30. [CrossRef]
36. Mochalski, P.; Unterkofler, K. Quantification of selected volatile organic compounds in human urine by gas chromatography selective reagent ionization time of flight mass spectrometry (GC-SRI-TOF-MS) coupled with head-space solid-phase microextraction (HS-SPME). *Analyst* **2016**, *141*, 4796–4803. [CrossRef]
37. Porto-Figueira, P.; Pereira, J.; Miekisch, W.; Camara, J.S. Exploring the potential of NTME/GC-MS, in the establishment of urinary volatomic profiles. Lung cancer patients as case study. *Sci. Rep.* **2018**, *8*, 13113. [CrossRef] [PubMed]
38. Tangerman, A. Measurement and biological significance of the volatile sulfur compounds hydrogen sulfide, methanethiol and dimethyl sulfide in various biological matrices. *J. Chromatogr. B* **2009**, *877*, 3366–3377. [CrossRef] [PubMed]
39. Blom, H.J.; Boers, G.H.J.; Vandenelzen, J.; Gahl, W.A.; Tangerman, A. Transamination of methionine in humans. *Clin. Sci.* **1989**, *76*, 43–49. [CrossRef]
40. Scholler, C.; Molin, S.; Wilkins, K. Volatile metabolites from some gram-negative bacteria. *Chemosphere* **1997**, *35*, 1487–1495. [CrossRef]
41. Bakhiya, N.; Appel, K.E. Toxicity and carcinogenicity of furan in human diet. *Arch. Toxicol.* **2010**, *84*, 563–578. [CrossRef] [PubMed]
42. Silva, C.L.; Passos, M.; Camara, J.S. Investigation of urinary volatile organic metabolites as potential cancer biomarkers by solid-phase microextraction in combination with gas chromatography-mass spectrometry. *Br. J. Cancer* **2011**, *105*, 1894–1904. [CrossRef] [PubMed]
43. Mochalski, P.; Diem, E.; Unterkofler, K.; Mundlein, A.; Drexel, H.; Mayhew, C.A.; Leiherer, A. In vitro profiling of volatile organic compounds released by Simpson-Golabi-Behmel syndrome adipocytes. *J. Chromatogr. B* **2019**, *1104*, 256–261. [CrossRef]
44. Silva, C.; Perestrelo, R.; Silva, P.; Tomas, H.; Camara, J.S. Breast Cancer Metabolomics: From Analytical Platforms to Multivariate Data Analysis. A Review. *Metabolites* **2019**, *9*, 102. [CrossRef] [PubMed]
45. Herman-Saffar, O.; Boger, Z.; Libson, S.; Lieberman, D.; Gonen, R.; Zeiri, Y. Early non-invasive detection of breast cancer using exhaled breath and urine analysis. *Comput. Biol. Med.* **2018**, *96*, 227–232. [CrossRef]
46. Giro Benet, J.; Seo, M.; Khine, M.; Guma Padro, J.; Pardo Martnez, A.; Kurdahi, F. Breast cancer detection by analyzing the volatile organic compound (VOC) signature in human urine. *Sci. Rep.* **2022**, *12*, 14873. [CrossRef]
47. Cala, M.; Aldana, J.; Sanchez, J.; Guio, J.; Meesters, R.J.W. Urinary metabolite and lipid alterations in Colombian Hispanic women with breast cancer: A pilot study. *J. Pharm. Biomed. Anal.* **2018**, *152*, 234–241. [CrossRef]
48. Zou, X.; Lu, Y.; Xia, L.; Zhang, Y.; Li, A.; Wang, H.; Huang, C.; Shen, C.; Chu, Y. Detection of volatile organic compounds in a drop of urine by ultrasonic nebulization extraction Proton Transfer Reaction Mass Spectrometry. *Anal. Chem.* **2018**, *90*, 2210–2215. [CrossRef]
49. Xu, W.; Zou, X.; Ding, H.W.; Ding, Y.T.; Zhang, J.; Liu, W.T.; Gong, T.T.; Nie, Z.C.; Yang, M.; Zhou, Q.; et al. Rapid and non-invasive diagnosis of type 2 diabetes through sniffing urinary acetone by a proton transfer reaction mass spectrometry. *Talanta* **2023**, *256*, 124265. [CrossRef]

Disclaimer/Publisher's Note: The statements, opinions and data contained in all publications are solely those of the individual author(s) and contributor(s) and not of MDPI and/or the editor(s). MDPI and/or the editor(s) disclaim responsibility for any injury to people or property resulting from any ideas, methods, instructions or products referred to in the content.

Article

The Integration of Metabolomics, Electronic Tongue, and Chromatic Difference Reveals the Correlations between the Critical Compounds and Flavor Characteristics of Two Grades of High-Quality Dianhong Congou Black Tea

Shan Zhang [1,2], Xujiang Shan [2,3], Linchi Niu [2], Le Chen [2,4], Jinjin Wang [2], Qinghua Zhou [4], Haibo Yuan [2,*], Jia Li [2,*] and Tian Wu [1,*]

1. School of Landscape Architecture and Horticulture Sciences, Southwest Forestry University, Kunming 650224, China; zhangshan@tricaas.com
2. Key Laboratory of Tea Biology and Resources Utilization, Ministry of Agriculture, Tea Research Institute, Chinese Academy of Agricultural Sciences, Hangzhou 310008, China; shanxujiang@tricaas.com (X.S.); niulinchi2001@163.com (L.N.); yjscl@hotmail.com (L.C.); jinjinwangtkzc@tricaas.com (J.W.)
3. State Key Laboratory of Tea Plant Biology and Utilization, Anhui Agricultural University, Hefei 230036, China
4. College of Environment, Zhejiang University of Technology, Hangzhou 310014, China; qhzhou@zjut.edu.cn
* Correspondence: 192168092@tricaas.com (H.Y.); jiali1986@tricaas.com (J.L.); wutianpotato@swfu.edu.cn (T.W.)

Abstract: Tea's biochemical compounds and flavor quality vary depending on its grade ranking. Dianhong Congou black tea (DCT) is a unique tea category produced using the large-leaf tea varieties from Yunnan, China. To date, the flavor characteristics and critical components of two grades of high-quality DCT, single-bud-grade DCT (BDCT), and special-grade DCT (SDCT) manufactured mainly with single buds and buds with one leaf, respectively, are far from clear. Herein, comparisons of two grades were performed by the integration of human sensory evaluation, an electronic tongue, chromatic differences, the quantification of major components, and metabolomics. The BDCT possessed a brisk, umami taste and a brighter infusion color, while the SDCT presented a comprehensive taste and redder liquor color. Quantification analysis showed that the levels of total polyphenols, catechins, and theaflavins (TFs) were significantly higher in the BDCT. Fifty-six different key compounds were screened by metabolomics, including catechins, flavone/flavonol glycosides, amino acids, phenolic acids, etc. Correlation analysis revealed that the sensory features of the BDCT and SDCT were attributed to their higher contents of catechins, TFs, theogallin, digalloylglucose, and accumulations of thearubigins (TRs), flavone/flavonol glycosides, and soluble sugars, respectively. This report is the first to focus on the comprehensive evaluation of the biochemical compositions and sensory characteristics of two grades of high-quality DCT, advancing the understanding of DCT from a multi-dimensional perspective.

Keywords: metabolomics; electronic tongue; flavor; taste; Dianhong Congou black tea; grades

1. Introduction

Black tea, as a recognized healthy beverage, has a long consumption history in the world, and it accounts for nearly 78% of the global tea consumption [1]. According to the processing technology used, black tea can be divided into three types, i.e., broken black tea, Congou black tea, and Souchoug black tea [2]. Among these, Congou black tea is popularly favored by consumers due to its elegant appearance, mellow taste, and sweet aroma [3]. Dianhong Congou black tea (DCT), which is a strip-shaped black tea, is a famous, geographically recognized brand produced by using the large-leaf tea varieties (*Camellia sinensis* L. var. assamica) from Yunnan province in the southwest of China [4] through the elaborate manufacturing steps of withering, rolling, fermentation, drying, and refining. Due to the special plant species and processing technology used, DCT is rich in tea polyphenols such as flavan-3-ols (catechins) and their derivatives such as theaflavins (TFs), thearubigins (TRs), and theasinensins (TBs), as well as phenolic acids and flavone/flavonol glycosides, etc., the contents of which are generally higher than in tea manufactured by small-/medium-leaf tea varieties [5]. On account of its higher content of tea polyphenols along with the other compounds such as amino acids, organic acids, and soluble sugars, DCT presents a strong, sweet-mellow, and umami taste profile in addition to its multiple health-protecting benefits [6,7]. In recent years, as an outstanding representative of premium Congou black tea, DCT has emerged as a popular herbal healthy beverage source and is exported to more than 30 countries around the world [5].

Grading is an important indicator for tea production as well as being a reference for consumers to estimate the tea quality. Based on the Group Standard of China Tea Science Society (T/CTSS 38-2021), DCT is divided into six grades, i.e., single-bud-, special-, first-, second-, third-, and fourth-grade, according to the tenderness of the young tea shoots, its sensory qualities, and its biochemical parameters, etc. Among these, single-bud-grade DCT (BDCT) and special-grade DCT (SDCT), which are produced by single buds with fresh leaf contents of no less than 90% and buds with one leaf, respectively, are commonly regarded as high-quality DCT according to the Group Standard of China Tea Science Society (T/CTSS 23-2021). Compared with middle-/low-quality DCT, high-quality DCT is favored by consumers, mostly due to its significant superiority in sensory characteristics [8]. However, it is difficult for average consumers to make a sensible choice between BDCT and SDCT, particularly due to the large gap in their prices in the market. In practice, consumers generally choose tea products based on the price, and more-expensive BDCT is often regarded as an optimal option. In fact, BDCT and SDCT both possess unique characteristics in terms of sensory quality and biochemical characterizations, which are suitable for various consumers' preferences [8]. Therefore, it is necessary to elucidate the differences between the two grades of high-quality DCT in order to afford theoretical support to improve the productivity of high-quality DCT and to provide a guide for the consumption of high-quality black tea in a rational manner.

At present, numerous studies have focused on the selection of tea varieties, processing technology improvement, chemical composition variation, and evaluating the flavor quality of black tea. For instance, the ratio of total polyphenols/total amino acids (P/A value) is often used as an indicator for the prediction of the manufacturing suitability of a variety [9] and as an index for the evaluation of black tea quality [10]. The ratio of TFs/TRs is also an important index to measure the processing technology and flavor characteristics of black tea [11]. In addition, there are some studies centered on the flavor features [5], aroma patterns [8], and quality evaluation methods [12,13] of various grades of DCT. However, few studies have been conducted to investigate the sensory and molecular differences between the two grades of high-quality DCT, i.e., BDCT and SDCT.

Currently, the sensory assessment approaches include human sensory evaluation, electronic tongue analysis, and chromatic difference measurement, etc. [4]. The analytical tools for chemical analysis to determine the biochemical compounds of tea include high-performance liquid chromatography (HPLC), gas chromatography–mass spectrometry (GC-MS), and high-resolution liquid chromatography–mass spectrometry (LC-MS), etc. HPLC is an effective analytical technique for the analysis of tea's major components. GC-MS and LC-MS are commonly performed for the qualitative and quantitative analysis of volatile and non-volatile compounds in tea in an untargeted pattern, respectively. A comprehensive combination of multiple sensory assessment approaches and chemical component analytical techniques is considered to be a more accurate method for the evaluation of tea quality. For example, Ma et al. characterized the key aroma-active compounds in high-grade Dianhong tea using GC-MS combined with sensory-directed flavor analysis [8]. Previous studies estimated the flavor quality of Keemun Congou black tea by a combination of sensory evaluation, HPLC, and LC-MS [14,15]. Therefore, it is advantageous to comprehensively study the two grades of high-quality DCT by integrating multiple approaches.

Herein, the aim of this study was to systematically compare the flavor characteristics of two grades of high-quality DCT, BDCT and SDCT, and to explore their correlations with the different key non-volatile compounds. To this end, a comprehensive sensory and molecular characterization on nine typical BDCT and nine typical SDCT samples was conducted by applying human sensory evaluation, an electronic tongue, chromatic differences, the quantification of main chemical components of the tea, and untargeted metabolomics profiling analysis. This study focused on the comparative assessment of two grades of high-quality DCT for the first time, intending to broaden the understanding about high-quality DCT with an objective, scientific, and global overview.

2. Materials and Methods

2.1. Chemicals and Reagents

High-purity solvents, including methanol, acetonitrile, formic acid, and acetic acid, were produced by either the Merck Company (Darmstadt, Germany) or the Sigma-Aldrich Company (St. Louis, MO, USA). Diagnostic solutions of the electronic tongue system used in this study including hydrochloric acid, sodium chloride, and monosodium glutamate (analytical grade) were purchased from the Evensen Biotechnology Company (Tianjin, China). Ultra-pure water was obtained using a Milli-Q System (Millipore, MA, USA).

2.2. Tea Samples

The 18 high-quality DCT samples, including 9 BDCT samples (mainly single-bud) and 9 SDCT samples (mainly one bud and one leaf), were collected from Fengqing county, Lincang city, Yunnan province, which is recognized as the core producing area of DCT [7].

2.3. Human Sensory Evaluation

Sensory evaluation of two groups of high-quality DCT (9 BDCT and 9 SDCT) was conducted following the Chinese National Standard (GB/T 23776-2018) by a group of professional tea tasters containing 2 females and 3 males from the Tea Research Institute, Chinese Academy of Agricultural Sciences. First, every tea sample (200 g) was placed in a white square plate for evaluating the dry tea's appearance. Next, each tea sample (3 g) was brewed using 150 mL of boiling pure water for 5 min in a clean porcelain cup, and then the tea infusion was filtered into a bowl for evaluating the liquor color, aroma, taste, and infused leaf, sequentially. The comments were given referring to the Tea Vocabulary for Sensory Evaluation (GB/T 14487-2017). The scoring was conducted using a 100-point grading system: total score (100%) = dry tea appearance (25%) + liquor color (10%) + aroma (25%) + taste (30%) + infused leaf appearance (10%). The tasters had no prior knowledge about the tea samples. All tea samples were served randomly. More detailed information is supplied in the supplementary information.

2.4. Electronic Tongue Measurement

An electronic tongue sensing system (α-Astree II, Alpha MOS company, Toulouse, France) was employed to acquire the taste fingerprints of the two groups (9 BDCT and 9 SDCT). The sensory array included a reference electrode made of Ag/AgCl and seven receptors of NMS, ANS, SCS, AHS, CTS, PKS, and CPS for discriminating the taste of umami, sweetness, bitterness, sourness, saltiness, and two comprehensive indexes, respectively. The preparation and detection methods of the tea infusions referred to an earlier study [3]. To ensure the accuracy of the data, every tea sample was brewed twice, as per the methods described in Section 2.3, and each tea infusion was measured four times. The average of eight repetitions was used as the final electronic tongue response.

2.5. Chromatic Difference Assessment

A colorimeter system (CM-5, Konica Minolta Investment Company, Shanghai, China) comprising indexes of L^* (luminance), $a^*(+)$ (redness), $a^*(-)$ (greenness), $b^*(+)$ (yellowness), $b^*(-)$ (blueness), and C^* (color saturation) was used to characterize the infusion color of the 18 tea samples (9 BDCT and 9 SDCT). The tea infusion was prepared and tested as previously described [3]. Briefly, every tea sample was brewed twice, and each tea infusion was measured three times by the colorimeter. Ultra-pure water was used as a blank. The final value was obtained from the average of six repetitions.

2.6. Quantitative Determination of the Major Tea Chemical Components

The levels of total free amino acids and total polyphenols in the 18 tea samples (9 BDCT and 9 SDCT) were determined following the Chinese National Standards of GB/T 8314-2013 and GB/8313-2018, respectively. The content of total soluble sugars was detected using the anthrone–sulfuric method [16]. The quantitative determination of caffeine, epigallocatechin gallate (EGCG), epigallocatechin (EGC), epicatechin gallate (ECG), epicatechin (EC), and catechin (C), was conducted using an HPLC system (Shimadzu, Kyoto, Japan). The content of total catechins was obtained by summing up the individual catechins of EGCG, EGC, ECG, EC, and C. The quantification of total TFs, TRs, and TBs was implemented using the systematic analysis reported in an early report [17]. Each sample was extracted and analyzed with three replicates.

2.7. Untargeted Metabolomics Based on LC-MS Analysis

The metabolomics analysis was conducted as per a previous study [18]. Briefly, the tea metabolites were extracted by adding 70% methanol (v/v) into finely ground tea powder. Each tea sample was extracted and analyzed with three replicates. An LC-MS run was conducted using an UHPLC apparatus (Dionex Ultimate 3000 system, Thermo Fisher, CA, USA) coupled to a Q Exactive Plus MS instrument (Thermo Fisher, CA, USA). LC separation was performed on an ACQUITY UPLC HSS T3 column (2.1 mm × 100 mm, 1.8 um, Waters, MA, USA) by gradient elution using 0.1% formic acid (v/v) in pure water and 0.1% formic acid (v/v) in acetonitrile as phase A and phase B, respectively. Negative modes were operated in the full-scan (m/z range of 100–1000) and HCD MS/MS modes. The capillary temperature, voltage, sheath gas flow, and auxiliary gas flow were set as 300 °C, 3.8 kV, 25 arb, and 5 arb, respectively. The quality control (QC) samples, prepared by mixing equal aliquots of all samples, were detected every eight injections during the whole run. The metabolite annotation was performed using database queries from the online systems of HMDB (https://hmdb.ca/ (accessed on 24 October 2022)) and Metline (https://metlin.scripps.edu (accessed on 6 November 2022)), MS/MS fragments, exact mass (within 5 ppm), retention time, and authentic standards validation.

2.8. Data Processing, Analysis, and Visualization

The raw data acquired from the LC-MS analysis were processed using the XCMS 3.4.1 R-package for generating a peak list containing the peak area intensity, charge-to-mass ratio (m/z), and retention time. The data pretreatment included normalization to the total ion intensity, the 80% rule, and QC evaluation, as previously described in [16]. The Mann–Whitney' U nonparametric test was used for the analysis of statistical differences between the two groups (9 BDCT and 9 SDCT). The Mann–Whitney U test, Bartlett's test of sphericity, and the Kaiser–Meyer–Olkin (KMO) test were performed using SPSS 26.0.0.0 (IBM, New York, NY, USA). The principal component analysis (PCA), partial least-squares analysis (PLS), partial least-squares discriminate analysis (PLS–DA), and the variable importance for the projection (VIP) plot were performed using SIMCA–P 14.1 (Umetrics, Umeå, Sweden). The heatmap analysis was performed using TBtools–II (v1.120, Toolbox for Biologists, Guangzhou, China). The graphs of the box plot were visualized using GraphPad Prism (GraphPad Software, San Diego, CA, USA). The pathway analysis was mapped by referring to the MetaboAnalyst (http://www.metaboanalyst.ca (accessed on 14 April 2023)) and KEGG (https://www.kegg.jp/ (accessed on 16 April 2023)) websites.

3. Results and Discussion

3.1. Human Sensory Evaluation

The sensory evaluation of the two grades of high-quality DCT, i.e., BDCT and SDCT, was performed by a group of experienced tea tasters. The results of the sensory features, including the dry tea appearance, liquor color, aroma, tea taste, infused leaf, and total score, are exhibited in Table 1. Detailed comments about and the scores of each tea sample are shown in Table S1. In terms of the dry tea appearance, the BDCT teas were tight and heavy, with a black bloom color and a golden and tippy appearance, while the SDCT teas were tight and heavy, bent, and with a black bloom color and a slightly golden and tippy appearance, which resulted in significant differences in the comments and scores (Figure 1A, Table 1). It was supposed that the tea made from a single bud was more likely to be shaped with a tight strip and present a more golden and tippy appearance, as compared with the tea made from a bud with one leaf. Regarding the other factors, the two groups had no significant differences in their scores of the liquor color, aroma, taste, and infused leaf, but they exhibited differentiation in the comments (Tables 1 and S1). Specifically, most of the BDCT samples presented a brisk taste with an umami, fruity-like sourness taste (Table S1) and a brighter liquor color (Figure 1B), while most of the SDCT samples presented a thick, mellow, and sweet taste (Table S1) and a redder liquor color (Figure 1B). In a word, the BDCT and SDCT showed no significant differences in their overall sensory scores, but they presented corresponding special sensory characteristics. To achieve an improved characterization, a more objective approach using an electronic tongue and chromatic difference analysis combined with a chemical investigation by quantifying the main chemical components and with untargeted metabolomics was conducted next.

Table 1. Results of the sensory indicators (dry tea appearance, liquor color, aroma, tea taste, infused leaf appearance, and total score) of the two grades of high-quality DCT samples as evaluated by human sensory evaluation.

Group	Sample Number	Dry Tea Appearance (25%)	Liquor Color (10%)	Aroma (25%)	Tea Taste (30%)	Infused Leaf (10%)	Total Score (100%)
BDCT	9	96.33 ± 0.5 [a]	90 ± 2.96	86.22 ± 4.82	84.56 ± 2.92	88.22 ± 1.64	88.83 ± 1.78
SDCT	9	93.56 ± 0.88 [b]	88.89 ± 2.93	85.67 ± 3.54	85 ± 2.60	87.89 ± 2.62	87.98 ± 1.71

[a,b] Different letters indicate significant differences between the mean scores of two groups ($p < 0.05$) as determined by the Mann–Whitney U test.

3.2. Electronic Tongue Profiles Measurement

The representative taste patterns, as characterized by an electronic tongue, of the BDCT and SDCT samples were visualized by a radar chart (Figure 1C). The BDCT tea samples showed an obvious umami response (NMS, 7.48) and a sourness response (AHS, 7.72). The SDCT tea samples exhibited a comprehensive taste including a strong sweetness (ANS, 6.66), an apparent bitterness (SCS, 6.52) integrated with umami (NMS, 5.57), a sourness (AHS, 5.38), and higher comprehensive index values of CPS (6.71) and PKS (6.56). An existing study has reported that the sourness, which is generally caused by organic acids, is responsible for the fruity-like taste in many foods [19]. A moderate sourness is thought to be conducive to a harmonious taste in black tea liquor, and an appropriate bitterness is beneficial to the mellow and thick sensation in tea infusions [20]. The results indicated that the BDCT infusion emerged as having an umami, fruity-like sour taste, while the SDCT infusion presented a complex and multi-dimensional taste formed by an appropriate interplay of sweetness, bitterness, umami, and sourness. The results of the electronic tongue analysis were consistent with the human sensory comments in terms of taste.

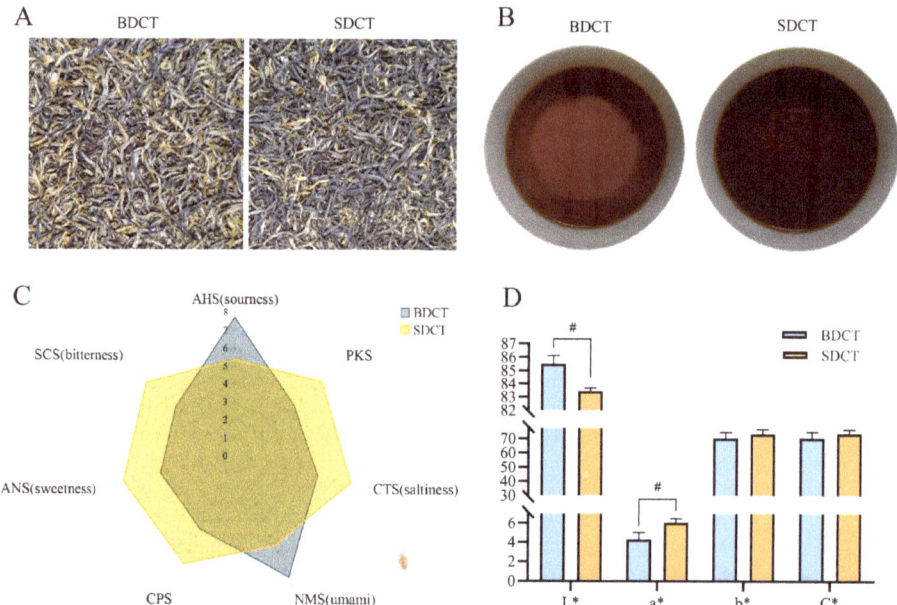

Figure 1. The representative tea samples' dry tea appearances (**A**), infusion colors (**B**), taste intensities, as evaluated by electronic tongue (**C**), and bar plots of color attributes, as evaluated by chromatic difference measurement (**D**), of the two grades of high-quality DCT. # Indicates mean values with significant differences between the two groups ($p < 0.05$).

3.3. Chromatic Difference Assessment

The color of a tea infusion is an important factor that affects the sensory quality of black tea. Premium black tea presents a bright and red liquor color, which is widely favored by consumers. Generally speaking, some consumers tend to like a redder liquor color, while others prefer a brighter liquor color. As shown in Figure 1D, the tea infusion in the BDCT presented a higher value of L* (luminance) ($p < 0.001$), reflecting a brighter liquor color compared with the SDCT. The a*(+) (redness) value in the SDCT group was significantly elevated ($p < 0.001$), indicating a redder liquor color in the SDCT. The values of b*(+) (yellowness) and C* (color saturation) were higher in the SDCT group, but they were not statistically significant ($p > 0.05$). The results were in agreement with the comments on the liquor color obtained by the human sensory evaluation, suggesting that each group of DCT had its own unique color that could satisfy different consumers' preferences.

3.4. Quantitative Determination of the Major Chemical Constituents

The total amount of polyphenols in tea, which mainly include flavan-3-ols, dimeric/polymeric catechins, flavonols and flavone/flavonol glycosides, and phenolic acids, accounts for 18~36% (w/w) of the dry weight of tea leaves [21]. The ratio of P/A has been regarded as an indicator for determining the suitability of tea cultivars [9] and an index for estimating the quality of black tea [10]. As shown in Table 2, the level of total polyphenols was significantly higher (BDCT vs. SDCT, fold change = 1.14, $p < 0.05$) and the total amino acid content was slightly lower (no significant difference) in the BDCT. Thus, a higher P/A value was observed in the BDCT compared with the SDCT (BDCT vs. SDCT, FC = 1.16, $p < 0.05$). A previous study has reported that the level of total polyphenols is generally negatively correlated with the maturity of young tea shoots [22], and the level of total amino acids reaches higher levels in moderate-maturity tea leaves such as those with one bud with one leaf [23]. It is believed that this result was largely due to the tenderness differences in the fresh leaves of the BDCT and SDCT.

Table 2. The contents of major biochemical components of tea in BDCT and SDCT.

Compounds	BDCT ($n = 9 \times 3$)	SDCT ($n = 9 \times 3$)	p Value	Fold Changes
Total polyphenols (%)	15.66 ± 2.22	13.62 ± 0.71	<0.05	1.14
Total amino acids (%)	3.37 ± 0.16	3.45 ± 0.18	n. s.	0.97
Total soluble sugars (%)	5.90 ± 0.46	6.09 ± 0.48	n. s.	0.99
Caffeine (%)	1.46 ± 0.42	1.72 ± 0.24	n. s.	0.88
Total catechins (%)	12.23 ± 1.95	9.20 ± 1.31	<0.05	1.34
EGCG (%)	8.78 ± 1.43	6.82 ± 0.78	<0.05	1.29
ECG (%)	1.52 ± 0.28	0.86 ± 0.18	<0.05	1.77
EGC (%)	1.30 ± 0.24	0.98 ± 0.17	<0.05	1.33
EC (%)	0.52 ± 0.22	0.30 ± 0.06	<0.05	1.76
C (%)	0.10 ± 0.06	0.13 ± 0.03	n. s.	0.83
TFs (%)	0.22 ± 0.05	0.17 ± 0.04	<0.001	1.36
TRs (%)	2.53 ± 0.32	2.64 ± 2.64	n. s.	0.96
TBs (%)	5.16 ± 0.70	5.30 ± 0.66	n. s.	1.03
TFs/TRs	0.09 ± 0.01	0.06 ± 0.01	<0.001	1.41
P/A value	4.64 ± 0.58	3.98 ± 0.36	<0.05	1.16

n. s., no significant difference; $p < 0.05$, significant difference; $p < 0.001$, extremely significant difference; the same below. Statistical significance was determined by the Mann–Whitney U test.

Total catechins, comprising galloylated catechins (EGCG, ECG) and non-galloylated catechins (EGC, EC, and C), account for 70~80% (w/w) of the amount of tea polyphenols [21]. The contents of total catechins and the individual compounds (EGCG, ECG, EGC, and EC) in the BDCT were significantly higher ($p < 0.05$) compared with the SDCT (Table 2). The changes in the concentrations of total catechins, ECG, and EGCG corresponded to the different tenderness of the young tea shoots. Catechin contents are negatively correlated with the growth of young shoots, with buds having a higher amount than first leaves [24]. As the important flavoring substances in tea, catechins are reported to impart a puckering astringency and a rough sensation in black tea infusions [1], which is generally described as "briskness" by sensory comments [25]. Therefore, we speculated that the higher amount of catechins may have been the factor contributing to the brisk taste in the BDCT.

Dimeric/polymeric catechins generated from oxidative condensation, i.e., TFs, TRs, and TBs, are the critical taste-active and colored substances in black tea. TFs impart a puckering, rough, and astringent sensation in tea infusions and are usually associated with the briskness taste and bright orange-yellow color in black tea infusion [26]. TRs are thought to be responsible for the redness color of tea liquors and the sweet taste of black tea [9]. An appropriate concentration of TBs has a positive effect on tea infusions, giving them a slightly sweet sensation, while an excessive amount of TBs is prone to cause a tea infusion to have a faint taste and dull color [21]. In addition, a moderately higher ratio of TFs/TRs is usually used as an index to measure the color and taste properties of black tea [11]. In this study, the contents of TRs and TBs were higher in the SDCT, though there were no significant differences ($p > 0.05$) (Table 2). Conversely, the concentration of TFs was significantly higher in the BDCT ($p < 0.001$), which was 1.36 times higher compared to the SDCT (Table 2). Furthermore, the ratio of TFs/TRs was obviously higher in the BDCT compared to the SDCT ($p < 0.001$) (Table 2). These results suggested that the BDCT teas were generally superior in terms of the features of a bright liquor color and a briskness taste due to the abundant accumulation of TFs, while the SDCT teas had advantages in terms of a redder liquor color and a sweet mellow taste due to the higher contents of TRs and TBs. In addition, the levels of total soluble sugars and caffeine showed no significant differences between the two groups.

In short, the results of the analysis of the main chemical constituents of the teas were basically in accordance with the aforementioned flavor features of the BDCT and SDCT. However, the tea flavor was influenced by the complex interplay of various quality-active compounds and was not limited to the several major components. Therefore, an untargeted metabolomics analysis was further conducted to investigate the global range of metabolites in the two grades of high-quality DCT.

3.5. Comprehensive Nontargeted Metabolomics Analysis

To comprehensively unravel the metabolic features of the BDCT and SDCT, an LC-MS-based untargeted metabolomics analysis was conducted. A total of 3104 ions were obtained. A typical chromatogram of the total ions is shown in Figure S1. To guarantee the reliability and repeatability of the data, the normalized intensities of the detected ions with replicate extractions were evaluated. The metabolite ions in the replicate extractions exhibited a high coefficient with $R^2 = 0.99$ (Figure S2). Bartlett's test of sphericity showed a high significance (chi–squared estimate of 527,322.793, $p < 0.001$), and the KMO value was observed to be 0.975, showing that the data were suitable for the factor analysis by PCA. A non-supervised PCA model including the BDCT, SDCT, and QC samples was used for a straightforward and global overview (Figure 2A). The QC samples were closely centered. The results suggested the reliable reproducibility of the present metabolomics analysis. In addition, the two groups were evidently separated in the PCA score plot, indicating a notable difference in metabolites in the two grades of tea samples. Next, a supervised PLS–DA model ($R2X = 66.6\%$, $R2Y = 98.7\%$, $Q2 = 97.4\%$) was established (Figure 2B), by which a more obvious distinction of the two grades was gained. The cross-validation using 200 permutation tests with the R2 (0.0, 0.421) and Q2 (0.0, −0.309) intercepts demonstrated that the PLS–DA model was reliable (Figure 2C). Subsequently, an S-plot, which visualized the metabolites and the classification patterns in a covariance matrix, was generated (Figure 2D). The potential critical metabolites with important contributions to the classification are highlighted with red squares.

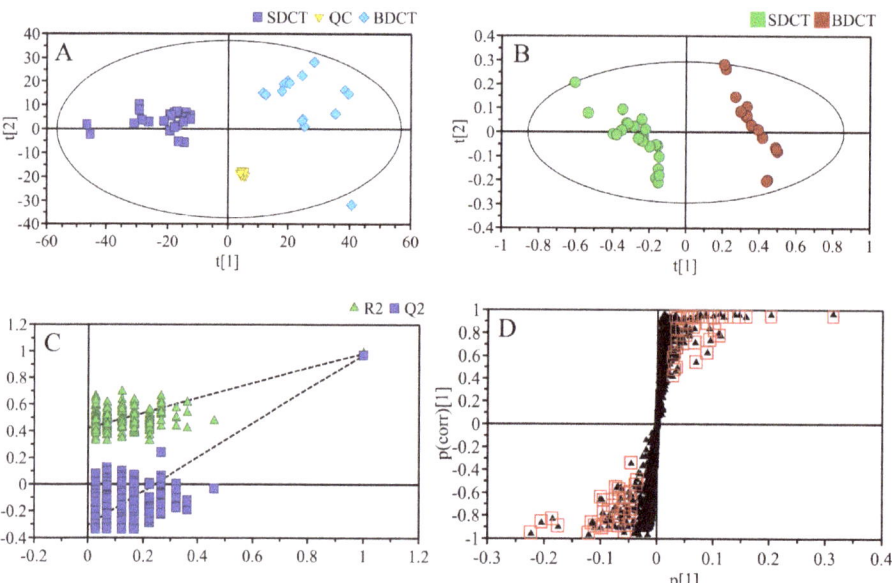

Figure 2. Multivariate statistical analysis of PCA score plot (**A**), PLS–DA score plot (**B**), cross-validation plot of PLS–DA model with 200 permutations (**C**), and S–plot of PLS–DA (**D**) for the two grades of high-quality DCT. The black triangles with red squares represent the potential important compounds.

3.6. The Key Metabolic Characteristics

A total of 56 prominently different compounds between the BDCT and SDCT were screened using the Mann–Whitney U test ($p < 0.05$) and VIP value (VIP > 1), and they consisted of three flavan-3-ols, six catechin dimers, twenty-six flavonols and flavone/flavonol glycosides, seven phenolic acids, six amino acids, four sugars, two organic acids, one flavone, and one nucleotide. Information about these compounds, including their ionization, m/z, RT, p value, VIP value, and MS/MS fragmentation, are shown in Table 3. A heat map was used to visualize these different compounds in the BDCT and SDCT (Figure 3). The yellow and blue color in the color scale represent the metabolite occurring at a higher or lower level than the average level of all the tea samples. Notable differences in the two grades of high-quality DCT were revealed. Generally speaking, catechins, dimeric/trimeric catechins including theasinensins, procyanidins, and theaflavin-3,3'-gallate (TF-3,3'-G), and a few phenolic acids with a galloyl group such as digalloylglucose, theogallin, etc., occurred at higher levels in the BDCT, while flavonols, flavone-C-glycosides, organic acids, soluble sugars, and most of the flavonol-O-glycosides, amino acids, and phenolic acids were at higher levels in the SDCT. To further elucidate the metabolite changes in the two grades of high-quality DCT manufactured with leaves with a different tenderness, the metabolic pathway involved with citric acid cycle (TCA cycle), phenylpropanoid metabolism, amino acid metabolism, flavone and flavonol metabolism, and flavonoid metabolism were mapped. The dynamic variations in the representative compounds between the two grades of high-quality DCT are shown in Figure 4.

Table 3. Detailed information of 56 key different compounds between BDCT and SDCT screened based on $p < 0.05$ and VIP > 1.

No.	Metabolite Identification	m/z	RT/min	p Value	VIP	MS/MS
	Flavan-3-ols and their derivatives					
1	Epiafzelechin [a]	273.0773	7.7	<0.001	1.6	187, 189, 229, 255
2	ECG [a]	305.0665	5.1	<0.001	1.1	125, 137, 165, 179, 219, 221, 261, 287
3	EGCG [a]	457.0767	6.6	<0.001	1.5	169, 193, 287, 305, 331
4	Procyanidin B1 [a]	577.1351	5.2	<0.001	1.1	125, 289, 407, 425, 451, 559
5	Procyanidin C1 [b]	865.1985	6.2	<0.001	1.2	125, 289, 407, 577, 695, 713, 739, 847
6	Theasinensin A [b]	913.1469	5.8	<0.001	1.5	285, 423, 573, 591, 743, 761
7	Theasinensin B [b]	761.1359	4.6	<0.001	1.2	423, 483, 575, 593, 609, 635, 743
8	Theasinensin F [b]	897.1520	7.2	<0.001	1.3	407, 727, 745
9	TF-3,3'-G [a]	867.1408	11.9	<0.001	1.3	125, 169, 241
	Flavonols and flavone/flavonol glycosides					
10	Apigenin 6-C-glucoside 8-C-arabinoside [b]	563.1406	7.5	<0.001	1.5	353, 383, 524, 443, 473, 503, 545
11	Apigenin-6,8-C-diglucoside [b]	593.1512	6.5	<0.001	1.4	473, 353, 503, 383, 575
12	Vitexin [a]	431.0983	8.6	<0.001	1.5	283, 311, 341
13	Vitexin-2-O-rhamnoside [a]	577.1563	8.6	<0.001	1.1	413, 293, 457
14	Vitexin-4″-O-glucoside [b]	593.1506	8.1	<0.001	1.5	293, 413
15	Kaempferol [a]	285.0414	12.2	<0.001	1.4	227, 239, 211
16	Kaempferol 3-O-galactosyl-rutinoside [b]	755.204	8.7	<0.001	1.4	285
17	Kaempferol 3-O-glucosyl-rutinoside [b]	755.204	9.1	<0.001	1.6	285

Table 3. Cont.

No.	Metabolite Identification	m/z	RT/min	p Value	VIP	MS/MS
18	Kaempferol 3-O-β-rutinoside [b]	593.1506	9.7	<0.001	1.7	285, 327
19	Kaempferol 7-(6″-galloylglucoside) [b]	599.1075	9.8	<0.001	1.6	125, 169, 313, 285, 447
20	Kaempferol-3-O-galactoside [b]	447.0933	9.7	<0.001	1.7	255, 284, 285, 327, 357
21	Dicoumaryl astragalin [b]	739.1675	12.2	<0.001	1.3	145, 285, 453, 593
22	p-Coumaroylastragalin [b]	593.1306	11.9	<0.001	1.3	285, 307, 447
23	Astragalin [a]	447.0933	10.2	<0.001	1.7	255, 284, 285, 327, 357
24	Myricetin 3-O-glucoside [b]	479.0825	7.7	<0.001	1.3	316, 317, 271
25	Quercetin [a]	301.0348	12.0	<0.001	1.6	107, 121, 151, 179
26	Isoquercitrin [a]	463.0882	9.0	<0.001	1.7	301, 300
27	Quercetin 3-arabinoside [b]	433.0799	9.7	<0.001	1.7	300, 271, 301, 255
28	Quercetin 3-O-galactosyl-rutinoside [b]	771.1989	8.0	<0.001	1.6	301, 343, 609
29	Quercetin 3-O-glucosyl-rutinoside [b]	771.1989	8.2	<0.001	1.5	301, 343, 609
30	Quercetin 7-(3-p-coumaroylglucoside) [b]	609.1279	11.8	<0.001	1.5	463, 300, 301
31	Quercetin-3-O-galactoside [b]	463.0882	8.8	<0.001	1.7	301, 300, 293
32	Quercetin-3-p-coumaroylrutinoside [b]	755.1873	11.8	<0.001	1.6	609, 591, 301, 271
33	3-Quercetin galloylglucoside [b]	615.1027	8.4	<0.001	1.4	463, 300, 301
34	7-Quercetin galloylglucoside [b]	615.1027	8.4	<0.001	1.4	463, 300, 301
35	Rutin [a]	609.1461	8.6	<0.001	1.7	301, 343
	Amino acids					
36	Aspartic acid [a]	132.0296	0.7	<0.001	1.3	88, 115
37	Glutamine [b]	146.0453	0.7	<0.001	1.4	109, 127
38	Histidine [a]	154.0616	0.6	<0.05	1.1	93, 137
39	Phenylalanine [a]	164.0711	2.4	<0.001	1.3	97, 137, 147
40	Theanine [a]	173.0926	1.1	<0.05	1.1	128, 155
41	Tyrosine [a]	180.066	1.2	<0.001	1.4	72, 93, 119, 163
	Phenolic acids					
42	Digalloylglucose [b]	483.078	5.2	<0.05	1.1	125, 169, 211, 271, 313, 331
43	Dihydroxy-benzoic acid [b]	153.0182	6.2	<0.05	1.6	109
44	Quinic acid [a]	191.0561	0.7	<0.001	1.6	85, 93, 127, 173
45	Shikimic acid [a]	173.0455	0.8	<0.001	1.1	73, 93, 111, 137
46	Theogallin [a]	343.0671	1.8	<0.001	1.5	191
47	p-Coumaric acid [a]	163.04	5.2	<0.001	1.4	119, 93
48	3-O-p-coumaroylquinic acid [b]	337.0929	6.2	<0.001	1.7	173
	Sugars					
49	Glucose [a]	179.0562	0.8	<0.001	1.6	59, 71, 89, 101, 113
50	Maltose [a]	341.1089	0.8	<0.001	1.3	113, 119, 143, 161, 179
51	Raffinose [a]	503.1612	0.7	<0.001	1.4	89, 101, 179, 221
52	ribonic acid [b]	165.0398	0.7	<0.001	1.1	75, 105, 129, 147
	Organic acids					
53	Citric acid [a]	191.0197	1.1	<0.001	1.2	85, 111, 173
54	Succinic acid [a]	117.0187	1.3	<0.001	1.1	73, 99
	Flavone					
55	Hydroxy trimethoxyflavone [b]	327.0893	8.0	<0.05	1.6	237, 211, 265
	Nucleotide					
56	UMP [b]	323.0286	0.8	<0.001	1.2	173, 211, 279, 305, 79, 193

[a] Confirmed by standards. [b] Identified based on exact mass and MS/MS. Statistical significance was determined by the Mann–Whitney U test.

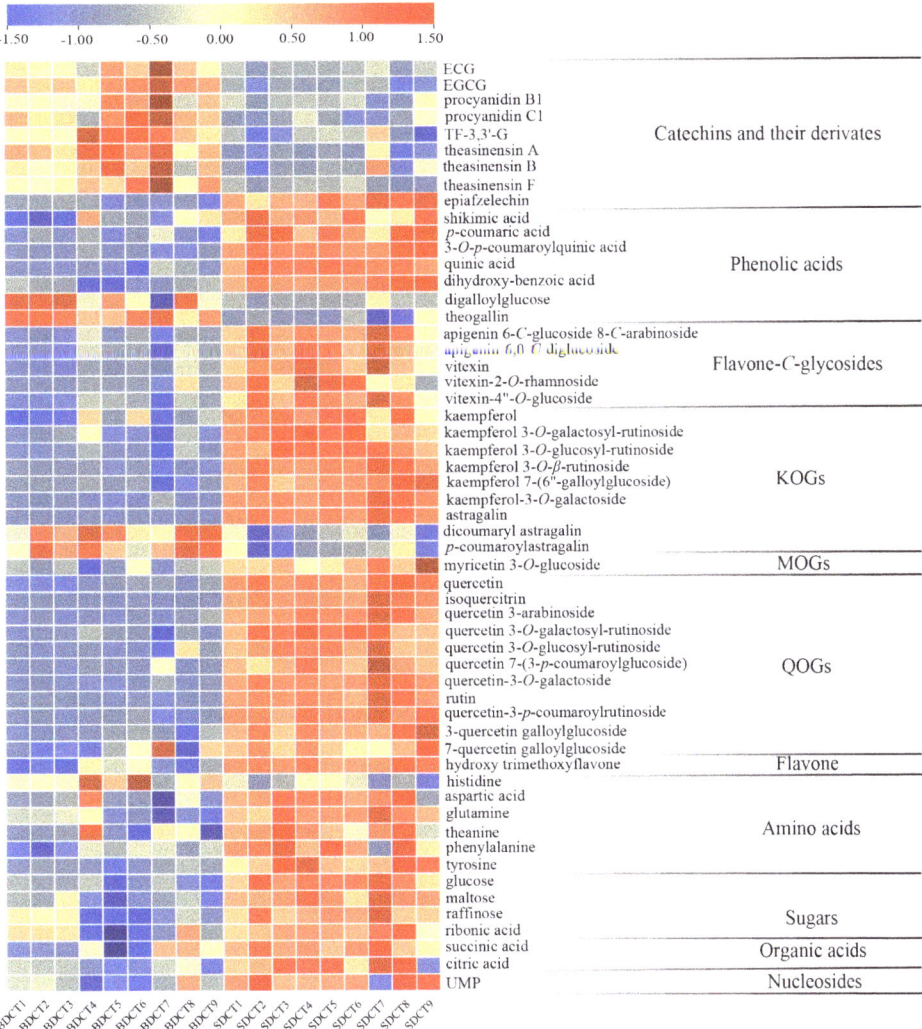

Figure 3. Heat map of the 56 different key compounds in the two grades of high-quality DCT. The data are shown as the mean of relative intensities of three replicates after being UV–scaled.

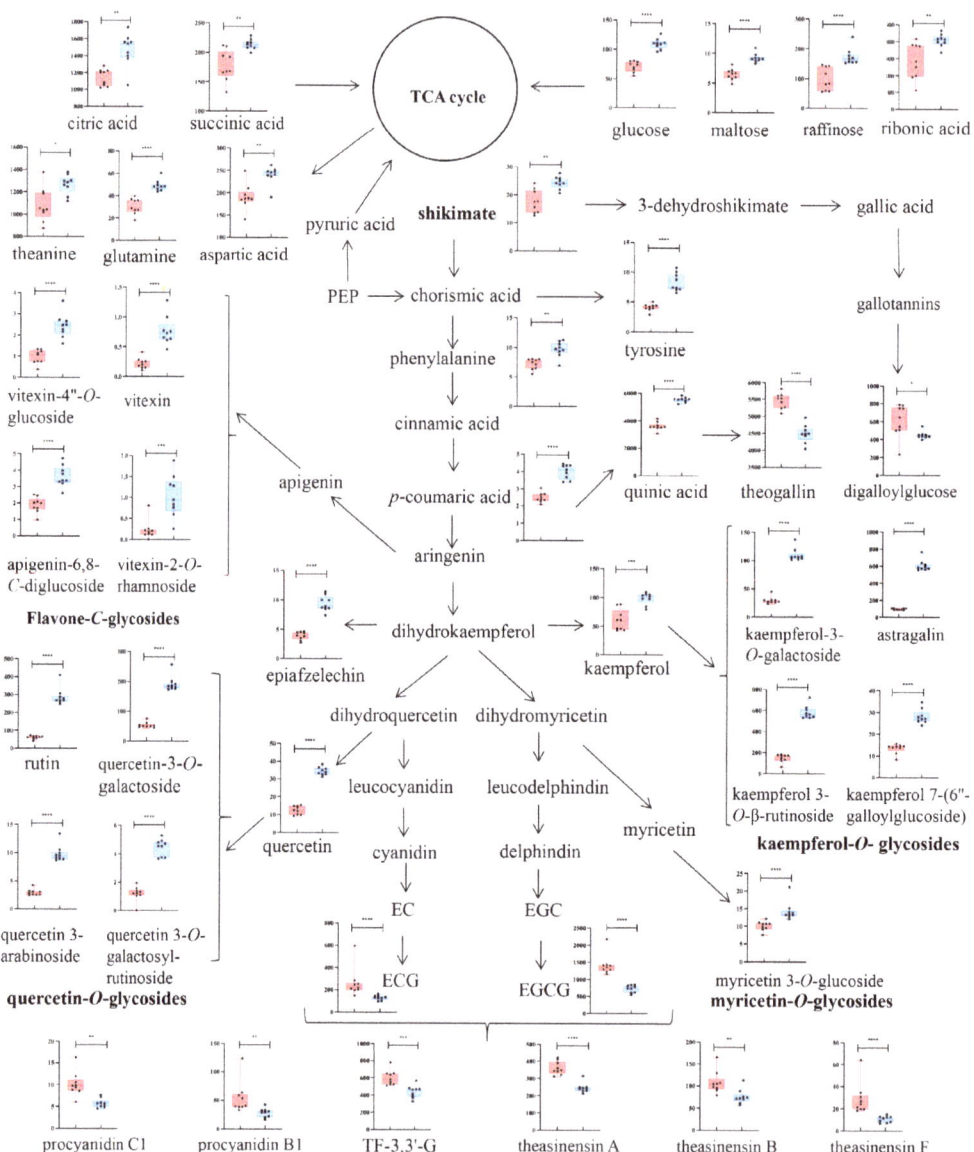

Figure 4. Mapping of the metabolic pathway and dynamic changes in the representative different key metabolites in two grades of high-quality DCT. Data are shown in scatter box plot as mean ± SD using relative abundance, as calculated by normalization to total ion intensity ($\times 10^5$). The orange and blue boxes represent metabolite intensities in BDCT and SDCT, respectively. * indicates $p < 0.05$, ** indicates $p < 0.001$, *** represents $p < 0.0005$, **** suggests $p < 0.0001$. Statistical significance was determined by the Mann–Whitney U test.

3.6.1. Flavan-3-ols and Their Derivatives

Flavan-3-ols, a group of the most abundant and characteristic metabolites in tea, have attracted much attention in biomedicine and food science due to their potent health-beneficial properties and are considered as the main contributors to the puckering astringency sensation in tea [1,27]. Among these, EGCG is the most abundant component affecting the taste of tea infusions and their bioactivity. An excessive intake of EGCG (734 mg/person/day) from green-tea-extract products is suspected to be related with liver toxicity, but daily tea consumption is considered to be safe since its EGCG content is much lower than the safe limit [28]. As shown in Figure 4, EGCG and ECG showed significantly higher levels in the BDCT than in the SDCT, which was in agreement with the results of the HPLC analysis (Table 2). Since ECG and EGCG belong to the group of galloylated catechins, this result was presumed to be possibly related with the higher galloylation level in buds than in leaves [29]. In addition, the downstream catechin derivatives, including procyanidins B1, procyanidins C1, theasinesins A, theasinesins B, theasinesins F, and TF-3,3′-G, also occurred at evidently higher levels in the BDCT (Figure 4). During black tea manufacturing, catechins undergo enzymatic oxidation to form various dimeric and oligomeric catechins, such as theasinesins, procyanidins, TFs, TRs, TBs, etc [30]. Both catechins and their water-soluble oxidation products have been reported as being the main contributors to the taste sensation of tea liquor [21]. A moderately higher content of theasinesins has been considered as a characteristic of high-quality black tea [19]. TF-3,3′-G, which is formed by the condensation of ECG and EGCG, is a major component of TFs and confers a briskness taste in tea [31]. It was speculated that the sufficient substrates of ECG and EGCG might have been responsible for the higher content of TF-3,3′-G in the BDCT tea samples [14,31]. The content of TF-3,3′-G was significantly higher in the BDCT, which was consistent with the trend of the TF content described in Section 3.4. Hence, we speculated that the higher contents of EGCG, ECG, theasinesins A, theasinesins B, theasinesins F, and TF-3,3′-G might be the important factors causing the briskness taste in the BDCT tea infusion and that the accumulation of these compounds came from the corresponding sufficient substrate supply in the tea buds. In addition, TFs largely contribute to the brightness of tea liquid color [3]. In the metabolomics analysis, TF-3,3′-G accumulated significantly, which might have been one of the factors responsible for high value of L* (luminance) in the BDCT, as mentioned above.

3.6.2. Phenolic Acids, Flavonols and Flavone/Flavonol Glycosides

Phenolic acids, as the precursors for the synthesis of catechins and flavonol glycosides, are important phenolic constituents with an antioxidative ability and contribute to the sourness, bitterness, and astringency taste in tea [27]. The representative different key compounds of phenolic acids, flavonols, and flavone/flavonol glycosides are exhibited in the metabolic pathway (Figure 4). Quinic acid and *p*-coumaric acid have been reported to be mainly responsible for the bitterness and astringency taste in tea liquid [32]. Theogallin, as a derivative of quinic acid, can enhance the umami taste in tea infusions [32]. As one of the hydrolysable tannins, digalloylglucose is considered to be correlated with the umami taste and higher quality of tea [33]. As manifested in Figure 4, *p*-coumaric acid and quinic acid demonstrated significantly higher levels in the SDCT and were beneficial in strengthening the bitterness and astringency taste in the tea infusions. On the contrary, the contents of theogallin and digalloylglucose were markedly higher in the BDCT, contributing to the umami taste and overall quality of the teas made from BDCT.

Flavonols and flavonol/flavone glycosides are also the main taste-active compounds in tea and present potential bioactivity in terms of antioxidant and cardiovascular-protective effects [27]. According to aglycones, flavonols and flavonol/flavone glycosides can be classified as apigenin-C-glycosides (ACGs), myricetin-O-glycosides (MOGs), quercetin-O-glycosides (QOGs), and kaempferol-O-glycosides (KOGs) [3]. As the downstream phenolic metabolites in the metabolic pathway, the contents of ACGs, MOGs, QOGs, and most of the KOGs were significantly higher in the SDCT than in the BDCT (Figure 4). These results were consistent with a previous report that showed that flavonol glycosides accumulated significantly more in White Peony tea (white tea processed using one bud with one leaf or two leaves) than in Silver Needle tea (white tea processed using only buds) [29]. As the second major phenolic metabolites in tea, flavonols and flavonol/flavone glycosides are mainly responsible for a mouth-drying astringency taste with an extremely low astringency taste threshold (0.001~19.8 μmol/L) [34], and they are responsible for the yellow liquor color of tea infusions [32]. In addition, it has been reported that flavonol glycosides can enhance the bitterness of caffeine [31]. As shown in Section 3.4, there were no significant differences in the content of caffeine between the two groups. However, the higher amount of flavonol glycosides was thought to enhance the bitterness taste of caffeine in the SDCT tea infusions. Therefore, the SDCT teas exhibited a higher intensity of SCS (bitterness), as reveled by the electronic tongue analysis (Figure 1C). Meanwhile, the higher value of b*(+) (yellowness) in the SDCT was considered to be related the yellowish color of flavonol glycosides (Figure 1D).

3.6.3. Soluble Sugars, Amino Acids, and Organic Acids

Soluble sugars are largely responsible for the sweetness in tea [1]. In this study, the mono-, di-, and oligosaccharides in the tea liquors, i.e., glucose, maltose, and raffinose, were in higher levels in the SDCT (Figure 4), which was thought to be related with the sweet taste in the SDCT.

Amino acids, as a group of important taste-active compound species of tea, can be divided into three types (i.e., bitter-, umami-, and sweet-tasting amino acids) according to their taste features [35]. The contents of bitter-tasting amino acids, such as phenylalanine and tyrosine, were significantly retained in the SDCT (Figure 4), which might have strengthened the bitterness taste of the tea infusions in this group. Meanwhile, umami-tasting amino acids, such as glutamine, aspartic acid, and theanine, also showed upward trends in the SDCT. Glutamine and aspartic acid are the main contributors to the umami taste in tea infusions [35]. Theanine exhibits a sweetness or freshness taste at different concentrations [1]. As shown in Figure 4, the significantly higher concentrations of glutamine, aspartic acid, and theanine in the SDCT were thought to largely enhance the umami taste in the SDCT.

Organic acids, as the crucial intermediate compounds of the TCA cycle and the phenylpropanoid metabolism pathway, contribute the most to the acidity of tea, which is often described as a "fruity-like" taste in black tea when the acidity degree is appropriate [19]. Succinic acid and citric acid are considered to be the highest contributors to the acidity among the organic acids in black tea [20]. As has been previously reported, high responses of bitterness and astringency suppress the sour taste in tea, which is thought to be beneficial to the overall taste of black tea [20]. As shown in Figure 4, the levels of the acidic compounds of citric acid and succinic acid were significantly higher in the SDCT than in the BDCT. However, the higher responses of bitterness and astringency in the SDCT might have suppressed its sour taste. Therefore, the response of AHS (sourness) was much lower in the SDCT compared to in the BDCT.

In summary, the BDCT teas showed a briskness, an umami, fruity-like taste, and a brighter liquor color due to the accumulation of EGCG, ECG, TF-3,3'-G, theasinesins, theogallin, digalloylglucose, etc. Meanwhile, the SDCT teas presented a redder liquor color and a multi-dimensional flavor integrated with an interplay of a moderate sweetness, bitterness, umami, and sourness due to the higher contents of soluble sugars (glucose, maltose, and raffinose), phenolic acids (p-coumaric acid and quinic acid), flavonol glycosides (most KOGs, ACGs, MOGs, and QOGs), umami-tasting amino acids (glutamine, aspartic acid, and theanine), bitter-tasting amino acids (phenylalanine and tyrosine), and organic acids (succinic acid and citric acid).

3.7. Correlation Analysis between the Different Key Metabolites and the Sensory Indicators

Aiming to further elucidate the correlations between the different key metabolites and the sensory indicators of tea, a PLS analysis was performed (Figure 5). In the figure, the further a metabolite was suited from the original point, the greater its contribution to the sensory variation. The X-variables represent the intensities of the compounds, while the Y-variables indicate the strength of the taste or color indicators of the tea infusion. Variables being located in the same region suggested a close positive correlation among them. As shown in Figure 5, higher contents of succinic acid, ribonic acid, and UMP were thought to be strongly related with the greater intensities of NMS (umami) and AHS (sourness). In addition, the total catechins, the flavan-3-ols of EGCG and ECG, the polymerized catechins of procyanidin B1, procyanidin C1, theasinensin F, and theasinensin B, and the phenolic acid of theogallin were found to be positively correlated with the strength of L* (luminance), which was evidently observed in the infusions of the BDCT teas. In contrast, most of the flavonols and flavonol/flavone glycosides (particularly ACGs and KOGs) and caffeine showed positive correlations with the sensory indicators of SCS (bitterness). Furthermore, TRs and TBs largely contributed to the intensities of ANS (sweetness), CPS and PKS (comprehensive sensory index), a*(+) (redness), b*(+) (yellowness), and C* (color saturation), which were obviously noticed in the SDCT teas. These compounds were considered to have contributed to the respective taste and liquor color quality features of the BDCT and SDCT.

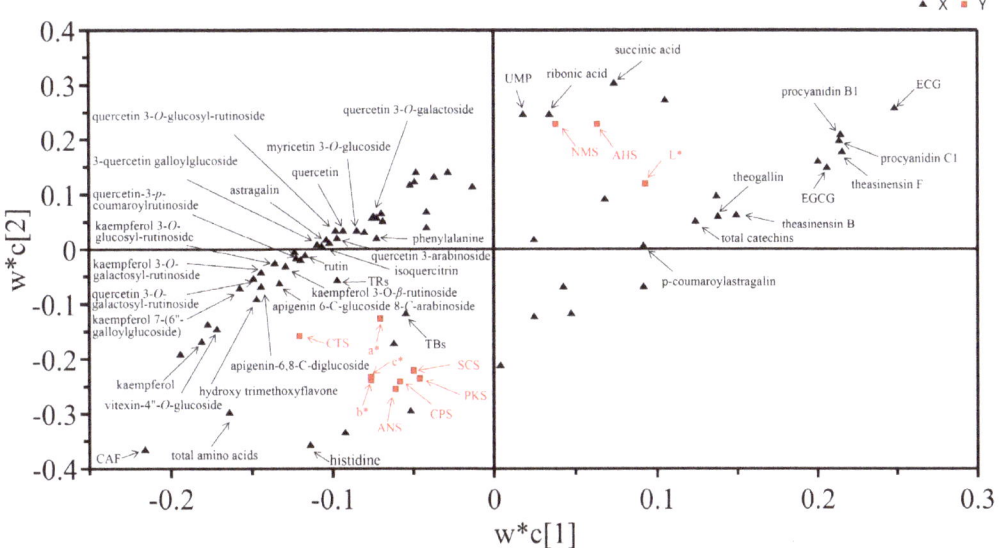

Figure 5. PLS analysis of the tea sensory indicators and the different key metabolites of the two grades of high-quality DCT. The metabolites were set as the X variables. The tea sensory indicators, including ANS, AHS, SCS, NMS, CTS, PKS, CPS, L*, a*, b*, and C*, were set as the Y variables.

4. Conclusions

In this study, the flavor characteristics and potential critical compounds of two grades of high-quality DCT, BDCT and SDCT, were revealed by using human sensory evaluation, an electronic tongue, chromatic differences, the quantification of the main components of the teas, and untargeted metabolomic analysis. The BDCT possessed an apparent briskness, an umami, fruity-like taste sensation, and a brighter infusion color, while the SDCT presented a multi-dimensional taste integrated with a moderate sweetness, bitterness, umami, and sourness and a redder tea infusion color. Flavan-3-ols of EGCG and ECG, polymerized catechins of theasinensins, TFs, TRs, and TBs, flavonols and flavone/flavonol glycosides of ACGs, KOGs, and the phenolic acid of theogallin, etc., were the compounds that contributed the most to the flavor characteristics of the two grades of high-quality DCT. To our knowledge, this is the first report focusing on the correlations between the biochemical compositions and sensory characteristics of two grades of high-quality DCT. These comprehensive comparisons are expected to provide an objective identification basis and a scientific guide for consumers' choices of high-quality DCT.

Supplementary Materials: The following supporting information can be downloaded at: https://www.mdpi.com/article/10.3390/metabo13070864/s1, Table S1: Detailed sensory comments and scores (dry tea appearance, liquor color, aroma, tea taste, infused leaf appearance, and total score) of two grades of high-quality DCT samples, as evaluated by human sensory evaluation; Figure S1: A typical chromatogram of total ions acquired from the metabolomics analysis; Figure S2: Scatter plot of normalized intensities of all detected ions in the two replicate extractions.

Author Contributions: Conceptualization, J.L. and T.W.; methodology, J.L., Q.Z. and H.Y.; software, X.S. and L.C.; validation, L.N. and L.C.; formal analysis, S.Z., L.N. and L.C.; investigation, S.Z., J.W., J.L. and X.S.; resources, X.S. and J.W.; data curation, J.L., J.W. and Q.Z.; writing—original draft preparation, S.Z.; writing—review and editing, J.L. and T.W.; visualization, S.Z., L.C. and L.N.; supervision, J.L., H.Y., T.W. and Q.Z.; project administration, J.L., T.W. and H.Y.; funding acquisition, J.L. and H.Y. All authors have read and agreed to the published version of the manuscript.

Funding: This work was supported by grants from the Science and Technology Innovation Project of the Chinese Academy of Agricultural Sciences (No. CAAS-ASTIP-TRICAAS), the China Agriculture Research System of MOF and MARA (CARS-19), and the Natural Science Foundation of China (No. 42277215).

Institutional Review Board Statement: Not applicable.

Informed Consent Statement: Not applicable.

Data Availability Statement: The data presented in this study are available within the article and in the Supplementary Materials.

Acknowledgments: The authors appreciate the generous provision of black tea samples from XiaoGuan Tea Co., Ltd. (Beijing, China).

Conflicts of Interest: The authors declare no potential conflicts of interest with respect to the research, authorship, and/or publication of this article.

References

1. Li, J.; Yao, Y.; Wang, J.; Hua, J.; Wang, J.; Yang, Y.; Dong, C.; Zhou, Q.; Jiang, Y.; Deng, Y.; et al. Rutin, gamma-aminobutyric acid, gallic acid and caffeine negatively affect the sweet-mellow taste of Congou black tea infusions. *Molecules* **2019**, *24*, 4221. [CrossRef] [PubMed]
2. Gao, C.; Huang, Y.; Li, J.; Lyu, S.; Wang, Z.; Xie, F.; Luo, Y.; Zhang, F.; Chen, Z.; Sun, W. Relationship between the grade and the characteristic flavor of PCT (Panyong Congou black tea). *Foods* **2022**, *11*, 2815. [CrossRef] [PubMed]
3. Wu, S.; Yu, Q.; Shen, S.; Shan, X.; Hua, J.; Zhu, J.; Qiu, J.; Deng, Y.; Zhou, Q.; Jiang, Y.; et al. Non-targeted metabolomics and electronic tongue analysis reveal the effect of rolling time on the sensory quality and nonvolatile metabolites of congou black tea. *LWT* **2022**, *169*, 113971. [CrossRef]
4. Ren, G.; Li, T.; Wei, Y.; Ning, J.; Zhang, Z. Estimation of Congou black tea quality by an electronic tongue technology combined with multivariate analysis. *Microchem. J.* **2021**, *163*, 105899. [CrossRef]
5. Wang, C.; Zhang, C.; Kong, Y.; Peng, X.; Li, C.; Liu, S.; Du, L.; Xiao, D.; Xu, Y. A comparative study of volatile components in Dianhong teas from fresh leaves of four tea cultivars by using chromatography-mass spectrometry, multivariate data analysis, and descriptive sensory analysis. *Food Res. Int.* **2017**, *100*, 267–275. [CrossRef] [PubMed]
6. Zhu, J.; Wang, J.; Yuan, H.; Ouyang, W.; Li, J.; Hua, J.; Jiang, Y. Effects of Fermentation Temperature and Time on the Color Attributes and Tea Pigments of Yunnan Congou Black Tea. *Foods* **2022**, *11*, 1845. [CrossRef]
7. Sun, Y.; Fu, Y.; Chen, R.; Zhang, Y.; Liao, T.; Xi, H.; Sun, S.; Cheng, Z. Profiling of volatile and non-volatile compounds in Dianhong by a combined approach of static headspace GC-MS and UPLC-MS. *CyTA J. Food* **2022**, *20*, 305–315. [CrossRef]
8. Ma, L.; Gao, M.; Zhang, L.; Qiao, Y.; Li, J.; Du, L.; Zhang, H.; Wang, H. Characterization of the key aroma-active compounds in high-grade Dianhong tea using GC-MS and GC-O combined with sensory-directed flavor analysis. *Food Chem.* **2022**, *378*, 132058. [CrossRef]
9. Chen, D.; Sun, Z.; Gao, J.; Peng, J.; Wang, Z.; Zhao, Y.; Lin, Z.; Dai, W. Metabolomics combined with proteomics provides a novel interpretation of the compound differences among Chinese tea cultivars (*Camellia sinensis* var. *sinensis*) with different manufacturing suitabilities. *Food Chem.* **2022**, *377*, 131976. [CrossRef]
10. Dong, C.; Zhu, H.; Wang, J.; Yuan, H.; Zhao, J.; Chen, Q. Prediction of black tea fermentation quality indices using NIRS and nonlinear tools. *Food Sci. Biotechnol.* **2017**, *26*, 853–860. [CrossRef]
11. Tanaka, T.; Yasumatsu, M.; Hirotani, M.; Matsuo, Y.; Li, N.; Zhu, H.T.; Saito, Y.; Ishimaru, K.; Zhang, Y.J. New degradation mechanism of black tea pigment theaflavin involving condensation with epigallocatechin-3-*O*-gallate. *Food Chem.* **2022**, *370*, 131326. [CrossRef]

12. Ren, G.; Ning, J.; Zhang, Z. Multi-variable selection strategy based on near-infrared spectra for the rapid description of dianhong black tea quality. *Spectrochim. Acta A Mol. Biomol. Spectrosc.* **2021**, *245*, 118918. [CrossRef]
13. Huang, J.; Ren, G.; Sun, Y.; Jin, S.; Li, L.; Wang, Y.; Ning, J.; Zhang, Z. Qualitative discrimination of Chinese dianhong black tea grades based on a handheld spectroscopy system coupled with chemometrics. *Food Sci. Nutr.* **2020**, *8*, 2015–2024. [CrossRef]
14. Guo, X.; Long, P.; Meng, Q.; Ho, C.T.; Zhang, L. An emerging strategy for evaluating the grades of Keemun black tea by combinatory liquid chromatography-Orbitrap mass spectrometry-based untargeted metabolomics and inhibition effects on alpha-glucosidase and alpha-amylase. *Food Chem.* **2018**, *246*, 74–81. [CrossRef]
15. Wen, M.; Han, Z.; Cui, Y.; Ho, C.T.; Wan, X.; Zhang, L. Identification of 4-O-p-coumaroylquinic acid as astringent compound of Keemun black tea by efficient integrated approaches of mass spectrometry, turbidity analysis and sensory evaluation. *Food Chem.* **2022**, *368*, 130803. [CrossRef]
16. Li, J.; Wu, S.; Yu, Q.; Wang, J.; Deng, Y.; Hua, J.; Zhou, Q.; Yuan, H.; Jiang, Y. Chemical profile of a novel ripened Pu-erh tea and its metabolic conversion during pile fermentation. *Food Chem.* **2022**, *378*, 132126. [CrossRef]
17. Wang, H.; Shen, S.; Wang, J.; Jiang, Y.; Li, J.; Yang, Y.; Hua, J.; Yuan, H. Novel insight into the effect of fermentation time on quality of Yunnan Congou black tea. *LWT* **2022**, *155*, 112939. [CrossRef]
18. Li, J.; Wang, J.; Yao, Y.; Hua, J.; Zhou, Q.; Jiang, Y.; Deng, Y.; Yang, Y.; Wang, J.; Yuan, H.; et al. Phytochemical comparison of different tea (*Camellia sinensis*) cultivars and its association with sensory quality of finished tea. *LWT* **2020**, *117*, 108595. [CrossRef]
19. Alasalvar, C.; Topal, B.; Serpen, A.; Bahar, B.; Pelvan, E.; Gokmen, V. Flavor characteristics of seven grades of black tea produced in Turkey. *J. Agric. Food Chem.* **2012**, *60*, 6323–6332. [CrossRef] [PubMed]
20. Zhang, X.; Du, X.; Li, Y.Z.; Nie, C.N.; Wang, C.M.; Bian, J.L.; Luo, F. Are organic acids really related to the sour taste difference between Chinese black tea and green tea. *Food Sci. Nutr.* **2022**, *10*, 2071–2081. [CrossRef] [PubMed]
21. Wan, X. *Biochemistry of Tea*, 3rd ed.; China Agriculture Publishing House: Beijing, China, 2003; pp. 9, 31, 191–192.
22. Liu, Z.; Bruins, M.E.; de Bruijn, W.J.C.; Vincken, J.-P. A comparison of the phenolic composition of old and young tea leaves reveals a decrease in flavanols and phenolic acids and an increase in flavonols upon tea leaf maturation. *J. Food Compos. Anal.* **2020**, *86*, 103385. [CrossRef]
23. Xu, C.; Liang, L.; Li, Y.; Yang, T.; Fan, Y.; Mao, X.; Wang, Y. Studies of quality development and major chemical composition of green tea processed from tea with different shoot maturity. *LWT* **2021**, *142*, 111055. [CrossRef]
24. Samanta, T.; Kotamreddy, J.N.R.; Ghosh, B.C.; Mitra, A. Changes in targeted metabolites, enzyme activities and transcripts at different developmental stages of tea leaves: A study for understanding the biochemical basis of tea shoot plucking. *Acta Physiol. Plant.* **2016**, *39*, 1–16. [CrossRef]
25. Singh, S.; Sud, R.K.; Gulati, A.; Joshi, R.; Yadav, A.K.; Sharma, R.K. Germplasm appraisal of western Himalayan tea: A breeding strategy for yield and quality improvement. *Genet. Resour. Crop Evol.* **2012**, *60*, 1501–1513. [CrossRef]
26. Obanda, M.; Owuor, P.O.; Mang'oka, R. Changes in the chemical and sensory quality parameters of black tea due to variations of fermentation time and temperature. *Food Chem.* **2001**, *75*, 395–404. [CrossRef]
27. Tan, J.; Dai, W.; Lu, M.; Lv, H.; Guo, L.; Zhang, Y.; Zhu, Y.; Peng, Q.; Lin, Z. Study of the dynamic changes in the non-volatile chemical constituents of black tea during fermentation processing by a non-targeted metabolomics approach. *Food Res. Int.* **2016**, *79*, 106–113. [CrossRef]
28. Dekant, W.; Fujii, K.; Shibata, E.; Morita, O.; Shimotoyodome, A. Safety assessment of green tea based beverages and dried green tea extracts as nutritional supplements. *Toxicol Lett.* **2017**, *277*, 104–108. [CrossRef]
29. Yang, C.; Hu, Z.; Lu, M.; Li, P.; Tan, J.; Chen, M.; Lv, H.; Zhu, Y.; Zhang, Y.; Guo, L.; et al. Application of metabolomics profiling in the analysis of metabolites and taste quality in different subtypes of white tea. *Food Res. Int.* **2018**, *106*, 909–919. [CrossRef]
30. Tanaka, T.; Matsuo, Y.; Kouno, I. Chemistry of secondary polyphenols produced during processing of tea and selected foods. *Int. J. Mol. Sci.* **2009**, *11*, 14–40. [CrossRef]
31. Scharbert, S.; Hofmann, T. Molecular definition of black tea taste by means of quantitative studies, taste reconstitution and omission experiments. *J. Agric. Food Chem.* **2005**, *53*, 5377–5384. [CrossRef]
32. Shan, X.; Yu, Q.; Chen, L.; Zhang, S.; Zhu, J.; Jiang, Y.; Yuan, H.; Zhou, Q.; Li, J.; Wang, Y.; et al. Analyzing the influence of withering degree on the dynamic changes in non-volatile metabolites and sensory quality of Longjing green tea by non-targeted metabolomics. *Front. Nutr.* **2023**, *10*, 1104926. [CrossRef] [PubMed]
33. Han, Z.; Wen, M.; Zhang, H.; Zhang, L.; Wan, X.; Ho, C.T. LC-MS based metabolomics and sensory evaluation reveal the critical compounds of different grades of Huangshan Maofeng green tea. *Food Chem.* **2022**, *374*, 131796. [CrossRef] [PubMed]
34. Scharbert, S.; Holzmann, N.; Hofmann, T. Identification of the astringent taste compounds in black tea infusions by combining instrumental analysis and human bioresponse. *J. Agric. Food Chem.* **2004**, *52*, 3498–3508. [CrossRef]
35. Yu, Z.; Yang, Z. Understanding different regulatory mechanisms of proteinaceous and non-proteinaceous amino acid formation in tea (*Camellia sinensis*) provides new insights into the safe and effective alteration of tea flavor and function. *Crit. Rev. Food Sci. Nutr.* **2020**, *60*, 844–858. [CrossRef] [PubMed]

Disclaimer/Publisher's Note: The statements, opinions and data contained in all publications are solely those of the individual author(s) and contributor(s) and not of MDPI and/or the editor(s). MDPI and/or the editor(s) disclaim responsibility for any injury to people or property resulting from any ideas, methods, instructions or products referred to in the content.

Article

Crosstalk between Breast Milk N-Acetylneuraminic Acid and Infant Growth in a Gut Microbiota-Dependent Manner

Runze Ouyang [1,2,3], Sijia Zheng [1,2,3], Xiaolin Wang [1,3], Qi Li [1,3], Juan Ding [4], Xiao Ma [5], Zhihong Zhuo [6], Zhen Li [7], Qi Xin [8], Xin Lu [1,2,3], Lina Zhou [1,2,3], Zhigang Ren [9], Surong Mei [10], Xinyu Liu [1,2,3,*] and Guowang Xu [1,2,3]

1. CAS Key Laboratory of Separation Science for Analytical Chemistry, Dalian Institute of Chemical Physics, Chinese Academy of Sciences, Dalian 116023, China
2. University of Chinese Academy of Sciences, Beijing 100049, China
3. Liaoning Province Key Laboratory of Metabolomics, Dalian 116023, China
4. Department of Quality Control, The First Affiliated Hospital of Zhengzhou University, Zhengzhou 450052, China
5. Department of Nursing, The First Affiliated Hospital of Zhengzhou University, Zhengzhou 450052, China
6. Department of Pediatric, The First Affiliated Hospital of Zhengzhou University, Zhengzhou 450052, China
7. Department of Interventional Radiology, The First Affiliated Hospital of Zhengzhou University, Zhengzhou 450052, China
8. Academy of Medical Sciences, Zhengzhou University, Zhengzhou 450052, China
9. Department of Infectious Diseases, The First Affiliated Hospital of Zhengzhou University, Zhengzhou 450052, China
10. State Key Laboratory of Environment Health (Incubation), Key Laboratory of Environment and Health, Ministry of Education, Key Laboratory of Environment and Health (Wuhan), Ministry of Environmental Protection, School of Public Health, Tongji Medical College, Huazhong University of Science and Technology, Wuhan 430030, China
* Correspondence: liuxy2012@dicp.ac.cn

Abstract: The healthy growth of infants during early life is associated with lifelong consequences. Breastfeeding has positive impacts on reducing obesity risk, which is likely due to the varied components of breast milk, such as N-acetylneuraminic acid (Neu5Ac). However, the effect of breast milk Neu5Ac on infant growth has not been well studied. In this study, targeted metabolomic and metagenomic analyses were performed to illustrate the association between breast milk Neu5Ac and infant growth. Results demonstrated that Neu5Ac was significantly abundant in breast milk from infants with low obesity risk in two independent Chinese cohorts. Neu5Ac from breast milk altered infant gut microbiota and bile acid metabolism, resulting in a distinct fecal bile acid profile in the high-Neu5Ac group, which was characterized by reduced levels of primary bile acids and elevated levels of secondary bile acids. Taurodeoxycholic acid 3-sulfate and taurochenodeoxycholic acid 3-sulfate were correlated with high breast milk Neu5Ac and low obesity risk in infants, and their associations with healthy growth were reproduced in mice colonized with infant-derived microbiota. *Parabacteroides* might be linked to bile acid metabolism and act as a mediator between Neu5Ac and infant growth. These results showed the gut microbiota-dependent crosstalk between breast milk Neu5Ac and infant growth.

Keywords: breast milk; N-acetylneuraminic acid; gut microbiota; bile acids; infant growth; germ-free mice

Citation: Ouyang, R.; Zheng, S.; Wang, X.; Li, Q.; Ding, J.; Ma, X.; Zhuo, Z.; Li, Z.; Xin, Q.; Lu, X.; et al. Crosstalk between Breast Milk N-Acetylneuraminic Acid and Infant Growth in a Gut Microbiota-Dependent Manner. *Metabolites* **2023**, *13*, 846. https://doi.org/10.3390/metabo13070846

Academic Editor: Cholsoon Jang

Received: 14 June 2023
Revised: 5 July 2023
Accepted: 12 July 2023
Published: 13 July 2023

Copyright: © 2023 by the authors. Licensee MDPI, Basel, Switzerland. This article is an open access article distributed under the terms and conditions of the Creative Commons Attribution (CC BY) license (https://creativecommons.org/licenses/by/4.0/).

1. Introduction

The healthy growth of infants during early life is associated with lifelong consequences. Being overweight or obese in childhood could increase the risk of adult adiposity and perhaps is associated with many health problems, such as type 1 diabetes and cardiovascular diseases [1,2]. It has been reported that infants with BMI z-scores greater than the 85th percentile of the World Health Organization standards were at risk of obesity [3,4].

Although genetics, early nutrition, lifestyle, and lack of physical activity are the direct factors leading to obesity, the potentially causal role of the early gut microbiota in childhood obesity has become increasingly prominent [5–7].

Sialylated oligosaccharides are some of the most important bioactive components in breast milk and could act as the prebiotics for the infant gut microbiota. In addition, sialylated oligosaccharides are associated with numerous benefits, such as promoting infant growth [8,9]. A previous study of two Malawian birth cohorts revealed that sialylated oligosaccharides could promote healthy infant growth in a microbiota-dependent manner in case of infant undernutrition [10]. Moreover, another study indicated that 3′-sialyllactose (3′-SL) plays a critical role in improving offspring's health [11].

Sialic acid could be released from sialylated oligosaccharides by sialidases of gut microbiota [10,12]. Although most of the sialic acid is combined with oligosaccharides and proteins in breast milk, about 3% of the sialic acid still exists in free form and plays a very important role [13]. N-acetylneuraminic acid (Neu5Ac) is the predominant form of free sialic acid in humans, and 3′-SL and 6′-sialyllactose (6′-SL) are two abundant Neu5Ac-binding oligosaccharides in breast milk [9,14]. Moreover, the free Neu5Ac might reflect the metabolic status of total sialic acid or its availability because concentrations dropped with decreases in the oligosaccharide-bound and protein-bound forms of sialic acid in breast milk [13]. At present, there are many studies illustrating that free Neu5Ac is crucial for improving infant brain development and enhancing immunity [15]. However, the relationship between free Neu5Ac and the obesity risk of infants has not been clarified. Importantly, a high level of free Neu5Ac is one of the most discriminative characteristics of breast milk, when compared with baby formula [13]. Understanding the role of Neu5Ac in the growth of infants could be beneficial for developing prebiotics or supplements to improve growth outcomes.

In this study, we first focused on the relationship between the level of Neu5Ac in early breast milk and infant growth in later infancy in two independent Chinese infant cohorts. Next, by combining metagenomics and targeted metabolomic analyses, we illustrated the influences of breast milk Neu5Ac on the infant gut microbiota and the related metabolites and explored the interaction of the Neu5Ac-related metabolites with infant growth. Finally, the mediator role of the gut microbiota and bacterial-derived metabolites in the link between Neu5Ac and growth was validated in a gnotobiotic mouse model.

2. Materials and Methods

2.1. Study Cohorts and Sample Collection

Breast milk and neonatal fecal samples were collected from Chinese mother–newborn dyads during the first week after delivery between May and December 2018 in the cohort in a previous study (Zhengzhou cohort, n = 58) [16]. The fecal samples were collected within 24 h of breast milk sampling. The newborns (25 males/33 females) were all full-term and healthy. Seventeen newborns were delivered through C-section. All the samples were kept at −80 °C and analyzed within 1 year. Detailed responses to questionnaires of maternal and neonatal characteristics including general information, feeding pattern, and antibiotic usage are recorded in Table 1.

For the Wuhan cohort, only breast milk samples were collected around the first month of lactation and growth data were collected when infants were around 3 years old (n = 201). All the infants were full-term, and 131 of them were delivered by C-section. All the samples were kept at −80 °C and analyzed within 1 year. Detailed responses to questionnaires of maternal and neonatal characteristics and the infant growth indicators at around 3 years are shown in Table 1. The study was approved by the Ethics Committee of the First Affiliated Hospital of Zhengzhou University. Written informed consent was received from all mothers.

Table 1. Characteristics of the mothers and infants included in the study.

	Zhengzhou Cohort (n = 58)	Wuhan Cohort (n = 201)
Sampling age of infant (days)		
Breast milk	4 (3–7)	41 (33–43)
Newborn feces	5 (3–7)	Not collected
Maternal characteristics		
Gestational age (day)	277 (272–281)	273 (266–280)
Delivery mode		
Natural delivery	41	70
C-section	17	131
Maternal antibiotic usage [1]	23	No information
Maternal BMI (kg/m^2) [2]		
Pre-pregnancy	20.26 (18.81–22.63)	20.06 (18.99–21.64)
Pre-delivery	27.07 (24.57–29.55)	27.32 (24.80–29.24)
Neonatal characteristics		
Infant sex [3]		
Male	25	111
Female	33	83
Feeding pattern [4]		
Mostly breastmilk feeding	18	102
Mixed feeding	40	98
Growth indicator [5]		
Infant age at time of BMI recording (months)	12 (12–13)	30 (24–36)
Infant BMI	16.98 (16.01–18.41)	16.78 (15.52–17.72)
Low obesity risk (No.)	22	118
High obesity risk (No.)	13	83

[1] There were 4 missing data in the Zhengzhou cohort; [2] 14 missing data in the Zhengzhou cohort and 18 missing data in the Wuhan cohort; [3] 4 missing data in the Wuhan cohort; [4] feeding pattern at the time of breast milk sampling. There were 1 missing datum in the Wuhan cohort; [5] 23 missing data in the Zhengzhou cohort. Growth indicator data from the Zhengzhou cohort and Wuhan cohort were collected when the infants were 1 year old and 3 years old, separately. All the continuous variables are shown in median and interquartile ranges.

2.2. Assessment of Infant Growth

Infant growth indicators from the Zhengzhou cohort were collected at around 1 year of age, while infant growth data from the Wuhan cohort were collected at around 3 years of age. The infant BMI z-scores adjusted for age and sex were calculated under the guidance of World Health Organization standards in both of the cohorts [3]. Infants with BMI z-score greater than the 85th percentile were considered as the high obesity risk group [4], and the rest were grouped as the low obesity risk group. Specifically, growth data on 35 infants from the Zhengzhou cohort were collected, of whom 13 were classified as the high obesity risk group and 22 as the low obesity risk group. For the 201 infants from the Wuhan cohort, 118 infants were grouped as low obesity risk, and 83 infants were grouped as high obesity risk (Table 1).

2.3. Quantification of Neu5Ac and Sialylated Oligosaccharides in Breast Milk

Breast milk Neu5Ac, 3′-SL, and 6′-SL were quantified by an online solid-phase extraction–hydrophilic interaction chromatography (SPE-HILIC-MS) platform according to a previously used method [16]. The compound identification is based on the comparison with t_R, m/z, and MS/MS fragments of the standards (Figure S1). In brief, 200 μL of breast milk samples was centrifuged at 8600× g for 20 min at 4 °C to eliminate lipids, and 200 μL of ethanol containing internal standards (Neu5Ac-^{13}C6) was added to the skim breast milk to remove proteins. After centrifugation at 15,000× g for 15 min at 4 °C, 50 μL of supernatant was lyophilized, re-solubilized using 200 μL acetonitrile/water (v/v = 1/1), and quantified on the SPE-HILIC-MS platform. Finally, we obtained the Neu5Ac, 3′-SL, and 6′-SL concentrations of 58 breast milk samples from the Zhengzhou cohort and 201 samples from the Wuhan cohort.

2.4. Fecal Microbiota Transplantation to Germ-Free Mice

The experiment was conducted under the protocols approved by the Cyagen Biological Animal Ethics Committee. Specifically, approximately 500 mg of frozen feces from one infant with matched clinical information and infant age at sampling (10 days post-partum) from each group was suspended in 5 mL of reduced phosphate-buffered saline (PBS) in Hungate tubes. Germ-free C57BL/6 wild-type mice were colonized with the respective donor microbiota after a 4 h fasting by oral gavage of 200 μL of fecal slurry per mouse. The fecal slurry of the same infant was transplanted to 6 mice in parallel as a group. The rest of the fecal slurry was stored at −80 °C for the second gavage. Mice that received fecal microbiota from the same infant were housed together in the same cage (two mice per cage and six mice per group). The colonization was repeated the week after using the same fecal slurry. During the colonization, mice were fed sterile fodder and sterile water. Mice were housed in gnotobiotic facilities in 12 h day/night cycles during the whole experiment. Body weight was recorded every 3 days and mice were euthanized 2 weeks after the first oral gavage. The feces and cecal content were sampled and immediately put in liquid nitrogen. Gut microbial colonization in the gnotobiotic mice was assessed with about 20 mg of mice feces collected the day before the mice were euthanized by 16S rRNA gene sequencing analysis as previously described [16].

2.5. Metagenomic Analysis of the Newborn Gut Microbiota

Total DNA of newborn fecal samples was extracted and quantified with a previously described method [16]. Then, the DNA extract was fragmented to an average size of roughly 400 bp using Covaris M220 (Gene Company Limited, Hong Kong, China). Paired-end libraries were generated with NEXTFLEX Rapid DNA-Seq (Bioo Scientific, Austin, TX, USA) and sequenced by HiSeq Reagent Kits on an Illumina Hiseq (Illumina, San Diego, CA, USA) in accordance with manufacturer instructions.

Fastp (version 0.20.0) was used to trim adaptors and remove low-quality reads from the raw paired-end reads [17]. Next, BWA (version 0.7.9a) was used to align the reads to the human genome [18], and any hits associated with the reads and their mated reads were removed. MEGAHIT (version 1.1.2) was used to assemble the metagenomics data [19]. Contigs that were at least 300 bp were chosen as the final assembled result. MetaGene was applied to predict the open reading frames (ORFs) from each assembled contig [20]. A non-redundant gene catalog was built using CD-HIT (version 4.6.1) with 90% sequence identity and 90% coverage [21]. SOAPaligner (version 2.21) was used to map the reads to the non-redundant gene catalog with 95% identity after quality control [22], and gene abundance was evaluated and the relative abundance was normalized with TPM value as described previously [23].

The NCBI NR database was used for taxonomic annotation. Diamond (version 0.8.35) was used to align the representative sequences of the non-redundant gene catalog to the database with an e-value cutoff of 1×10^{-5} [24]. The genes that occurred in more than 5% of the samples were included in the following analysis [25]. The relative abundance of genes of the same genus was summed up to assess the relative abundance of the gut microbiota at the genus level. The Kyoto Encyclopedia of Genes and Genomes (KEGG) database was used for functional annotation by Diamond with an e-value cutoff of 1×10^{-5}. The relative abundance of KEGG orthology (KO) was estimated by summing the relative abundance of genes of the same KO and renormalizing to one.

2.6. Bile Acid Analysis

For each sample, 100 mg of frozen infant fecal samples or 10 mg of frozen mice cecum content, a zirconia bead, and 1 mL of extraction solvent were mixed and homogenized with a mixed grinding apparatus (MM400, Retsch, Germany). The extraction solvent was composed of methanol with 0.3 μg/mL of cholic acid (CA)-d5, 0.9 μg/mL of chenodeoxycholic acid (CDCA)-d4, 0.6 μg/mL of glycocholic acid (GCA)-d5, 0.6 μg/mL of glycochenodeoxycholic acid (GCDCA)-d4, 0.3 μg/mL of taurocholic acid (TCA)-d5, and

0.3 µg/mL of taurodeoxycholic acid (TDCA)-d5. Next, the mixture was centrifuged at 4 °C, 14,000× *g* for 10 min to remove proteins. Then, 800 µL of the supernatant was lyophilized and reconstituted in 300 µL of acetonitrile/water ($v/v = 1/1$) for the following analysis.

An ultra-high-performance liquid chromatograph (UHPLC) coupled to a Shimadzu 8050 Triple Quad mass spectrometer (Shimadzu, Kyoto, Japan) with the electrospray ionization (ESI) source in negative ion mode was used for the targeted bile acid analysis. A Waters ACQUITY UPLC C8 column (1.7 µm, 2.1 × 100 mm) was used for separation. The column temperature was set as 40 °C and the flow rate was 0.2 mL/min. Mobile phase A was 10 mM NH_4HCO_3 aqueous solution and mobile phase B was acetonitrile. Multiple reaction monitoring (MRM) was used for mass analysis. The mass parameters were set as follows: nebulizing gas flow 3 L/min, heating gas flow 10 L/min, interface temperature 300 °C, heat block temperature 400 °C, and drying gas flow 10 L/min.

LabSolutions (version 5.89, Shimadzu, Kyoto, Japan) was used for instrument control and peak extraction. All the peak areas were corrected by internal standards and sample weight. Then, the absolute concentrations of bile acids were calculated by the external standard method and normalized by total sum scaling.

2.7. Statistical Analysis

All the box plots, bar plots, and scatter plots were visualized in GraphPad Prism 9.0 (GraphPad Software Inc., Boston, MA, USA). A Mann–Whitney U test was used for the significance test of continuous variables.

Beta diversity of the neonatal gut microbiota was calculated to evaluate the correlation between newborn gut genera and breast milk Neu5Ac content based on Bray–Curtis distance (RStudio 4.1.0, vegan package 2.5.7) and visualized in a principal coordinate analysis (PCoA) plot with the ggplot2 package (version 3.4.0), and adonis was conducted to evaluate the significant difference. Alpha diversity was calculated based on Simpson and Pielou indexes (RStudio 4.1.0, vegan package 2.5.7) to evaluate the gut microbial richness and evenness. Linear discriminant analysis effect size (LefSe) analyses were performed to assess the discriminative bacteria between different groups, and an LDR score above 2.0 was considered as significant [26]. Redundancy analysis (RDA) was conducted on the basis of the infant gut microbiota at the genus level on an online platform (http://cloud.biomicroclass.com/CloudPlatform) (accessed on 12 June 2023). Difference analysis of metabolic pathways was conducted with Welch's *t*-test followed by FDR correction with the Benjamini–Hochberg method in STAMP [27], and a corrected *p*-value less than 0.05 was considered significant. Spearman correlation analysis was applied to estimate associations between genera and bacterial metabolic pathways in GraphPad Prism 9.0.

Levels of bile acids between different groups were compared by the logarithmic ratio of relative contents and a significant difference was defined as a combination of the fold change (FC) and *p*-value ($|Log_2 FC| > 0.6$ and $p < 0.05$) and visualized in a volcano plot (ggpubr package 0.5.0, ggthemes package 4.2.4). Regression analysis was performed to evaluate the correlations between breast milk Neu5Ac level and infant growth and the correlation between the level of bile acids and infant growth or mouse growth in RStudio (lm function), and $p < 0.05$ was considered to show a significant difference. Random forest models were conducted as a feature selection technique to evaluate which bile acids were most important to differentiate samples based on infant growth (randomForest package, 4.7.1).

3. Results

3.1. Correlation between Breast Milk Neu5Ac Concentrations and Obesity Risk in Chinese Infants

We first characterized the relationship between Neu5Ac/3′-SL/6′-SL in breast milk samples collected around 1 week post-partum and infant obesity risk at 1 year of age in the Zhengzhou cohort ($n = 58$). The results showed that high breast milk Neu5Ac and 3′-SL were both correlated with low obesity risk of the infants at the age of 1 year; that is, infants in the low obesity risk group were exposed to significantly higher levels of breast milk Neu5Ac and 3′-SL during the neonatal period (Figure 1a). Further adjusting for the confounders, including infant sex, delivery mode, feeding, and infant age at breast milk sampling, did not change the positive association between Neu5Ac/3′-SL and low obesity risk (Figure 1c).

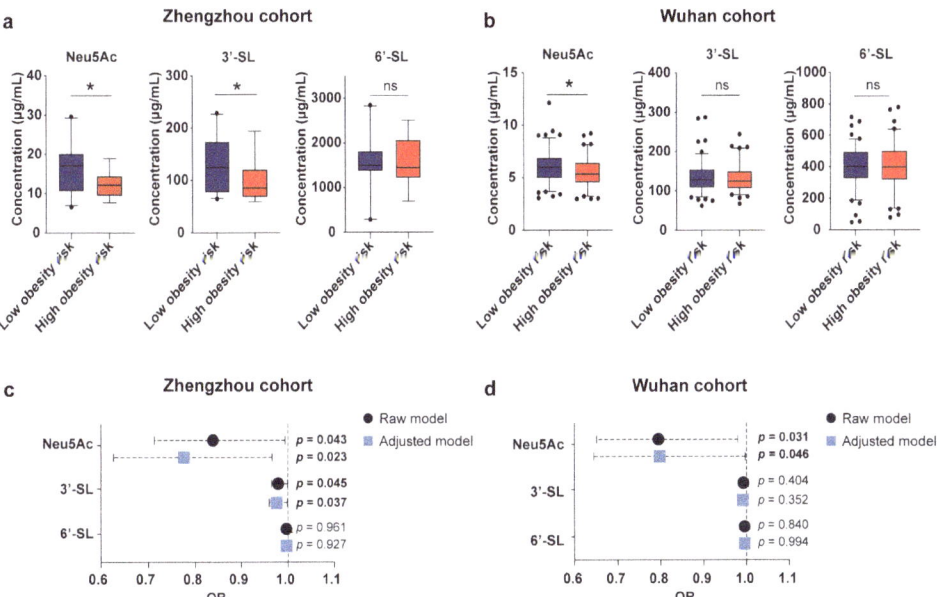

Figure 1. Correlation between breast milk Neu5Ac concentrations and the obesity risk of Chinese infants. (**a,b**) Concentrations of Neu5Ac, 3′-SL, and 6′-SL in breast milk of Chinese mothers collected from the Zhengzhou cohort ($n = 35$) (**a**) and Wuhan cohort ($n = 201$) (**b**), binned by the growth of their infants. Concentrations of Neu5Ac, 3′-SL, and 6′-SL are shown as mean ± SEM (* $p < 0.05$, ns $p > 0.05$, Mann–Whitney U test). (**c,d**) Regression analysis of breast milk Neu5Ac, 3′-SL, and 6′-SL concentrations and infant obesity risk in the Zhengzhou cohort (**c**) and Wuhan cohort (**d**). Adjusted confounders: infant sex, delivery mode, feeding pattern, and infant age at the time of breast milk sampling. OR, odds ratio; Neu5Ac, N-acetylneuraminic acid; 3′-SL, 3′-sialyllactose; 6′-SL, 6′-sialyllactose.

Similar to the Zhengzhou cohort, early Neu5Ac content was significantly elevated in the breast milk of low obesity risk infants aged 3 years old in the Wuhan cohort (Figure 1b). Correction for confounding factors did not alter the correlation between Neu5Ac and infants' obesity risk (Figure 1d). Meanwhile, the association between 3′-SL and infant obesity risk was not observed in the Wuhan cohort. Altogether, these results from the two independent Chinese cohorts suggested that early breast milk Neu5Ac is important for reducing the obesity risk of infants later in infancy.

3.2. Gut Microbiota-Dependent Association between Breast Milk Neu5Ac and Growth

Neu5Ac is involved in the establishment of newborn gut microbiota, which is closely related to the growth of infants later. To further explore the role of gut microbiota in the correlation between Neu5Ac and infant growth, fecal samples of the infants from the Zhengzhou cohort were collected within 24 h of breast milk sampling ($n = 58$) and a fecal microbiota transplantation (FMT) experiment was conducted in germ-free mice to verify the mediating role of microbiota between Neu5Ac concentration and growth.

First, the infants in the Zhengzhou cohort were divided into the low-Neu5Ac (LN) group ($n = 29$) and high-Neu5Ac (HN) group ($n = 29$) according to the median concentration of Neu5Ac in their breast milk (Table 2). The LN group included those infants with breast milk Neu5Ac concentrations below the median, and the HN group included those infants with breast milk Neu5Ac concentrations above the median. The HN group was exposed to breast milk Neu5Ac concentrations 2.5 times higher than that of the LN group ($p < 0.0001$, Mann–Whitney U test) (Figure 2a). Moreover, the proportion of low obesity risk infants was higher in the HN group (Figure 2b).

Table 2. Clinical characteristics of the infants in the LN group and HN group in the Zhengzhou cohort.

	LN Group ($n = 29$)	HN Group ($n = 29$)
Sampling age of infant (days)		
Breast milk	6 (4–13)	4 (3–5)
Newborn feces	5 (4–13)	4 (3–5)
Maternal characteristics		
Gestational age (day)	277 (274–280)	273 (270–284)
Delivery mode		
Natural delivery	25	16
C-section	4	13
Maternal antibiotic usage [1]	8	15
Maternal BMI (kg/m^2) [2]		
Pre-pregnancy	19.53 (18.78–22.03)	21.57 (19.53–23.42)
Pre-delivery	26.30 (24.45–28.40)	27.55 (26.02–30.20)
Neonatal characteristics		
Infant sex		
Male	13	12
Female	16	17
Feeding pattern [3]		
Mostly breastmilk feeding	10	8
Mixed feeding	19	21
Growth indicator [4]		
Infant age at time of BMI recording (months)	13 (12–13)	12 (12–13)
Infant BMI	17.97 (16.01–18.66)	16.64 (16.00–17.12)
Low obesity risk (No.)	9	13
High obesity risk (No.)	10	3

[1] There were 1 missing datum in LN group and 3 missing data in HN group; [2] 4 missing data in LN group and 10 missing data in HN group; [3] feeding pattern at the time of breast milk sampling; [4] 10 missing data in LN group and 13 missing data in HN group. Growth indicator data from the Zhengzhou cohort were collected when the infants were 1 year old. All the continuous variables are shown in median and interquartile ranges.

Next, a fecal sample from one infant with matched clinical information from each of the LN and HN groups was taken separately to make a fecal slurry (Table S1), then transplanted into young (5-week-old) male germ-free C57BL/6 mice by oral gavage. The fecal slurry of the same infant was transplanted into 6 mice in parallel as a group (Figure 2c). After the oral gavage of fecal slurry was administered twice, and two weeks of growth monitoring was carried out, we found that mice colonized with fecal microbiota from the HN infants had stable weight gain, while mice colonized with the fecal bacteria from the LN group showed a dramatic weight loss followed by a slow weight gain until the mice were sacrificed (Figure 2d). Additionally, two of the LN fecal bacteria-colonized mice exhibited ruffled back fur after the first colonization, suggesting a sub-healthy state of the mice. Collectively, we speculated that breast milk Neu5Ac-related gut microbiota is involved in the regulation of healthy growth.

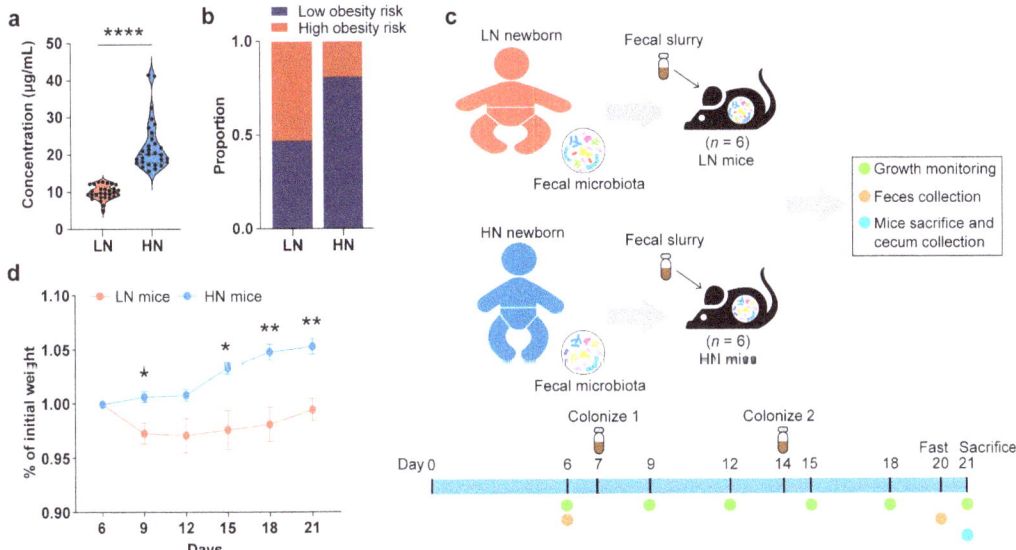

Figure 2. Effect of gut microbiota in infants fed with breast milk with different Neu5Ac levels on the growth of gnotobiotic mice. (**a**) The concentration of breast milk Neu5Ac in the LN group and HN group. Concentration is shown as mean ± SEM (**** $p < 0.0001$, Mann–Whitney U test). (**b**) Ratio of low obesity risk infants to high obesity risk infants between LN group and HN group. (**c**) Design of gnotobiotic mouse experiments. (**d**) Weight gain normalized to body weight of LN mice and HN mice (** $p < 0.01$, * $p < 0.05$, Mann–Whitney U test). LN mice: mice colonized with the feces from the infants in the LN group; HN mice: mice colonized with the feces from the infants in the HN group.

3.3. Effect of Breast Milk Neu5Ac on the Composition and Function of Gut Microbiota

To further explore the specific role of breast milk Neu5Ac in the colonization of infant gut microbiota from early life, metagenomic sequencing analysis was performed on the fecal samples collected in parallel with the breast milk to profile the gut microbiota of infants from the Zhengzhou cohort, and the effects of breast milk Neu5Ac on gut microbiota were characterized. In total, we generated 243.28 GB of paired-end reads of high-quality sequences (average 4.19 Gb per sample).

Clear separation of the gut microbiota was observed between the LN group and HN group (Bray–Curtis distance, adonis, permutations = 999, $p = 0.008$) (Figure 3a). However, breast milk Neu5Ac levels did not cause a change in alpha diversity between the two groups (Figure S2). In terms of the gut microbiota composition, only a few bacteria were associated with Neu5Ac content. For instance, *Bifidobacterium* was abundant in the LN group and *Klebsiella* was enriched in the HN group (Figure 3b,c). Similarly, this difference in microbial composition was also present in germ-free mice colonized with fecal microbiota from the LN and HN groups, although the gut microbial richness and evenness were higher in the LN group (Figure S3a–d). Notably, breast milk Neu5Ac was not the unique impact factor on the infant gut microbiota, delivery mode also had an impact on the gut microbial community according to the RDA (Figure 4a). Therefore, the effect of breast milk Neu5Ac levels on the composition of gut microbiota in vaginally delivered infants of the Zhengzhou cohort was further analyzed, and the enrichment of *Klebsiella* in the HN group could also be observed (Figure 4b).

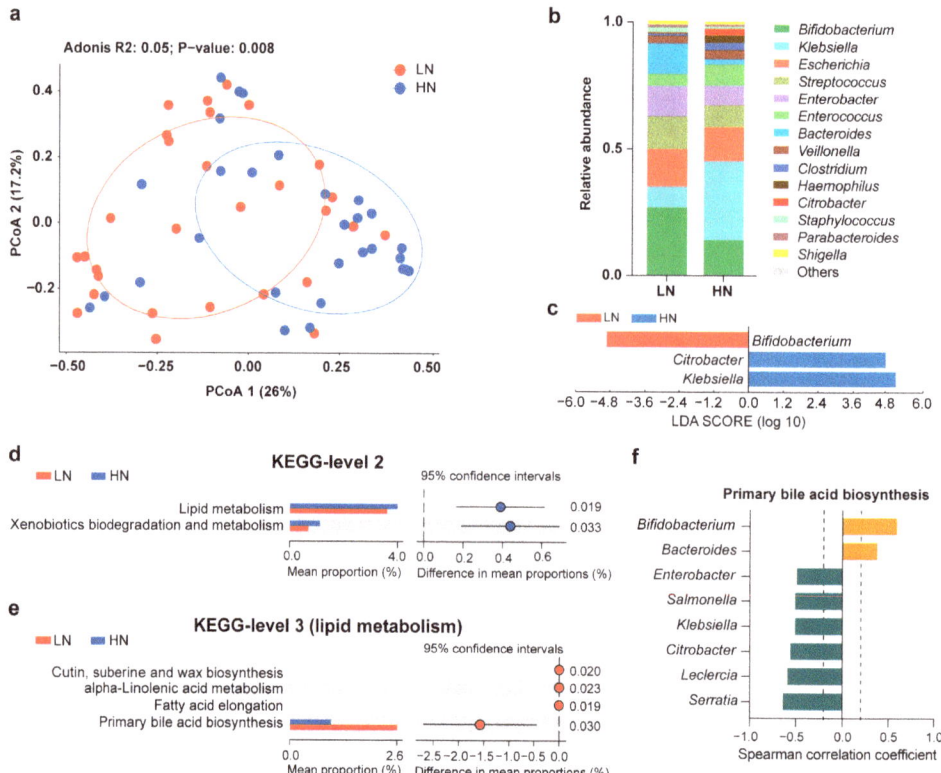

Figure 3. Effects of breast milk Neu5Ac on the composition and function of the infant gut microbiota in the Zhengzhou cohort. (**a**) Principal coordinate analysis (PCoA1 and PCoA2) of the infant gut microbiota at the genus level based on Bray–Curtis distance, with community structure differences tested by adonis analysis of variance with 999 permutations. (**b**) Comparisons of the mean relative abundance of gut microbiota at the genus level between the LN group and HN group. Genera with relative abundance above 1% are shown in the bar plot. "Others" indicates sum of the bacteria with relative abundance less than 1%. (**c**) Discriminative bacteria between LN group and HN group. Genera with relative abundance above 1% are included in the analysis. (**d**) Differences in the microbial metabolic pathway based on level 2 of the KEGG database. (**e**) Differences in microbial lipid metabolic pathway of the KEGG database (level 3). (**f**) Spearman correlation coefficients (r) between gut genera and primary bile acid biosynthesis pathway. Orange indicates a positive correlation and green indicates a negative correlation. Genera with relative abundance above 1% are included in the analysis. Significant correlations with |r| > 0.2 and $p < 0.05$ are shown.

Further analysis based on the functional capacity of the infant gut microbiota indicated that lipid metabolism, especially bile acid metabolism, was highly correlated with breast milk Neu5Ac (Figure 3d,e). Moreover, the discriminative bacteria between HN and LN groups exhibited different levels of involvement in primary bile acid biosynthesis (Figure 3f). For instance, *Bifidobacterium* and *Bacteroides*, the LN group's abundant bacteria, were positively associated with primary bile acid biosynthesis, while *Klebsiella*, the bacterium enriched in the HN group, was negatively correlated with primary bile acid biosynthesis. Altogether, these findings suggested the presence of an interaction between Neu5Ac and gut microbial bile acid metabolism.

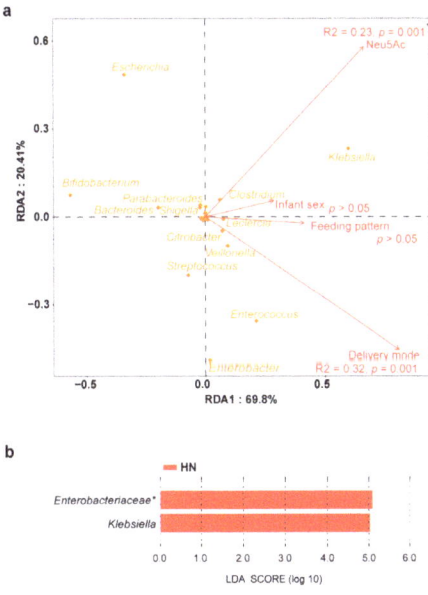

Figure 4. Influencing factors of gut microbiota in the Zhengzhou cohort. (**a**) Redundancy analysis of impact factors on the infant gut microbiota. (**b**) Discriminative bacteria between LN group and HN group in the vaginally delivered infants. Genera with relative abundance above 1% are included in the analysis. Asterisks (*) indicates the unclassified bacteria at the genus level.

3.4. Conjugated Bile Acids in the Correlations between Breast Milk Neu5Ac and Infant Obesity Risk

Next, targeted tandem mass spectrometry was used to measure the levels of 19 bile acids of fecal samples in the Zhengzhou cohort, and the distinct bile acid profile was characterized between the LN group and HN group. In general, a higher proportion of primary unconjugated bile acids, namely, CA and CDCA, was found in the LN group (Figure 5a). Primary unconjugated bile acids can be converted into secondary or conjugated bile acids by gut microbiota. We observed that the HN group contained a greater abundance of conjugated bile acids, especially taurine- and sulfo-conjugated deoxycholic acid (DCA), CDCA, and lithocholic acid (LCA) (Figure 5b and Table S2).

Notably, many of the discriminative bile acids between the LN group and HN group were significantly correlated with infant obesity risk later in life. In regression models adjusting for the confounders (infant sex, delivery mode, feeding pattern, and infant age at breast milk sampling), the total proportions of sulfo-, taurine-, and glycine-conjugated bile acids all inversely contributed to a high BMI in infants at 12 months of age (Figure 5c). Looking at the individual bile acids, glycodeoxycholic acid (GDCA), taurolithocholic acid (TLCA), glycolithocholic acid (GLCA), and taurochenodeoxycholic acid 3-sulfate (TCDCS) were the main bile acids that were related to infant growth. Consistently, the random forest models using the newborn bile acid content to predict infant obesity risk had an out-of-bag error rate of 13.79%, and the most important bile acids contributing to the model were TLCA, GDCA, and glycolithocholic acid 3-sulfate (GLCAS) (Figure S4).

To further explore the association of infant obesity risk with the bile acid levels in newborns, the above Neu5Ac-related bile acids were classified in quartiles and logistic regression analyses were performed. A higher relative abundance of TLCA was associated with a 91.5% decrease in high obesity risk (above vs. below median, OR = 0.085, 95% CI, 0.013–0.542, p = 0.007). An increased level of GDCA was correlated with a 93.7% decrease in high obesity risk (above vs. below median, OR = 0.063, 95% CI, 0.010–0.421, p = 0.004).

Elevation of TCDCS was associated with an 85.9% decrease in high obesity risk (above vs. below median, OR = 0.141, 95% CI, 0.023–0.857, p = 0.033). Additionally, the enrichment of taurodeoxycholic acid 3-sulfate (TDCS) or GLCAS tended to be correlated with high obesity risk (above vs. below median, OR = 0.239, 95% CI, 0.047–1.219, p = 0.085). Taken together, breast milk Neu5Ac may be correlated with gut microbial bile acid metabolism, which in turn is associated with infant obesity risk.

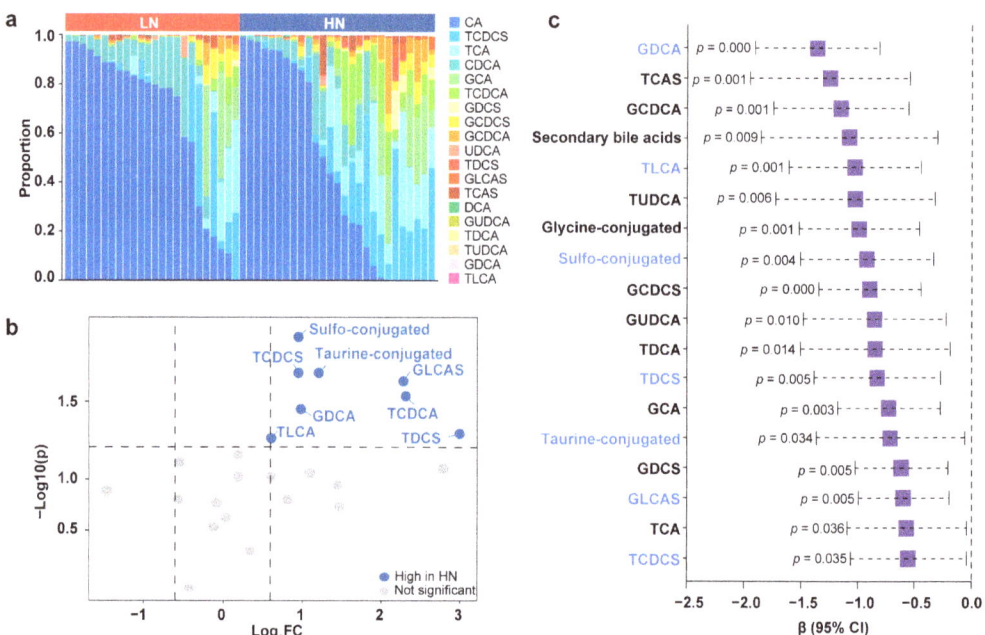

Figure 5. Linkage of conjugated bile acids in the correlations between breast milk Neu5Ac and infant growth. (**a**) Fecal bile acid composition in newborns of LN group and HN group (GDCS, glycodeoxycholic acid 3-sulfate; GCDCS, glycochenodeoxycholic acid 3-sulfate; UDCA, ursodeoxycholic acid; TCAS, taurocholic acid 3-sulfate; GUDCA, glycoursodeoxycholic acid; TUDCA, tauroursodeoxycholic acid). (**b**) Volcano plot of different bile acids between LN group and HN group. Fold changes were calculated as the ratio of each bile acid between the HN group and LN group and converted logarithmically. Blue indicates Log$_2$ FC > 0.6 and p < 0.05 (Mann–Whitney U test), and grey indicates no significant difference. See also Table S2. (**c**) Regression analysis to estimate the associations between the newborn fecal bile acids and infant BMI aged 12 months. Linear regression models were controlled for the following confounders: infant sex, delivery mode, feeding pattern, and infant age at sampling. Bile acids marked with blue are those breast milk Neu5Ac-related bile acids identified in (**b**).

3.5. Validation of the Interactions among Neu5Ac, Gut Microbiota, Bile Acid Metabolism, and Healthy Growth in Gnotobiotic Mice

We next examined whether the above observed correlations of breast milk Neu5Ac with bile acid metabolism and healthy growth in infants also occurred in the FMT mice. Similar to the Zhengzhou cohort, two sulfo-conjugated bile acids, TDCS and TCDCS, were enriched in the mice colonized with the fecal bacteria from the HN group (Figure 6a and Table S3), suggesting that the differences in bile acids were related to gut microbiota. Just as mentioned above, mice colonized with the bacteria from the HN infants exhibited a healthier growth, and our results showed that the HN group's abundant bile acids, TDCS and TCDCS, may have contributed to this (Figure 6b).

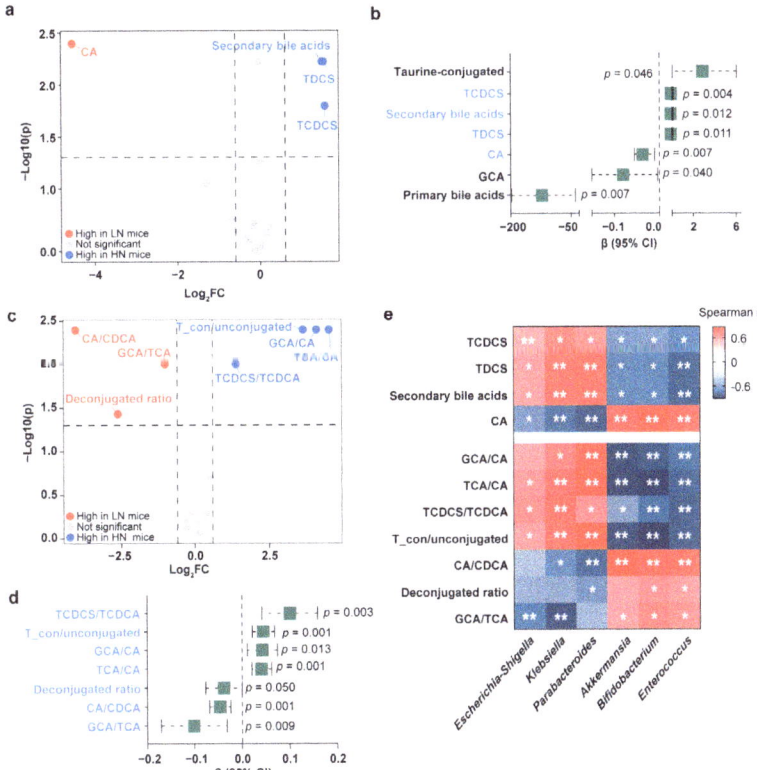

Figure 6. Validation of the association between bile acid and healthy weight gain in mice. (**a**) Volcano plot of different bile acids between LN mice and HN mice. (**b**) Regression analysis to estimate the associations between the mouse cecal bile acid content and weight gain. (**c**) Volcano plot of bile acid conversion to evaluate the enzymatic activity of bile acid metabolism between LN mice and HN mice. (**d**) Regression analysis to estimate the associations between the mouse cecal bile acid conversion and weight gain. (**e**) Heatmap of Spearman correlation coefficients between mouse cecal bile acids and gut microbiota at the genus level (* $p < 0.05$ and ** $p < 0.01$). Fold changes were calculated as the ratio of each bile acid between HN mice and LN mice and converted logarithmically. Blue indicates Log_2 FC > 0.6 and $p < 0.05$ and red indicates Log_2 FC < −0.6 and $p < 0.05$, and grey indicates no significant difference (Mann–Whitney U test). See also Tables S3 and S4. Bile acids marked with blue are discriminative bile acids between LN mice and HN mice identified in (**a**) or (**c**).

Additionally, by calculating the ratios of bile acids, we analyzed the activities of enzymes involved in bile acid metabolism and found a significant elevation of GCA/CA, TCA/CA, and TCDCS/taurochenodeoxycholic acid (TCDCA) in the HN mice (Figure 6c and Table S4), suggesting that enzymes involved in the conjugation of bile acids were elevated in the HN mice. Notably, GCA/CA, TCA/CA, and TCDCS/TCDCA were all found to be associated with favorable growth in the mice (Figure 6d).

Finally, we found that the HN group-related bacteria, *Klebsiella* and *Parabacteroides*, were positively associated with the healthy growth-correlated bile acids (Figure 6e). Specifically, *Klebsiella* and *Parabacteroides* were both positively correlated with TDCS and TCDCS and negatively correlated with CA. The positive associations of *Klebsiella* and *Parabacteroides* with conjugated bile acids further indicated that the enzymes that facilitates bile acid conjugation may be highly expressed in these bacteria.

4. Discussion

Based on the two independent cohorts comprising different lactation periods of breast milk samples from Chinese mothers, our study indicated that the milk from mothers of children exhibiting low obesity risk in later infancy contained a higher level of Neu5Ac in early lactation. Approximately 73% of the Neu5Ac in breast milk is conjugated with oligosaccharides to generate sialylated oligosaccharides, while there is still small but significant amount of Neu5Ac existing in free form, much higher than in formula milk [13]. Previous studies have proposed that breast milk sialylated oligosaccharides were related to infant growth in both full-term and pre-term infants [10,28,29]. Consistently, our study showed that the Neu5Ac in breast milk was also associated with reduced obesity risk in infants. The concentration of Neu5Ac in breast milk changed across lactation [29], and our study further indicated that Neu5Ac levels in breast milk collected one week post-partum (colostrum and transitional milk) and one month post-partum (mature milk) might both affect growth later in infancy. Infant sex, delivery mode, feeding, and infant age at breast milk sampling might be the impact factors of both breast milk Neu5Ac/3′-SL concentration and infant growth [9,30–32], and we showed that adjusting for the above confounders did not change the positive association between Neu5Ac/3′-SL and low obesity risk. The Wuhan cohort and the Zhengzhou cohort are two independent cohorts composed of breast milk samples from different periods, and the similar associations between Neu5Ac and infant obesity risk from the above two cohorts raised the possibility that Neu5Ac plays an important role in infant growth. Although the liver has the ability to de novo synthesize sialic acid from glucose, the activity of the rate-limiting enzyme (UDP-N-acetylglucosamine-2-epimerase) is low during the neonatal period [33], indicating that breast milk Neu5Ac is an important source for newborns, and dietary supplementation with Neu5Ac might be beneficial for infants who cannot be breastfed.

Neu5Ac is a nine-carbon monosaccharide that plays a role in shaping infant gut microbiota. *Bifidobacterium* expresses sialidase that could liberate sialic acids from sialyloligosaccharides, gangliosides, and glycoproteins [34]. The enrichment of *Bifidobacterium* in the LN group might make up the deficiency of sialic acid content in breast milk. *Klebsiella* could utilize sialic acid as a carbon source as, consistently, high breast milk Neu5Ac was related to abundant *Klebsiella* [35]. A previous study indicated that sialyllactose, a conjugated form of Neu5Ac, tended to interplay with the gut bacterial transcriptional function more than composition [10]. As a microbial metabolite of sialyllactose [10], we noted that Neu5Ac was also intensely correlated with the metabolic function of gut microbiota. Gut bacterial lipid metabolism, especially bile acid metabolism, varied considerably with the level of breast milk Neu5Ac. Bile acids are some of the most important kinds of gut microbial metabolites. They are synthesized in the liver from cholesterol through two different metabolic pathways, namely, the classic pathway and the alternative pathway. Primary bile acids comprising CA and CDCA are products of the above two pathways, respectively [36]. CA and CDCA are conjugated to either taurine or glycine in the liver, secreted into the bile, and released into the duodenum after ingestion of food. Once in the gut, bile acids are transformed by gut microbiota to produce a wide variety of secondary bile acids [37]. The wide array of secondary bile acids in the HN group suggested that the gut microbiota of that group possessed more abundant bile acid metabolic function. Sulfonation of bile acids increases their solubility and enhances their fecal excretion, and elevated levels of sulfo-conjugated bile acids have been detected in fecal samples of breastfed infants [38,39]. The content of Neu5Ac in breast milk is much higher than that in formula milk [13], and the positive associations between breast milk Neu5Ac and sulfo-conjugated bile acids identified in our study suggested that higher Neu5Ac content in breast milk might contribute to the elevated levels of sulfo-conjugated bile acids in breastfed infants. Collectively, reduced levels of primary bile acids and elevated levels of secondary bile acids have been reported as a feature of healthy individuals [40,41], which was consistent with the characteristics of the fecal bile acid pool in the HN group infants, who displayed lower obesity risk.

Bile acids are key regulators in maintaining the energy balance and metabolic homeostasis of the host [42,43]. It has been reported that dysregulated bile acid metabolism is associated with growth faltering or obesity [44,45], and our study further proposed that the microbiota-derived bile acids might be the mediators in the complicated interaction between breast milk Neu5Ac and infant growth. In our study, several bile acids belonging to conjugated CDCA, DCA, and LCA were elevated in the HN group and were potentially related to reducing the obesity risk of the infants from that group. Consistently, a previous study has reported that levels of CDCA and DCA were correlated with activities of energy metabolism enzymes, such as gastrointestinal hormones, pancreatic peptide YY, and glucagon-like peptide-1 (GLP-1) [46]. Moreover, CDCA, DCA, and LCA have been identified as signaling molecules for the activation of farnesoid X receptor (FXR) [47]. By binding to the intestinal FXR, bile acids can induce the expression of endocrine hormone fibroblast growth factor 19 (FGF19), which suppresses lipogenesis and increases fatty acid oxidation in the liver, thereby regulating body weight gain and reducing the risk of obesity [48].

Although humans and mice metabolize Neu5Ac in different ways, and its main derivatives are also different [12,49], the aim of the study was to explore the association between Neu5Ac-related gut microbiota and infant growth, thus germ-free mice were used for colonization of infant gut microbiota whose breast milk has different levels of Neu5Ac. Germ-free mice are leaner than conventionally raised mice [50]. The healthy germ-free mice would gain weight after colonization with gut microbiota from infants [51]. Thus, the stable weight gain of HN mice indicated that mice transplanted with fecal bacteria from the HN infants exhibited a healthier growth, which is in line with the fact that the HN infants presented healthier growth due to a low risk of obesity. Therefore, the similar healthy growth of HN infants and HN mice provided additional evidence that the crosstalk between breast milk Neu5Ac and infant growth is gut microbiota-dependent. Further bile acid analysis of the gnotobiotic mice validated the idea that gut microbial bile acid metabolism is important in the weight regulation of the mice after FMT. As bile acid composition in mice is somewhat distinct from that in humans [52], a total of 14 bile acids identified in the infants were focused on in the FMT mice. A previous report showed that the cecum bile acid pool in mice was mainly composed of primary bile acids [53], and the majority of the bile acids identified in the mouse models of our study were taurine-primary bile acids. Conventionalization of germ-free C57BL/6 mice with a normal microbiota produces an increase in both body weight and body fat [51,54]. Our study consistently suggested that two taurine-conjugated bile acids in HN infants, TDCS and TCDCS, were linked with healthy weight gain in the mice after FMT. Bile acid regulation pathways in humans and mice are remarkably similar, that is, CDCA and DCA could engage FXR and activate expression of FGF15, thus having a myriad of other effects including regulation of lipogenesis and metabolic rate in mice [55]. Next, we found that two discriminative bacteria in the HN group, *Klebsiella* and *Parabacteroides*, were correlated with the level of the growth-regulating bile acids. It has been well established that *Parabacteroides* is capable of producing secondary bile acids [56] and thus may be able to alleviate obesity and metabolic dysfunctions [57]. However, no reports on the metabolism of bile acids by *Klebsiella* have been published so far. Therefore, we speculated that the correlation of *Klebsiella* with bile acids reflects that certain bacteria interacting with *Klebsiella* might be the primary metabolizers of bile acids, while *Klebsiella* benefits secondarily. Clearly, further investigation into the roles of *Klebsiella* and its related bacteria in bile acid metabolism is warranted.

However, this study still has several limitations. First, some of the infants included in this study were delivered by cesarean section or were subjected to mixed feeding, both of which are important factors affecting infant growth. Although adjusting for these factors in the regression model did not affect the correlation between breast milk Neu5Ac and infant growth in the two cohorts, future studies should validate the results of this study using vaginally delivered and breastfed infants. Second, our study focused only on Chinese cohorts, so it is unclear whether our findings can be generalized to infants in different countries. Next, despite our analysis of breast milk specimens from Chinese mothers revealing an association between Neu5Ac abundance and infant growth, additional time-series studies of mother–infant dyads from other cohorts are needed to determine the extent to which breast milk Neu5Ac-related variations in infant microbiota composition and function correlate with infant growth outcomes.

5. Conclusions

Our study demonstrated that breast milk Neu5Ac was associated with infant obesity risk in a gut microbiota-dependent manner. Breast milk Neu5Ac altered gut microbiota and reprogrammed bile acid metabolism, resulting in a distinct fecal bile acid profile in the HN group, which was characterized by reduced levels of primary bile acids and elevated levels of secondary bile acids. The conjugated DCA and CDCA were elevated in the HN group and positively correlated with reducing infant obesity risk. Especially, two sulfo- and taurine-conjugated bile acids, TDCS and TCDCS, were Neu5Ac-related and they were also helpful for healthy growth promotion, and the associations with healthy growth were reproduced in mice colonized with infant-derived microbiota. Finally, we proposed that *Parabacteroides* might be involved in bile acid metabolism and act as the mediator between Neu5Ac and infant growth. Additional studies are needed to clarify the mechanism of specific gut bacterial species and the correlated bile acids that contribute to early life growth and development. Our study might help to identify strategies to develop prebiotics to improve the growth outcomes of infants.

Supplementary Materials: The following supporting information can be downloaded at: https://www.mdpi.com/article/10.3390/metabo13070846/s1, Figure S1: The MS/MS spectra of Neu5Ac in standard sample and breast milk sample; Figure S2: Box plots of alpha diversity evaluated by Simpson and Pielou indexes between LN and HN group; Figure S3: Gut microbial composition of the gnotobiotic mice 2 weeks after fecal microbiota transplantation; Figure S4: Important bile acids from random forests for classifying low obesity risk and high obesity risk infants in the Zhengzhou cohort; Table S1: Clinical information of the donors for fecal microbiota transplantation; Table S2: Fold change of fecal bile acids between LN and HN infants; Table S3: Fold change of cecal bile acids between LN and HN mice; Table S4: Fold change of cecal bile acid conversion between LN and HN mice.

Author Contributions: Conceptualization, R.O., X.L. (Xinyu Liu) and G.X.; Data curation, R.O.; Formal analysis, R.O.; Funding acquisition, L.Z. and X.L. (Xinyu Liu); Investigation, J.D. and X.M.; Methodology, S.Z., X.W. and Q.L.; Resources, J.D., Z.Z., Z.L., Q.X., S.M. and Z.R.; Supervision, X.L. (Xin Lu), X.L. (Xinyu Liu) and G.X.; Visualization, R.O. and S.Z.; Writing—original draft, R.O.; Writing—review and editing, X.L. (Xinyu Liu) and G.X. All authors have read and agreed to the published version of the manuscript.

Funding: This research was funded by the Youth Science and Technology Star Project Support Program (No. 2020RQ067) from the Scientific and Technological Office of Dalian, the Youth Innovation Promotion Association of Chinese Academy of Sciences (No. 2021186), and the Innovation Program (DICP I202019) of Science and Research from Dalian Institute of Chemical Physics, Chinese Academy of Sciences.

Institutional Review Board Statement: The study was approved by the Ethics Committee of the First Affiliated Hospital of Zhengzhou University (2017-KY-12 and 26 September 2017).

Informed Consent Statement: Informed consent was obtained from all subjects involved in the study.

Data Availability Statement: The data for this study have been deposited in the European Nucleotide Archive (ENA) at EMBL-EBI under accession number PRJEB 57911.

Acknowledgments: We thank all the mothers and infants who participated in this study. We thank Petia Kovatcheva-Datchary for helpful suggestions on designing research and analyzing data.

Conflicts of Interest: The authors declare no conflict of interest.

References

1. Zucker, I.; Zloof, Y.; Bardugo, A.; Tsur, A.M.; Lutski, M.; Cohen, Y.; Cukierman-Yaffe, T.; Minsky, N.; Derazne, E.; Tzur, D.; et al. Obesity in late adolescence and incident type 1 diabetes in young adulthood. *Diabetologia* **2022**, *65*, 1473–1482. [CrossRef] [PubMed]
2. Umer, A.; Kelley, G.A.; Cottrell, L.E.; Giacobbi, P., Jr.; Innes, K.E.; Lilly, C.L. Childhood obesity and adult cardiovascular disease risk factors: A systematic review with meta-analysis. *BMC Public Health* **2017**, *17*, 683. [CrossRef]
3. World Health Organization. *WHO Child Growth Standards*; World Health Organization: Geneva, Switzerland, 2006.
4. Forbes, J.D.; Azad, M.B.; Vehling, L.; Tun, H.M.; Konya, T.B.; Guttman, D.S.; Field, C.J.; Lefebvre, D.; Sears, M.R.; Becker, A.B.; et al. Association of exposure to formula in the hospital and subsequent infant feeding practices with gut microbiota and risk of overweight in the first year of life. *JAMA Pediatr.* **2018**, *172*, e181161. [CrossRef] [PubMed]
5. NCD Risk Factor Collaboration (NCD-RisC). Worldwide trends in body-mass index, underweight, overweight, and obesity from 1975 to 2016: A pooled analysis of 2416 population-based measurement studies in 128·9 million children, adolescents, and adults. *Lancet* **2017**, *390*, 2627–2642. [CrossRef]
6. Stanislawski, M.A.; Dabelea, D.; Wagner, B.D.; Iszatt, N.; Dahl, C.; Sontag, M.K.; Knight, R.; Lozupone, C.A.; Eggesbo, M. Gut microbiota in the first 2 years of life and the association with body mass index at age 12 in a norwegian birth cohort. *mBio* **2018**, *9*, e01751-18. [CrossRef] [PubMed]
7. Alderete, T.L.; Jones, R.B.; Shaffer, J.P.; Holzhausen, E.A.; Patterson, W.B.; Kazemian, E.; Chatzi, L.; Knight, R.; Plows, J.F.; Berger, P.K.; et al. Early life gut microbiota is associated with rapid infant growth in Hispanics from Southern California. *Gut Microbes* **2021**, *13*, 1961203. [CrossRef]
8. Fehr, K.; Moossavi, S.; Sbihi, H.; Boutin, R.C.T.; Bode, L.; Robertson, B.; Yonemitsu, C.; Field, C.J.; Becker, A.B.; Mandhane, P.J.; et al. Breastmilk feeding practices are associated with the co-occurrence of bacteria in mothers' milk and the infant gut: The CHILD cohort study. *Cell Host Microbe* **2020**, *28*, 285–297.e284. [CrossRef]
9. Lis-Kuberka, J.; Orczyk-Pawilowicz, M. Sialylated oligosaccharides and glycoconjugates of human milk. The impact on infant and newborn protection, development and well-being. *Nutrients* **2019**, *11*, 306. [CrossRef]
10. Charbonneau, M.R.; O'Donnell, D.; Blanton, L.V.; Totten, S.M.; Davis, J.C.; Barratt, M.J.; Cheng, J.; Guruge, J.; Talcott, M.; Bain, J.R.; et al. Sialylated milk oligosaccharides promote microbiota-dependent growth in models of infant undernutrition. *Cell* **2016**, *164*, 859–871. [CrossRef]
11. Harris, J.E.; Pinckard, K.M.; Wright, K.R.; Baer, L.A.; Arts, P.J.; Abay, E.; Shettigar, V.K.; Lehnig, A.C.; Robertson, B.; Madaris, K.; et al. Exercise-induced 3'-sialyllactose in breast milk is a critical mediator to improve metabolic health and cardiac function in mouse offspring. *Nat. Metab.* **2020**, *2*, 678–687. [CrossRef]
12. Juge, N.; Tailford, L.; Owen, C.D. Sialidases from gut bacteria: A mini-review. *Biochem. Soc. Trans.* **2016**, *44*, 166–175. [CrossRef] [PubMed]
13. Wang, B.; Brand-Miller, J.; McVeagh, P.; Petocz, P. Concentration and distribution of sialic acid in human milk and infant formulas. *Am. J. Clin. Nutr.* **2001**, *74*, 510–515. [CrossRef] [PubMed]
14. Claumarchirant, L.; Sanchez-Siles, L.M.; Matencio, E.; Alegria, A.; Lagarda, M.J. Evaluation of sialic acid in infant feeding: Contents and bioavailability. *J. Agric. Food Chem.* **2016**, *64*, 8333–8342. [CrossRef] [PubMed]
15. Varki, A. Sialic acids in human health and disease. *Trends Mol. Med.* **2008**, *14*, 351–360. [CrossRef] [PubMed]
16. Ding, J.; Ouyang, R.; Zheng, S.; Wang, Y.; Huang, Y.; Ma, X.; Zou, Y.; Chen, R.; Zhuo, Z.; Li, Z.; et al. Effect of breastmilk microbiota and sialylated oligosaccharides on the colonization of infant gut microbial community and fecal metabolome. *Metabolites* **2022**, *12*, 1136. [CrossRef] [PubMed]
17. Chen, S.; Zhou, Y.; Chen, Y.; Gu, J. Fastp: An ultra-fast all-in-one fastq preprocessor. *Bioinformatics* **2018**, *34*, i884–i890. [CrossRef]
18. Li, H.; Durbin, R. Fast and accurate short read alignment with burrows-wheeler transform. *Bioinformatics* **2009**, *25*, 1754–1760. [CrossRef]
19. Li, D.; Liu, C.-M.; Luo, R.; Sadakane, K.; Lam, T.-W. Megahit: An ultra-fast single-node solution for large and complex metagenomics assembly via succinct de bruijn graph. *Bioinformatics* **2015**, *31*, 1674–1676. [CrossRef]
20. Noguchi, H.; Park, J.; Takagi, T. Metagene: Prokaryotic gene finding from environmental genome shotgun sequences. *Nucleic Acids Res.* **2006**, *34*, 5623–5630. [CrossRef]
21. Fu, L.; Niu, B.; Zhu, Z.; Wu, S.; Li, W. CD-HIT: Accelerated for clustering the next-generation sequencing data. *Bioinformatics* **2012**, *28*, 3150–3152. [CrossRef]
22. Li, R.; Li, Y.; Kristiansen, K.; Wang, J. SOAP: Short oligonucleotide alignment program. *Bioinformatics* **2008**, *24*, 713–714. [CrossRef]
23. Qin, J.; Li, Y.; Cai, Z.; Li, S.; Zhu, J.; Zhang, F.; Liang, S.; Zhang, W.; Guan, Y.; Shen, D.; et al. A metagenome-wide association study of gut microbiota in type 2 diabetes. *Nature* **2012**, *490*, 55–60. [CrossRef]
24. Buchfink, B.; Xie, C.; Huson, D.H. Fast and sensitive protein alignment using diamond. *Nat. Methods* **2015**, *12*, 59–60. [CrossRef]
25. Bäckhed, F.; Roswall, J.; Peng, Y.; Feng, Q.; Jia, H.; Kovatcheva-Datchary, P.; Li, Y.; Xia, Y.; Xie, H.; Zhong, H.; et al. Dynamics and stabilization of the human gut microbiome during the first year of life. *Cell Host Microbe* **2015**, *17*, 690–703. [CrossRef]
26. Segata, N.; Izard, J.; Waldron, L.; Gevers, D.; Miropolsky, L.; Garrett, W.S.; Huttenhower, C. Metagenomic biomarker discovery and explanation. *Genome Biol.* **2011**, *12*, R60. [CrossRef]
27. Parks, D.H.; Tyson, G.W.; Hugenholtz, P.; Beiko, R.G. STAMP: Statistical analysis of taxonomic and functional profiles. *Bioinformatics* **2014**, *30*, 3123–3124. [CrossRef]

28. Saben, J.L.; Sims, C.R.; Abraham, A.; Bode, L.; Andres, A. Human milk oligosaccharide concentrations and infant intakes are associated with maternal overweight and obesity and predict infant growth. *Nutrients* **2021**, *13*, 446. [CrossRef]
29. Chen, Z.; Chang, Y.; Liu, H.; You, Y.; Liu, Y.; Yu, X.; Dou, Y.; Ma, D.; Chen, L.; Tong, X.; et al. Distribution and influencing factors of the sialic acid content in the breast milk of preterm mothers at different stages. *Front. Nutr.* **2022**, *9*, 753919. [CrossRef]
30. Han, S.M.; Derraik, J.G.B.; Binia, A.; Sprenger, N.; Vickers, M.H.; Cutfield, W.S. Maternal and infant factors influencing human milk oligosaccharide composition: Beyond maternal genetics. *J. Nutr.* **2021**, *151*, 1383–1393. [CrossRef]
31. Armstrong, J.; Reilly, J.J. Breastfeeding and lowering the risk of childhood obesity. *Lancet* **2002**, *359*, 2003–2004. [CrossRef]
32. Quecke, B.; Graf, Y.; Epure, A.-M.; Santschi, V.; Chiolero, A.; Carmeli, C.; Cullati, S. Caesarean section and obesity in young adult offspring: Update of a systematic review with meta-analysis. *Obes. Rev.* **2022**, *23*, e13368. [CrossRef] [PubMed]
33. Gal, B.; Ruano, M.J.; Puente, R.; García-Pardo, L.A.; Rueda, R.; Gil, A.; Hueso, P. Developmental changes in UDP-N-acetylglucosamine 2-epimerase activity of rat and guinea-pig liver. *Comp. Biochem. Physiol. B Biochem. Mol. Biol.* **1997**, *118*, 13–15. [CrossRef] [PubMed]
34. Kiyohara, M.; Tanigawa, K.; Chaiwangsri, T.; Katayama, T.; Ashida, H.; Yamamoto, K. An exo-alpha-sialidase from *bifidobacteria* involved in the degradation of sialyloligosaccharides in human milk and intestinal glycoconjugates. *Glycobiology* **2011**, *21*, 437–447. [CrossRef] [PubMed]
35. McDonald, N.D.; Lubin, J.B.; Chowdhury, N.; Boyd, E.F. Host-derived sialic acids are an important nutrient source required for optimal bacterial fitness in vivo. *mBio* **2016**, *7*, e02237-15. [CrossRef] [PubMed]
36. Hamilton, J.P.; Xie, G.; Raufman, J.-P.; Hogan, S.; Griffin, T.L.; Packard, C.A.; Chatfield, D.A.; Hagey, L.R.; Steinbach, J.H.; Hofmann, A.F. Human cecal bile acids: Concentration and spectrum. *Am. J. Physiol. Gastrointest. Liver Physiol.* **2007**, *293*, G256–G263. [CrossRef]
37. Guzior, D.V.; Quinn, R.A. Review: Microbial transformations of human bile acids. *Microbiome* **2021**, *9*, 140. [CrossRef]
38. Alnouti, Y. Bile acid sulfation: A pathway of bile acid elimination and detoxification. *Toxicol. Sci.* **2009**, *108*, 225–246. [CrossRef]
39. Sillner, N.; Walker, A.; Lucio, M.; Maier, T.V.; Bazanella, M.; Rychlik, M.; Haller, D.; Schmitt-Kopplin, P. Longitudinal profiles of dietary and microbial metabolites in formula- and breastfed infants. *Front. Mol. Biosci.* **2021**, *8*, 660456. [CrossRef]
40. Franzosa, E.A.; Sirota-Madi, A.; Avila-Pacheco, J.; Fornelos, N.; Haiser, H.J.; Reinker, S.; Vatanen, T.; Hall, A.B.; Mallick, H.; McIver, L.J.; et al. Gut microbiome structure and metabolic activity in inflammatory bowel disease. *Nat. Microbiol.* **2019**, *4*, 293–305. [CrossRef]
41. Wang, Y.; Gao, X.; Zhang, X.; Xiao, Y.; Huang, J.; Yu, D.; Li, X.; Hu, H.; Ge, T.; Li, D.; et al. Gut microbiota dysbiosis is associated with altered bile acid metabolism in infantile cholestasis. *mSystems* **2019**, *4*, e00463-19. [CrossRef]
42. Jia, W.; Xie, G.; Jia, W. Bile acid-microbiota crosstalk in gastrointestinal inflammation and carcinogenesis. *Nat. Rev. Gastroenterol. Hepatol.* **2017**, *15*, 111–128. [CrossRef]
43. Perino, A.; Schoonjans, K. Metabolic messengers: Bile acids. *Nat. Metab.* **2022**, *4*, 416–423. [CrossRef] [PubMed]
44. Zhao, X.; Setchell, K.D.R.; Huang, R.; Mallawaarachchi, I.; Ehsan, L.; Dobrzykowski, E., III; Zhao, J.; Syed, S.; Ma, J.Z.; Iqbal, N.T.; et al. Bile acid profiling reveals distinct signatures in undernourished children with environmental enteric dysfunction. *J. Nutr.* **2021**, *151*, 3689–3700. [CrossRef] [PubMed]
45. So, S.S.Y.; Yeung, C.H.C.; Schooling, C.M.; El-Nezami, H. Targeting bile acid metabolism in obesity reduction: A systematic review and meta-analysis. *Obes. Rev.* **2020**, *21*, e13017. [CrossRef]
46. Roberts, R.E.; Glicksman, C.; Alaghband-Zadeh, J.; Sherwood, R.A.; Akuji, N.; le Roux, C.W. The relationship between postprandial bile acid concentration, GLP-1, PYY and ghrelin. *Clin. Endocrinol.* **2011**, *74*, 67–72. [CrossRef]
47. de Aguiar Vallim, T.Q.; Tarling, E.J.; Edwards, P.A. Pleiotropic roles of bile acids in metabolism. *Cell Metab.* **2013**, *17*, 657–669. [CrossRef]
48. Fang, S.; Suh, J.M.; Reilly, S.M.; Yu, E.; Osborn, O.; Lackey, D.; Yoshihara, E.; Perino, A.; Jacinto, S.; Lukasheva, Y.; et al. Intestinal FXR agonism promotes adipose tissue browning and reduces obesity and insulin resistance. *Nat. Med.* **2015**, *21*, 159–165. [CrossRef] [PubMed]
49. Varki, A. Uniquely human evolution of sialic acid genetics and biology. *Proc. Natl. Acad. Sci. USA* **2010**, *107* (Suppl. 2), 8939–8946. [CrossRef]
50. Moretti, C.H.; Schiffer, T.A.; Li, X.; Weitzberg, E.; Carlstrom, M.; Lundberg, J.O. Germ-free mice are not protected against diet-induced obesity and metabolic dysfunction. *Acta Physiol.* **2021**, *231*, e13581. [CrossRef]
51. Hiltunen, H.; Hanani, H.; Luoto, R.; Turjeman, S.; Ziv, O.; Isolauri, E.; Salminen, S.; Koren, O.; Rautava, S. Preterm infant meconium microbiota transplant induces growth failure, inflammatory activation, and metabolic disturbances in germ-free mice. *Cell Rep. Med.* **2021**, *2*, 100447. [CrossRef]
52. Honda, A.; Miyazaki, T.; Iwamoto, J.; Hirayama, T.; Morishita, Y.; Monma, T.; Ueda, H.; Mizuno, S.; Sugiyama, F.; Takahashi, S.; et al. Regulation of bile acid metabolism in mouse models with hydrophobic bile acid composition. *J. Lipid Res.* **2020**, *61*, 54–69. [CrossRef]
53. Collins, S.L.; Stine, J.G.; Bisanz, J.E.; Okafor, C.D.; Patterson, A.D. Bile acids and the gut microbiota: Metabolic interactions and impacts on disease. *Nat. Rev. Microbiol.* **2023**, *21*, 236–247. [CrossRef]
54. Bäckhed, F.; Ding, H.; Wang, T.; Hooper, L.V.; Koh, G.Y.; Nagy, A.; Semenkovich, C.F.; Gordon, J.I. The gut microbiota as an environmental factor that regulates fat storage. *Proc. Natl. Acad. Sci. USA* **2004**, *101*, 15718–15723. [CrossRef]
55. Li, T.; Chiang, J.Y.L. Bile acid signaling in metabolic disease and drug therapy. *Pharmacol. Rev.* **2014**, *66*, 948–983. [CrossRef]

56. Li, M.; Wang, S.; Li, Y.; Zhao, M.; Kuang, J.; Liang, D.; Wang, J.; Wei, M.; Rajani, C.; Ma, X.; et al. Gut microbiota-bile acid crosstalk contributes to the rebound weight gain after calorie restriction in mice. *Nat. Commun.* **2022**, *13*, 2060. [CrossRef]
57. Wang, K.; Liao, M.; Zhou, N.; Bao, L.; Ma, K.; Zheng, Z.; Wang, Y.; Liu, C.; Wang, W.; Wang, J.; et al. *Parabacteroides distasonis* alleviates obesity and metabolic dysfunctions via production of succinate and secondary bile acids. *Cell Rep.* **2019**, *26*, 222–235. [CrossRef]

Disclaimer/Publisher's Note: The statements, opinions and data contained in all publications are solely those of the individual author(s) and contributor(s) and not of MDPI and/or the editor(s). MDPI and/or the editor(s) disclaim responsibility for any injury to people or property resulting from any ideas, methods, instructions or products referred to in the content.

Article

Anti-Allergic Effect of Dietary Polyphenols Curcumin and Epigallocatechin Gallate via Anti-Degranulation in IgE/Antigen-Stimulated Mast Cell Model: A Lipidomics Perspective

Jun Zeng [1,2,*,†], Jingwen Hao [1,†], Zhiqiang Yang [1,†], Chunyu Ma [1], Longhua Gao [1], Yue Chen [3], Guiling Li [1,2] and Jia Li [4,*]

1. College of Ocean Food and Biological Engineering, Jimei University, Xiamen 361021, China
2. Xiamen Key Laboratory of Marine Functional Food, Xiamen 361021, China
3. The Affiliated Stomatology Hospital, School of Medicine, Zhejiang University, Hangzhou 310000, China
4. Key Laboratory of Tea Biology and Resources Utilization, Ministry of Agriculture, Tea Research Institute, Chinese Academy of Agricultural Sciences, Hangzhou 310008, China
* Correspondence: junzeng@jmu.edu.cn (J.Z.); jiali1986@tricaas.com (J.L.); Tel.: +86-592-6181487 (J.Z.); +86-571-86650637 (J.L.)
† These authors contributed equally to this work.

Abstract: Polyphenol-rich foods exhibit anti-allergic/-inflammatory properties. As major effector cells of allergies, mast cells undergo degranulation after activation and then initiate inflammatory responses. Key immune phenomena could be regulated by the production and metabolism of lipid mediators by mast cells. Here, we analyzed the antiallergic activities of two representative dietary polyphenols, curcumin and epigallocatechin gallate (EGCG), and traced their effects on cellular lipidome rewiring in the progression of degranulation. Both curcumin and EGCG significantly inhibited degranulation as they suppressed the release of β-hexosaminidase, interleukin-4, and tumor necrosis factor-α from the IgE/antigen-stimulated mast cell model. A comprehensive lipidomics study involving 957 identified lipid species revealed that although the lipidome remodeling patterns (lipid response and composition) of curcumin intervention were considerably similar to those of EGCG, lipid metabolism was more potently disturbed by curcumin. Seventy-eight percent of significant differential lipids upon IgE/antigen stimulation could be regulated by curcumin/EGCG. LPC-O 22:0 was defined as a potential biomarker for its sensitivity to IgE/antigen stimulation and curcumin/EGCG intervention. The key changes in diacylglycerols, fatty acids, and bismonoacylglycerophosphates provided clues that cell signaling disturbances could be associated with curcumin/EGCG intervention. Our work supplies a novel perspective for understanding curcumin/EGCG involvement in antianaphylaxis and helps guide future attempts to use dietary polyphenols.

Keywords: allergy; dietary polyphenols; curcumin; EGCG; lipidomics

1. Introduction

Allergic diseases, including food allergies, allergic asthma, rhinitis, and dermatitis, cause significant morbidity worldwide [1,2]. Food allergies, for instance, occur in up to 10% of the worldwide population and are associated with an increasing prevalence every year [2]. Allergy is a serious public health and food safety concern.

Dietary components have gained increasing attention in recent years for their ability to prevent and alleviate allergic responses [2,3]. As the main component in fruits, vegetables, and other edible plant parts, polyphenols and polyphenol-rich foods have been reported to exhibit anti-allergic/-inflammatory properties [2,3]. Dietary polyphenols, such as curcumin (turmeric) and epigallocatechin gallate (EGCG, green tea), are emerging topics of interest and research, as these compounds exhibit potential benefits for human health [4]. Polyphenols may interfere with allergic reactions by inhibiting the release of chemical mediators (histamine, hexosaminidase, or leukotrienes), cytokine production, signal transduction, and gene expression in mast cells, basophils, or T cells [2]. Furthermore, the formation of complexes between polyphenols and proteins, such as peanut and cashew proteins, has been shown to prevent antibody recognition of allergens via allergen precipitation and reduce IgE binding to allergens [5–7]. Plant-based proteins (PP)–phenolic compounds (PC) conjugates and complexes have been reported to exhibit potential allergy-reducing activities [8]. The regulation of gut microbiota by polyphenols might also contribute to their anti-allergic/-inflammatory properties [2]. An increasing number of clinical and epidemiological studies have provided evidence that dietary polyphenol consumption and a reduction in risk factors for chronic allergic diseases are correlated [3]. Although in vivo and in vitro studies have indicated polyphenols' anti-allergic/-inflammatory effects, knowledge of their mechanism of action remains incomplete. A deeper understanding of pathogenic mechanisms is needed to explore promising biomarkers of allergic diseases as well as identify the involvement of polyphenols in cellular and molecular events central to antianaphylaxis.

The immune mechanism underlying allergic disorders encompasses an adaptive Th2-type response [2]. Antigen-specific immunoglobulin E (IgE) antibodies, together with major effector cells of allergies (i.e., the mast cell), are crucial for the development of allergic disorders [9]. Mast cells express the high-affinity IgE receptor FcεRI on their surface [10]. Antigen stimulation activates mast cells sensitized with IgE [1]. Mast cells undergo degranulation after activation, initiate an acute inflammatory response, and contribute to the progression of chronic diseases [1]. The degranulation of mast cells results in the release of β-hexosaminidase, a common degranulation marker, histamine, inflammatory cytokines, and lipid-derived mediators [10]. Mast cells are well-known producers of different lipid mediators [9]. Currently, the production and metabolism of these lipid mediators have in turn been shown to regulate mast cell functions in an increasing number of studies [9,11]. Bioactive lipids of leukotrienes, prostanoids, platelet-activating factor (PAF), and sphingolipids have been reported to influence cell signaling via multiple mechanisms [12], including by formatting structural support platforms (lipid rafts) for receptor signaling complexes, by transducing signals as primary/secondary messengers, and by serving as kinase/phosphatase cofactors [12]. Then, bioactive lipids could remodel key immune phenomena (degranulation, chemotaxis, and sensitization). In rapidly emerging research, the modulation of mast cell reactivity by lipid metabolism, in addition to proteins, is revealing novel and unprecedented targets [9]. These targets may serve to preclude mast cell effects in allergic reactions. Thus, it is important to gain more comprehensive insights into the effects of lipidome remodeling on mast cell degranulation, changes in cellular lipid composition induced by allergens, and changes in lipid transport and metabolism in mast cells. High-throughput lipidomics is an emerging analytical strategy that enables a wide range of lipids to be explored on the scale of individual lipid molecular species, supplying a global and detailed map of the lipidome response to external stimulation [13]. To the best of our knowledge, on the scale of the lipidome, the characterization of mast cell lipid composition and stimulus-specific changes upon polyphenol interventions remains limited.

Turmeric, the major source of curcumin, is a spice that has been traditionally used in Asian countries for culinary purposes. As a natural polyphenol derivative, curcumin got approved to be "generally recognized as safe" (GRAS) by the US Food and Drug Administration (FDA) [14]. The tea plant (*Camellia sinensis*) is native to East Asia and has traditionally been consumed worldwide as "tea". Green tea is also a rich source of the natural polyphenol EGCG, which is present most abundantly [15]. Curcumin and EGCG were suggested to exhibit anti-allergic/-inflammatory potential in previous reports [3,14] and our preliminary studies. The compounds were defined as two typical dietary polyphenols in this study to analyze and compare their mechanisms of action underlying the progression of mast cell degranulation. As an in vitro mast cell model, the basophilic leukemia (RBL-2H3) cell line has been successfully applied in previous studies to investigate IgE-FcεRI interactions and degranulation, as well as screen antiallergy drug candidates [16,17]. According to previous studies that traced the release of mediators from mast cells, the allergic reaction progresses along the following time course: the immediate phase (within 1 h of allergen challenge) and the later phase (after 3–48 h) [18]. Hence, in this study, curcumin/EGCG intervention in the progression of degranulation was explored at the interfacial stage of the allergic reaction (i.e., 1 h and 3 h). Specially, we investigated comprehensive lipidome rewiring in an IgE/antigen (i.e., dinitrophenyl-bovine serum albumin, DNP-BSA)-stimulated RBL-2H3 degranulation model using a nontargeted lipidomics approach based on ultra-high-performance liquid chromatography coupled to mass spectrometry (UPLC—MS). A sample set of cells from the control groups and curcumin/EGCG intervention groups was traced at both 1 h and 3 h for lipidomics investigation. Our work provides a novel perspective on understanding the action of antigen stimulation and curcumin/EGCG involvement in antianaphylaxis.

2. Materials and Methods

2.1. Cell Culture and Cell Viability

RBL-2H3 cells were purchased from Procell Life Science & Technology Co., Ltd. (Wuhan, China) and cultured in minimal essential medium (MEM) with 15% heat-inactivated fetal bovine serum (FBS), 100 units/mL penicillin, and 100 μg/mL streptomycin at 37 °C in a humidified incubator (5% CO_2). The cytotoxic effects of curcumin and EGCG were evaluated by MTT assay (CellTiter 96 Aqueous One Solution Cell Proliferation Assay; Promega, Solarbio, Beijing, China) (n = 6).

2.2. Sample Collection

Chemicals and reagents are provided in the Supporting Information. First, RBL-2H3 cells were incubated with 200 ng/mL anti-DNP-IgE for 18 h. After washing with phosphate-buffered saline (PBS) three times, the IgE-sensitized cells were exposed to 10 μM curcumin (i.e., Cur group) or 200 μM EGCG (i.e., EGCG group) and then stimulated with 500 ng/mL DNP-BSA simultaneously. The coincubation of IgE-sensitized cells with curcumin/EGCG and DNP-BSA was performed for 1 h or 3 h. A vehicle control group without both IgE and DNP-BSA was set up in parallel with the experiment (i.e., Veh group), as well as a positive control group prepared by incubation of IgE-sensitized cells with the DNP-BSA antigen (i.e., AG group).

The structures of curcumin and EGCG and the scheme of this experimental design are shown in Figure 1A–C. To monitor the curcumin/EGCG intervention on IgE-mediated degranulation, cell specimens from control groups (incl. Veh and AG) and intervention groups (incl. Cur and EGCG) were collected at both 1 h and 3 h. Four independent biological replicates of RBL-2H3 cells were prepared for each group at each time point for the lipidomics study.

2.3. Nontargeted Lipidomics Study

Each sample was placed in a 10 cm Petri dish with 8 mL of cell media (about 10^7 cells). The cell medium was completely aspirated after sample collection, followed by washing with Dulbecco's phosphate-buffered saline (DPBS) solution and inactivation with liquid nitrogen immediately.

After lipidome extraction, freeze-dried lipid extracts were reconstituted in 10 μL of dichlormethane:methanol (2:1) and then diluted five times in ACN:isopropanol:water (65:30:5). Finally, 50 μL of cell lipid extracts was analyzed using an UltiMate 3000 UPLC system (Thermo, Waltham, MA, USA) coupled with a quadrupole Orbitrap mass spectrometer (Q-Exactive, Thermo, USA). LC separation was performed using a BEH C8 column (2.1 × 100 mm, 1.7 μm) (Waters, Milford, MA, USA). Full-scan MS for lipid profiling and data-dependent MS/MS (ddMS2) for lipid identification were performed in both positive and negative electrospray (ESI) ion modes.

Details of the lipidomics study, including lipidome extraction and lipidomics analysis by UPLC-Q-Exactive MS, are provided in the Supporting Information and were adopted from our previously published method [19–21]. Quality control (QC) samples, which were generated by pooling equal aliquots of lipid extracts from each sample, were prepared as real samples and regularly inserted into the analysis sequence to monitor the robustness of lipidomic analysis.

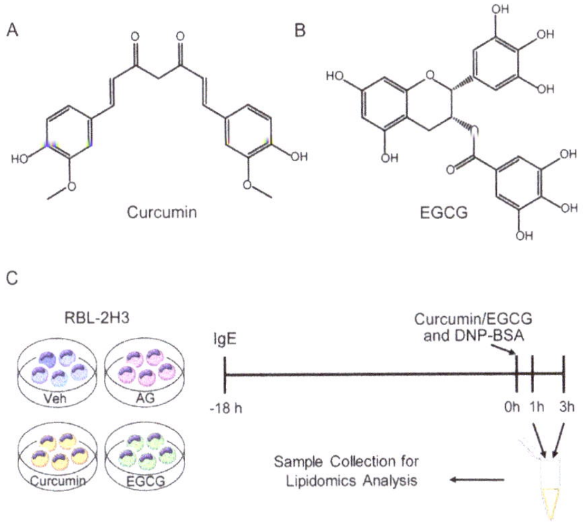

Figure 1. Characteristics of the experimental model. (**A**) Structure of curcumin. (**B**) Structure of EGCG. (**C**) Scheme of the experimental design. Both curcumin and EGCG inhibit IgE-mediated degranulation in the RBL-2H3 cell model. Veh: vehicle control group; AG: IgE/antigen stimulation group; curcumin: curcumin intervention group; EGCG: EGCG intervention group.

2.4. Data Processing and Statistics

Lipid species were identified according to accurate m/z, tandem mass spectrometry (MS/MS) fragmentation patterns, and retention behavior. The LIPID MAPS database (http://www.lipidmaps.org/, accessed on 1 January 2023) and MS-DIAL software (http://prime.psc.riken.jp/compms/msdial/main.html, accessed on 1 January 2023) were used for lipid queries. For the quantification of these identified lipids, peak areas were obtained by high-resolution extracted ion chromatogram using Trace Finder software (Thermo, USA). Two thresholds, both m/z and retention time, were applied to the extraction process of peak area (an m/z tolerance of ± 5 ppm and a retention time extraction window of ± 15 s). Peak checking and noise removal were carried out to reduce errors.

To eliminate systematic bias, the peak area of each lipid species was normalized to the total intensity of all lipid species in a given sample. Prior to statistical analysis, lipids with a percentage relative standard deviation (%RSD) higher than 30% in all QCs were removed from the dataset. Then, the dataset was subjected to SIMCA-P 11.0.0.0 software (Umetrics, Malmö, Sweden) for principal component analysis (PCA) and partial least squares discriminant analysis (PLS-DA) with unit variance (UV) scaling. To assess the univariate statistical significance, two-way analysis of variance (ANOVA), Wilcoxon Mann–Whitney test, and false discovery rate (FDR) correction (Benjamini–Hochberg method) were employed using Multi Experiment Viewer (MeV) software (open-source genomic analysis software, version 4.9.0) and an in-house-developed MATLAB program (The MathWorks, Natick, MA, USA). On the basis of hierarchical cluster analysis (HCA), a heatmap was also generated with MeV to visualize the relative levels of lipids. Receiver operating characteristic curve (ROC) and binary logistic regression were performed using SPSS Statistics software (SPSS Inc., Chicago, DE, USA).

The abbreviations used in this study for lipid classes are as follows: (Hex)Cer, (hexosyl)ceramide; SM, sphingomyelin; CE, cholesterol esters; (L)PC, (lyso)phosphatidylcholine; (L)PE, (lyso)phosphatidylethanolamine; OxPE, oxidized phosphatidylethanolamine; (L)PG, (lyso)phosphatidylglycerol; (L)PI, (lyso)phosphatidylinositol; (L)PS, (lyso)phosphatidylserine; (L)PA, (lyso)phosphatidic acid; PEtOH, phosphatidylethanol; DG, diacylglycerol; TG, triacylglycerol; CL, cardiolipin; CoQ, coenzyme Q; ASM, acylsphingomyelin; NAE, N-acyl ethanolamines; GM3, ganglioside GM3; FA, fatty acid; OxFA, oxidized fatty acid; CAR, acylcarnitine; (H)BMP, (hemi)bismonoacylglycerophosphate; ST, sterol; o/p-, ether and plasmalogen.

2.5. β-Hexosaminidase Release Assay

β-Hexosaminidase activity in culture supernatants was measured as an indicator of degranulation [10]. The amount of β-hexosaminidase released from RBL-2H3 cells was quantified according to previous reports with slight modifications (n = 3) [22].

Briefly, after the coincubation of IgE-sensitized cells with curcumin/EGCG and DNP-BSA, the supernatant was collected and centrifuged at 1500 rpm for 5 min, while the cells were incubated in Tyrode's buffer containing 1% Triton X-100 for 5 min. The supernatant and cell lysate were transferred to 96-well black microplates (25 µL/well) and then incubated with 1.2 mM 4-methylumbelliferyl-N-acetyl-β-D-glucosamincide dissolved in 0.1 M citrate buffer (pH 4.5) at 37 °C for 30 min (100 µL/well). The fluorescence intensity was measured at 450 nm with a microplate reader. The β-hexosaminidase release (%) and inhibition of release (%) were calculated as follows:

$$\beta\text{hexosaminidase release}(\%) = \frac{F_{\text{supernatant}}}{F_{\text{supernatant}} + F_{\text{cell lysate}}} \times 100 \quad (1)$$

$$\text{Inhibition of release }(\%) = \frac{AG - \text{Intervention}}{AG - \text{Veh}} \times 100 \quad (2)$$

where F in Equation (1) is the fluorescence intensity. AG, Veh, and intervention in Equation (2) refer to the β-hexosaminidase release (%) of the AG, Veh, and intervention groups.

2.6. TNF-α and IL-4 Release Assay

To determine the tumor necrosis factor-α (TNF-α) and interleukin-4 (IL-4) concentrations in the culture media, all samples were centrifuged (17,000× g, 10 min) at 4 °C and stored at −80 °C until analysis. Then, the levels of TNF-α and IL-4 were measured using ELISA kits (Elabscience, Wuhan, China), in accordance with the manufacturers' instructions (n = 3).

3. Results

3.1. Inhibitory Effect of Curcumin/EGCG on IgE-Mediated Degranulation

The release of β-hexosaminidase was first measured as a general indicator of degranulation and a hallmark characteristic of allergic reactions upon allergen stimulation [10]. The modeling with 200 ng/mL anti-DNP-IgE for sensitization and the coincubation of DNP-BSA and curcumin/EGCG for stimulation was defined in our preliminary studies for better inhibition of β-hexosaminidase release (Figures 2A,B and S1A–C).

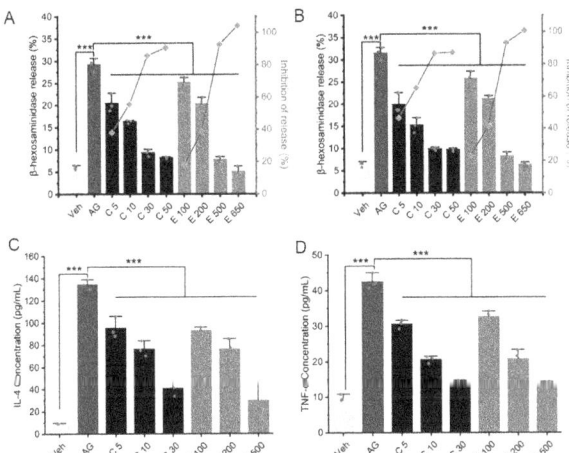

Figure 2. Both curcumin and EGCG inhibit IgE-mediated degranulation in the RBL-2H3 cell model. (**A**,**B**) exhibit β-hexosaminidase release at 1 h and 3 h, respectively. The β-hexosaminidase release (%) is represented by each column, and the line reveals the inhibition of release (%). (**C**) IL-4 release at 3 h. (**D**) TNF-α release at 3 h. All data are presented as the mean ± SD. Individuals in each group are represented by the red dot in each column. C5, C10, C30, and C50 denote intervention groups in which IgE/antigen-stimulated cells were treated with 5, 10, 30 and 50 μM curcumin; E100, E200, E500, and E650 denote intervention groups in which IgE/antigen-stimulated cells were treated with 100, 200, 500, and 650 μM EGCG. ***: $p < 0.001$.

To assess the effect of curcumin/EGCG on the IgE-mediated allergic response, cell morphology and the release of two representative proinflammatory cytokines (TNF-α and IL-4) were also analyzed, in accordance with the evaluation of β-hexosaminidase release. RBL-2H3 cells from the Veh group displayed fibroblastic morphology (Figure S2). Activating RBL-2H3 cells by an IgE–antigen complex induced cell swelling, and significantly improved the levels of β-hexosaminidase, IL-4, and TNF-α (Figures S2 and 2). After curcumin/EGCG treatment, the activated cells exhibited improved morphology, and the release of β-hexosaminidase, IL-4, and TNF-α was significantly suppressed in a dose-dependent manner (Figures S2 and 2).

Cell viability was not obviously affected at concentrations less than 10 μM curcumin and 200 μM EGCG (Figure S1D,E). Then, 10 μM curcumin and 200 μM EGCG were used in subsequent lipidomics studies. We observed that the suppression of β-hexosaminidase release by 10 μM curcumin (i.e., 55.72% and 65.27% inhibition of release for 1 h and 3 h, respectively) was more potent than that by 200 μM EGCG (i.e., 38.61% and 41.24% inhibition of release for 1 h and 3 h, respectively; Figure 2A,B). Compared with that at 1 h, the percent inhibition of β-hexosaminidase release at 3 h increased, indicating the progression of curcumin/EGCG intervention. Interestingly, the decrease in proinflammatory cytokine production induced by 10 μM curcumin and 200 μM EGCG was similar. These results confirmed that the inhibition of IgE-mediated RBL-2H3 degranulation by curcumin/EGCG was successfully implemented in this study.

3.2. Lipidome of RBL-2H3 Cells

To trace curcumin/EGCG action in the progression of degranulation, RBL-2H3 cells from the control groups (incl. Veh and AG) and intervention groups (incl. Cur and EGCG) were analyzed at both 1 h and 3 h for lipidomics investigation. The large-scale lipidomics profiling of RBL-2H3 cells revealed approximately 1800 lipid features in a nontargeted pattern (Figure S3). A total of 957 lipid species were finally identified, including 75 fatty acyls, 161 glycerolipids, 582 glycerophospholipids, 123 sphingolipids, 14 sterols, and

2 prenols (Tables S1, S2 and Figure S4). This profiling revealed that the chemical structures, compositions, and polarities of the cellular lipidome were largely diverse and complex.

The reliability and robustness of the acquired lipidomics data were investigated by evaluating QC samples and confirmed to be satisfactory for complex biological samples (Figure S5). Detailed information on the identified lipids and QC evaluation is described in the Supporting Information.

3.3. Global Profiling of Lipidome Disturbance

Global profiling of the cellular lipidome was visualized by unsupervised PCA. Two types of metabolic disturbance, time-related and treatment-induced changes, were clearly visible on the score plot (Figure 3A). Cell specimens collected at 1 h and 3 h were presented on the two sides of the PCA score plot. The control (i.e., Veh and AG) and different intervention groups (i.e., Cur and EGCG) showed a clear trend of discrimination along the second principal component. An overview of these lipidome differences was further quantified by analyzing the Euclidean distance between the AG group and other groups at each time point (Figure 3B).

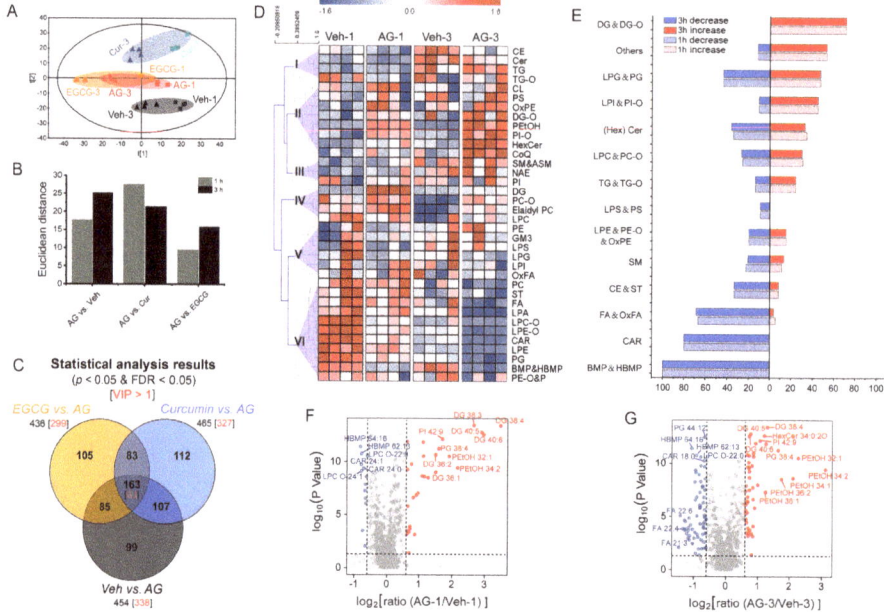

Figure 3. Global profiling of lipidome disturbance and lipidome changes associated with IgE-mediated degranulation. In lipidomics sampling, 10 μM curcumin and 200 μM EGCG were defined based on cell viability results. (**A**) PCA score plot for the classification of vehicle (Veh-1 and Veh-3), IgE/antigen stimulation (AG-1 and AG-3), curcumin (Cur-1 and Cur-3), and EGCG (EGCG-1 and EGCG-3) groups at both 1 h and 3 h. (**B**) Euclidean distance. (**C**) Venn diagram for an overview of the statistical results. The black numbers mean lipids with $p < 0.05$ and FDR < 0.05. The red numbers mean lipids with $p < 0.05$, FDR < 0.05, and VIP > 1. (**D**) Heatmap of each lipid class. The sum of the relative responses from each lipid class was UV scaled and subjected to hierarchical clustering. (**E**) Percentage of significantly differential lipids in response to IgE-mediated degranulation (Veh vs. AG, $p < 0.05$ and FDR < 0.05). The number of lipid species that were either significantly up- or downregulated was normalized to the total number of lipids detected in each family. (**F**,**G**) volcano plots for the comparisons at 1 h and 3 h, respectively. The red (or blue) dot denotes the lipid with $p < 0.05$ and a ratio more than 3/2 (or less than 2/3).

To specify significant differential lipids in response to allergen stimulation and curcumin/EGCG intervention, the univariate statistical significance of lipids was evaluated ($p < 0.05$ and FDR < 0.05). Three PLS-DA models (i.e., Veh vs. AG, curcumin vs. AG, and EGCG vs. AG, Figures S6–S8) were developed to screen significant differential lipids based on variable importance in the projection (VIP) values (VIP > 1). Lipids with multivariate and univariate statistical importance in the classification were cross-refined (i.e., the intersection, Figure 3C) and then assumed to be representative differential characteristics.

3.4. Lipidome Changes Associated with IgE-Mediated Degranulation

The comparison between the AG and Veh groups indicated that the cellular lipid metabolism was rewired during the progression of IgE-mediated degranulation (Figures 3A and S6). The classification of PCA showed robust lipidome disturbances in response to allergen stimulation at both 1 h and 3 h (Figure 3A), and the progression of degranulation was supported by the more evident metabolic changes at 3 h when compared with 1 h (Figure 3D). A total of 454 significant differential lipids changed by degranulation were discovered (Veh vs. AG, $p < 0.05$ and FDR < 0.05, Figure 3C, Table S3).

The sum of responses in each lipid class was analyzed to cluster into six major groups in a heatmap according to the similarity of variation tendencies (Figure 3D). Enrichment of changes in lipid classes was pinpointed by normalizing the number of significant differential lipids changed by degranulation to the total number of lipids detected in each family (Figure 3E). DG(-O) (incl. DG and DG-O), manifesting a greater response in AG at both 1 h and 3 h, were found to be the top lipid category associated with significant upregulation upon allergen stimulation (Figure 3E). In contrast, (H)BMP (incl. BMP and HBMP), CAR, and (Ox)FA (incl. FA and OxFA) were discovered to be the top lipid classes associated with significant downregulation in AG (Figure 3D,E). Notably, the percentage of either increased or decreased lipids with significant differences was similar between 1 h and 3 h (Figure 3E), indicating that common lipidome change patterns are shared during the progression of degranulation.

Unique metabolic disturbances were also observed at different stages of degranulation. Volcano plots indicated that at the later stage of 3 h, the fold change of significant differential lipids increased and the range of fold changes broadened, suggesting that the metabolic disturbances progressed (Figure 3F,G). Specifically, DGs were observed with an obvious increase at 1 h, followed by a callback at 3 h (Figure 3F). With the progression of degranulation, FAs underwent downregulation from the immediate phase to the later phase and exhibited a prominent decrease upon 3 h of antigen stimulation (Figure 3G). The emergence of these response patterns may involve degranulation dynamics.

3.5. Comparison between Curcumin and EGCG Intervention

Global profiling of the cellular lipidome was analyzed to depict the intervention effects of curcumin/EGCG from the perspective of lipidome remodeling (Cur vs. AG and EGCG vs. AG, Figures S7 and S8).

Multivariable differences were pinpointed by Euclidean distances in PCA (Figure 3B). Unlike the EGCG group, curcumin intervention was evidenced by the more evident metabolic changes at 1 h compared with 3 h, highlighting the active intervention by curcumin at the early stage of allergen stimulation. Univariate statistical tests identified 454 significant differential lipids in response to allergen actions (Table S3). According to the Venn diagram (Figure 3C), a total of 270 (59%) and 248 (55%) of these differential lipids exhibited significant quantitative alterations when the curcumin and EGCG groups were compared with the AG group ($p < 0.05$ and FDR < 0.05). Together, 355 (78%) of the significant differential lipids could be changed upon either curcumin or EGCG intervention. Both multivariable and univariate statistical results implied that the lipidome intervention in Cur was more prominent than that in EGCG, which is consistent with their differences in the suppression of IgE-mediated RBL-2H3 degranulation (Figure 2).

To specify the lipidome regulation by curcumin/EGCG, the representative differential lipids associated with allergen stimulation were further subjected to a heatmap (Figure 4). The baseline level was defined as the average readings from time-matched AG groups. The lipid contents of each sample from the Veh, Curcumin, and EGCG groups were then divided by the average of time-matched AG groups to produce the ratio. As compared to the alteration pattern associated with allergen stimulation (Veh vs. AG), the lipidome modulation by curcumin/EGCG could be identified (Cur vs. AG and EGCG vs. AG). Both curcumin and EGCG regulate the response pattern of CAR, CE, CoQ, glycerophospholipids [(H)BMP, LPC(-O) (incl. LPC and LPC-O), PC(-O) (incl. PC and PC-O), PG, (L)PE (incl. PE and LPE) and PE(-O/-p)], and sphingolipids [(Hex)Cer (incl. Cer and HexCer) and SM] at either 1 h or 3 h. These lipid species exhibited similar response patterns between the intervention and Veh groups, implying their sensitivity to both curcumin and EGCG interventions. Aside from those common features, curcumin exhibits a greater potency than that of EGCG in the upregulation of FA metabolism. Curcumin recovered the abundance of DG and PEtOH with an immediate decrease at 1 h. In contrast, the abundance of CAR and TG was improved by EGCG at 3 h. Special nonrecovered response patterns were also discovered for the curcumin/EGCG intervention groups, in which the changes in lipids were different from those of the control group. The inherent biological variation could at least partially contribute to the emergence of these response patterns. Another explanation might involve the functional diversity of polyphenols.

We further determined whether there were any changes associated with the lipid composition upon curcumin/EGCG intervention (Figures 5A and S9). At the immediate phase of allergen stimulation, no significant alterations in FA acyl chain composition were observed (Figure 5A). After 3 h of allergen stimulation (i.e., the later phase), there was a significantly lower abundance of unsaturated fatty acids [incl. monounsaturated fatty acids (MUFAs) and polyunsaturated fatty acids (PUFAs)] than saturated fatty acids (SFAs) (Figure 5B). Three hours of curcumin intervention significantly improved the depletion of MUFA/PUFA ($p < 0.05$, Figure 5C,D). Likewise, the MUFA/PUFA content was greatly replenished by 3 h of EGCG treatment (Figure 5E,F).

Then, from the perspective of lipidome remodeling, the activity of curcumin/EGCG in the suppression of IgE-mediated degranulation is involved in the comprehensive regulation of both lipid response and composition.

3.6. Defining Potential Biomarkers

Considering the association between phenotype and metabolic changes, we further examined potential biomarkers to discriminate degranulation and validate curcumin/EGCG inhibition.

Compared with the control group, allergen stimulation induced a total of 338 representative significant differential lipids that were cross-refined by screening for univariate statistical significance ($p < 0.05$, FDR < 0.05), multivariate VIP values (VIP > 1), and covariance $p(corr)$ values ($|p(corr)| > 0.3$). These significant lipids were further picked via the criteria of change magnitude (fold change $>3/2$ or $<2/3$) and stricter analysis quality (within-group variation $< 15\%$). Then, these candidates were subjected to ROC analysis. A total of 19 lipids were finally designated as biomarker candidates with the best discrimination ability (AUC = 1), spanning the lipid categories of CAR, DG, HBMP, PEtOH, PG, PI(-O), and LPC-O (Table 1).

The biomarker candidates that were sensitive to allergen stimulation were subsequently validated in the evaluation of the curcumin/EGCG intervention. Curcumin significantly recovered the abundance of LPC-O 22:0, LPC-O 24:1, HBMP 58:8, CAR 24:0, CAR 24:1, and PG 38:4, while LPC-O 22:0 could be significantly improved by EGCG. LPC-O 22:0 was discovered in the overlap.

LPC-O 22:0 was defined as a potential biomarker (Figures 5G and S10). The exploitation of LPC-O 22:0 achieved an AUC value of 1 in the discrimination of degranulation (Veh vs. AG), and both the sensitivity and specificity were 100% (Figure 5H). Furthermore, for the comparison between all intervention individuals (incl. curcumin and EGCG intervention groups) and AG groups, satisfactory discrimination results were also acquired in the evaluation of inhibition manner, resulting in an AUC value of 0.836 and sensitivity and specificity of 68.8% and 87.5%, respectively (Figure 5I). These evaluations confirmed the indicator function of LPC-O 22:0, implying its key significance in the degranulation process of RBL-2H3 cells.

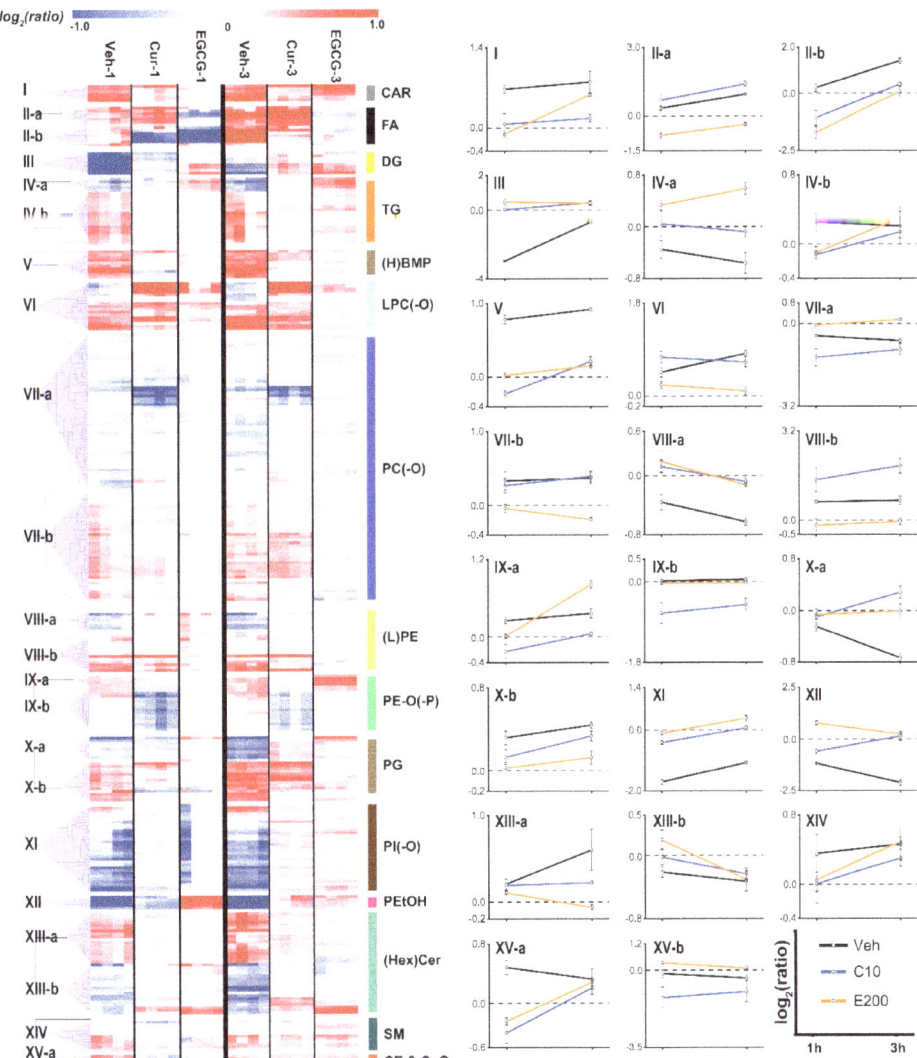

Figure 4. Comparison between curcumin and EGCG intervention. Representative differential lipids associated with allergen stimulation were subjected to heatmap to specify the regulation by curcumin/EGCG. The conversion dataset with relative contents of lipids (i.e., the contents of lipids for each sample divided by the average values from time-matched AG individuals) was logarithmically scaled, and then categorized in the tree of hierarchical clustering analysis based on the similarity of the regulation-response pattern (**left** panel). Representative metabolites were selected from each cluster to present the response trajectory (**right** panel). Each point in the trajectory was presented as the average relative content ± SD.

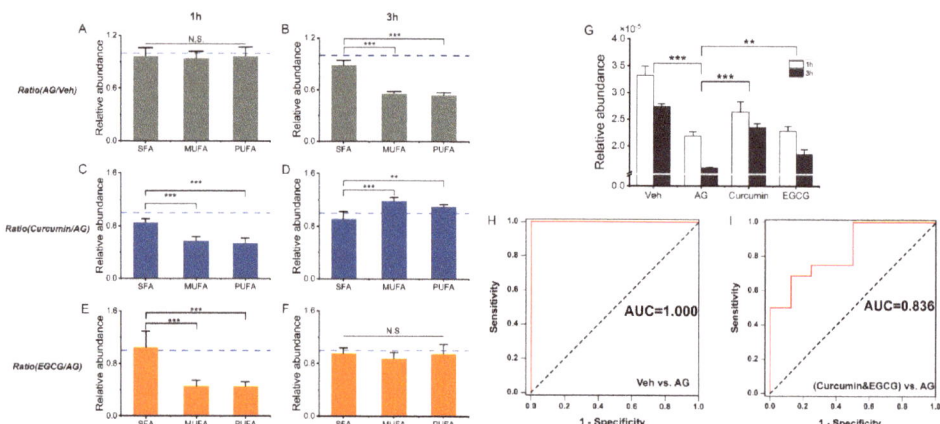

Figure 5. Changes in the content and composition of important lipids. (**A–F**) present FA changes associated with acyl chain composition. (**G**) Relative content of LPC-O 22:0. (**H,I**) ROC curves of LPC-O 22:0. Diagnostic potential was evaluated based on binary logistic regression. Each column is presented as the mean ± SD. **: $0.001 < p < 0.01$, ***: $p < 0.001$, N.S.: no significance.

Table 1. Statistical information of biomarker candidates.

Compound	Fatty Acids	Veh vs. AG			AG/Veh		Curcumin vs. AG			Cur/AG		EGCG vs. AG			EGCG/AG	
		p	FDR	Ratio-1 h	Ratio-3 h	p	FDR	Ratio-1 h	Ratio-3 h	p	FDR	Ratio-1 h	Ratio-3 h			
CAR 24:0	24:0	4.51×10^{-10}	1.27×10^{-8}	0.611	0.522	9.38×10^{-4}	2.64×10^{-3}	1.159	1.240	3.52×10^{-1}	4.63×10^{-1}	0.849	1.241
CAR 24:1	24:1	1.77×10^{-10}	6.10×10^{-9}	0.580	0.582	1.48×10^{-4}	5.12×10^{-4}	1.103	1.350	6.81×10^{-2}	1.27×10^{-1}	0.861	1.220
DG 38:4	18:0_20:4	3.90×10^{-14}	1.82×10^{-11}	11.091	2.447	5.23×10^{-1}	6.16×10^{-1}	0.913	1.229	1.44×10^{-5}	1.11×10^{-4}	1.495	1.743
DG 40:5	18:0_22:5	1.64×10^{-13}	4.88×10^{-11}	7.501	1.702	1.56×10^{-1}	2.37×10^{-1}	0.969	1.320	1.16×10^{-4}	6.10×10^{-4}	1.372	1.397
DG 40:6	18:0_22:6	2.62×10^{-13}	4.88×10^{-11}	7.822	1.649	1.54×10^{-2}	3.22×10^{-2}	1.025	1.351	1.08×10^{-4}	5.81×10^{-4}	1.399	1.324
HBMP 58:8	18:1_18:1_22:6	2.76×10^{-9}	5.47×10^{-8}	0.648	0.643	3.19×10^{-4}	1.00×10^{-3}	1.081	1.278	2.01×10^{-1}	3.00×10^{-1}	1.080	0.992
HBMP 62:13	18:1_22:6_22:6	1.72×10^{-10}	1.11×10^{-9}	0.582	0.529	7.29×10^{-1}	7.93×10^{-1}	0.855	1.158	6.06×10^{-2}	1.14×10^{-1}	1.018	1.111
HBMP 64:16	20:4_22:6_22:6	3.66×10^{-12}	3.41×10^{-10}	0.569	0.478	2.99×10^{-3}	7.68×10^{-3}	0.725	1.041	5.32×10^{-2}	1.02×10^{-1}	0.999	1.140
LPC-O 22:0	O-22:0	2.52×10^{-11}	1.30×10^{-9}	0.658	0.581	1.35×10^{-7}	2.42×10^{-6}	1.209	1.477	5.04×10^{-4}	1.90×10^{-3}	1.045	1.163
LPC-O 24:1	O-24:1	7.71×10^{-10}	1.99×10^{-8}	0.584	0.524	8.44×10^{-9}	7.14×10^{-7}	1.287	1.517	2.89×10^{-2}	6.04×10^{-2}	1.060	1.052
PEtOH 34:1	16:0_18:1	2.64×10^{-9}	5.34×10^{-8}	2.239	4.310	1.86×10^{-1}	2.71×10^{-1}	0.670	1.121	1.98×10^{-6}	2.49×10^{-5}	1.728	1.197
PEtOH 36:1	18:0_18:1	5.14×10^{-8}	6.47×10^{-7}	1.511	2.358	1.09×10^{-1}	1.77×10^{-1}	0.696	1.047	8.64×10^{-5}	4.82×10^{-4}	1.524	1.156
PG 36:4	16:0_20:4	1.79×10^{-11}	1.11×10^{-9}	1.505	1.650	3.32×10^{-4}	1.03×10^{-3}	0.795	1.007	3.62×10^{-7}	6.02×10^{-6}	1.088	1.328
PG 38:4	18:0_20:4	5.71×10^{-12}	4.83×10^{-10}	3.010	3.106	9.49×10^{-4}	2.66×10^{-3}	0.723	0.993	3.01×10^{-7}	5.22×10^{-6}	1.341	1.634
PI 36:5	16:0_20:5	1.42×10^{-10}	1.47×10^{-10}	2.175	2.374	2.15×10^{-4}	1.02×10^{-3}	1.043	1.197	9.42×10^{-1}	9.68×10^{-1}	0.864	1.135
PI 42:9	20:4_22:5	7.28×10^{-13}	9.68×10^{-11}	3.257	2.087	9.68×10^{-5}	3.62×10^{-4}	0.759	1.061	1.86×10^{-2}	4.14×10^{-2}	0.939	1.321
PI-O 36:4	O-16:0_20:4	1.72×10^{-10}	6.10×10^{-9}	1.691	1.759	9.41×10^{-2}	1.57×10^{-1}	0.905	1.012	3.72×10^{-1}	4.83×10^{-1}	0.859	1.225
PI-O 38:6	O-16:0_22:6	5.88×10^{-10}	1.61×10^{-8}	1.556	1.658	2.83×10^{-3}	7.32×10^{-3}	1.035	1.127	9.71×10^{-2}	1.69×10^{-1}	0.868	0.999
PI-O 40:7	O-18:1_22:6	1.26×10^{-12}	1.47×10^{-10}	1.546	1.894	4.01×10^{-1}	5.09×10^{-1}	1.013	1.010	3.18×10^{-2}	6.56×10^{-2}	0.945	0.924

4. Discussion

Due to their potential benefits for human health, dietary polyphenols, such as curcumin (turmeric) and EGCG (green tea), have been a major topic of interest. An increasing number of trials have shown a correlation between dietary polyphenol consumption and a reduction in risk factors for chronic diseases [2,3]. Two typical dietary polyphenols, curcumin and EGCG, were confirmed to show anti-allergic potential in our study. Both curcumin and EGCG significantly suppressed the release of the indicator of degranulation (β-hexosaminidase) and representative pro-inflammatory cytokines (IL-4 and TNF-α) on IgE/antigen-stimulated RBL-2H3 cells.

Lipids can act as vital intermediates in various cellular communication processes. In this study, the key lipidome remodeling of antigen-stimulated RBL-2H3 cells was further investigated to understand the curcumin/EGCG intervention that underlies the progression of degranulation. Global disturbances in the cellular lipidome were discovered upon IgE/allergen stimulation. Enrichment of 454 significant differential lipids ($p < 0.05$, FDR < 0.05, AG vs. Veh) pinpointed the top lipid categories associated with significant upregulation [i.e., DG(-O)] and downregulation [i.e., (H)BMP, FA, and CAR] after stimulation. Although the progression of IgE-mediated degranulation was revealed by the more evident metabolic changes at 3 h than at 1 h, similar lipidome change patterns were shared during the progression of degranulation.

Notably, 78% of those significant differential lipids could be regulated upon either curcumin or EGCG intervention. Aside from common features in the improvement of the cellular lipidome by these two typical dietary polyphenols, special lipidome patterns were also found for different intervention groups. Both multivariable and univariate statistical results implied that lipidome regulation in the Cur group (i.e., 10 μM curcumin) was more prominent than that in the EGCG group (i.e., 200 μM EGCG). These lipidomics changes were consistent with the suppression of β-hexosaminidase release by curcumin, which was enhanced compared with EGCG despite their similar performance in decreasing proinflammatory cytokine production. Furthermore, when compared with EGCG, the active intervention by curcumin was highlighted in the immediate phase of the allergic reaction.

4.1. DG Metabolism

As the key secondary lipid messengers, DGs were the top lipid categories that underwent significant changes upon degranulation (Figure 3D,E). DGs were significantly increased at the immediate phase of IgE/allergen stimulation, followed by callback at the later phase (Figure 3F).

The interaction of allergens with IgE–FcεRI complexes results in the formation of signaling complexes that converge on the activation of phospholipase C (PLC) [23]. PLC activation leads to the enzymatic cleavage of phosphoinositol 4,5-bisphosphate (PIP2) into DG and inositol 1,4,5-triphosphate (IP3) [23,24]. IP3 mediates the release of intracellular Ca^{2+} [10,23,24]. DG targets, such as protein kinase C (PKC), Ras guanyl nucleotide-releasing proteins (RasGRP), and the canonical transient receptor potential (TRPC) channel protein, have been shown to be critical in controlling mast cell degranulation [10,23,24]. Consequently, DGs act as the key secondary lipid messengers for transducing signals downstream of receptors. The levels of DGs are tightly associated with the magnitude and duration of the degranulation responses generated. Then, in the AG group, the upregulation of DGs might indicate that antigen–IgE–FcεRI complexes successfully initiated the signaling cascade to activate the process of mast cell degranulation.

The abundance of DGs was regulated by curcumin with an immediate decrease at the early phase of allergen stimulation, indicating that curcumin could inhibit DG-related signal transduction to partially block the process of degranulation. In contrast, DGs could not be recovered by EGCG, suggesting that EGCG was absent from the inhibition of DG-related signal transduction, i.e., antigen stimulation. This was in accordance with the better suppression of mast cell degranulation obtained by curcumin than by EGCG.

4.2. FA Metabolism

With the progression of degranulation, FAs were downregulated from the immediate phase to the later phase and exhibited a prominent decrease upon 3 h of antigen stimulation (Figure 3G). When compared with controls, no significant alterations in FA acyl chain composition in AG were observed at 1 h (Figure 5A), while the abundance of unsaturated fatty acids (incl. MUFA and PUFA) was significantly lower than that of SFA at 3 h (Figure 5B). PUFA metabolism is recognized as an important factor in immune regulation and disease control. The depletion of n-6 PUFAs leads to the production of highly proinflammatory mediators, such as prostaglandins (PGs), thromboxanes (TXs), leukotrienes (LTs), and lipoxins (LXs) [12,25]. As one of the most significant differential lipid classes upon allergen stimulation, the comprehensive regulation of FA abundance and composition may involve degranulation dynamics.

We found that compared to EGCG, curcumin is more potent in upregulating FA metabolism (Figure 4). Three hours of curcumin intervention significantly improved the depletion of MUFA/PUFA ($p < 0.05$, Figure 5C,D). Likewise, the MUFA/PUFA content was replenished by 3 h of EGCG treatment (Figure 5E,F). These changes in the FA profile caused by curcumin/EGCG might contribute to the modification of mast cell gene expression [26]. PPAR-β and -γ have been reported to be expressed in human and murine mast cells and involved in the suppression of mast cell maturation and IgE/antigen-induced production of proinflammatory cytokines [27,28]. At the later phase, the replenishment of PUFAs has been suggested to activate PPAR-γ and change the expression of the antigen response machinery (e.g., FcεRI, Fyn, Lyn, Syk) and degranulation machinery (e.g., calcium channels, vesicle docking molecules) [26]. As a result, mast cells become less susceptible to antigen activation. On the other hand, as the building blocks for lipids, FAs participate in forming membrane phospholipid bilayers and facilitating protein acylation, which is important for the structure and function of membranes/lipid rafts. It has been reported that lipid rafts are especially vital for FcεRI-mediated signal transduction [26]. In IgE/antigen-stimulated cells, FcεRI is quickly recruited to lipid rafts to initiate signaling [26,29]. FcεRI signaling could also be regulated by many lipid raft components [26]. When compared to nonraft regions, the phospholipids in lipid rafts prefer higher levels of saturated fatty acids [30]. Thus, the modification of the FA profile by curcumin/EGCG in the improvement of MUFA/PUFA levels might influence lipid raft function and then reduce FcεRI signaling induced by IgE/antigen, followed by suppression of mediator release from mast cells.

In addition, CARs, which play a critical role in transporting FAs into mitochondria so they can be oxidized to produce energy, were also defined as one of the top lipid classes with significant changes (Figure 3E). In this study, CARs were found to be decreased in response to IgE/antigen stimulation. The downregulation of CARs may lead to the inhibition of FA β-oxidative, and then promote more FA flow to the pathway of eicosanoic acid synthesis to produce proinflammatory mediators (Figure 6). Decreased FA oxidation and mitochondrial dysfunction have been reported in sensitized mice to support our discoveries [31]. As expected, CARs could be greatly improved by curcumin/EGCG intervention.

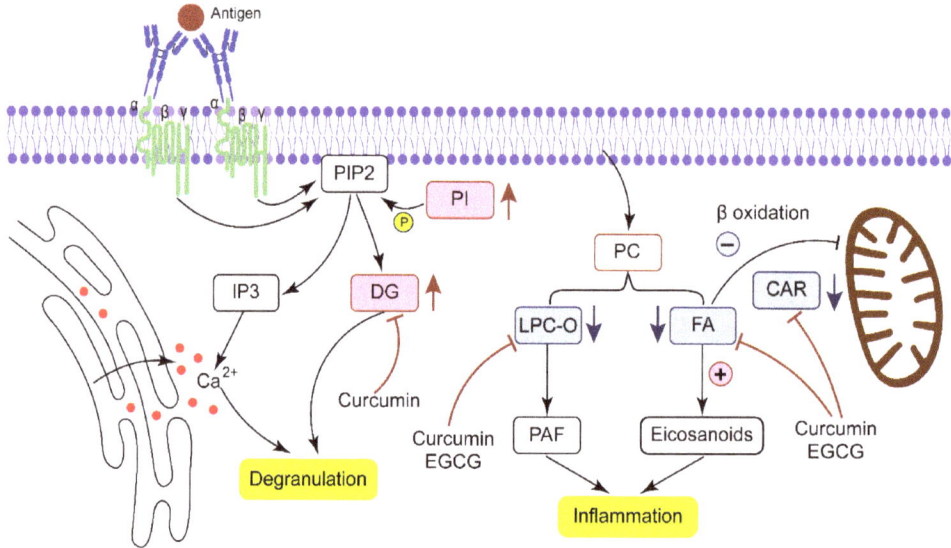

Figure 6. Schematic diagram showing the major rewiring of lipid metabolism.

4.3. BMP Metabolism

In this study, (H)BMPs were found to be decreased upon IgE/antigen stimulation. BMPs are markedly enriched in the inner membranes of late endosomes, particularly lysosomes, and play a key role in lysosomal integrity and function [32,33].

BMP-enriched vesicles serve in endosomal-lysosomal tracking and function as docking structures for the activation of lysosomal hydrolytic enzymes [32]. The unique *sn-1:sn-1'* stereoconfiguration of BMP confers its higher resistance to the hydrolytic lysosomal environment [32]. Then, BMP's negative charge could be retained, facilitating its role as the docking site and essential cofactor for some lysosomal proteins that contain positively charged domains [32,34]. Indeed, the modulation of ABHD6 (i.e., BMP hydrolase) activity has been found to alter the immune response in a murine model of lung inflammation [35].

In addition, BMP was reported to be a relatively abundant phospholipid in mast cell-derived extracellular vesicles (EVs), especially degranulated mast cells [36]. Thus, such EVs are derived not only from the plasma membrane or multivesicular bodies but probably also from secretory lysosomes. Mast cells can release EVs constitutively and after IgE-mediated degranulation [36]. In addition to transferring RNA species to other mast cells and containing lipid mediators [37], mast cell-derived EVs exert immune-stimulatory effects on dendritic cells and T/B cells [38,39]. Then, the decrease in (H)BMPs upon IgE/antigen stimulation might also be involved in the release of EVs after IgE-mediated degranulation. Curcumin/EGCG intervention produced favorable effects on the improvement of (H)BMP-related metabolism.

LPC-O 22:0 was defined as a potential biomarker for its sensitivity to IgE/antigen stimulation and curcumin/EGCG intervention. The remodeling of membrane phospholipids PC by phospholipase A2 (PLA2) generates arachidonic acid (AA) and LPC [12]. Furthermore, LPC is converted to PAF via LPC acetyltransferase (LPCAT) [40]. As one of the key lipid mediators that mast cells abundantly synthesize, PAF signals via the G-protein coupled receptor (GPCR) (PAF receptor, PAFR), which initiates a signaling cascade [12]; then degranulation and an enhancement in inflammation are triggered [12]. The recovery of the potential biomarker LPC by curcumin/EGCG thus indicated that their effective intervention may involve the inhibition of PAF generation.

Certainly, these lipidome changes might be the result of the development of protein–polyphenol complexes with possibly lower allergenic potential [5,6], as well as more sophisticated and comprehensive anti-allergy mechanisms. Following that, a growing number of studies have demonstrated that mast cell synthesis and metabolism of lipid mediators, in turn, influence cellular processes [9,11]. The role of lipids in the pathogenesis of allergic disease has long been studied. Our study indicated that a potency was observed with curcumin than with EGCG in the disturbance of lipid metabolism, in accordance with the superior effects of curcumin observed when compared with EGCG in the suppression of the degranulation process. A considerable similarity between curcumin intervention and EGCG was discovered in their lipidome remodeling patterns. Our study confirmed the significance of DG, FA, BMP, and LPC metabolism for IgE/antigen stimulation and subsequent curcumin/EGCG intervention. Both changes in lipid response and composition patterns indicated that lipids influence the degranulation process via multiple mechanisms, including (i) producing highly proinflammatory mediators, (ii) mediating intracellular signaling cascades by acting as second messengers, (iii) activating a diverse family of receptors, and (iv) forming structural support platforms (lipid rafts) and extracellular vesicles.

The goal of this study is to describe curcumin/EGCG-induced lipidome change in order to give a novel perspective on curcumin/EGCG participation in anaphylaxis. It should be noted, however, that research into the process is currently restricted. The underlying mechanisms that influence lipid alterations by putative protein–polyphenol complexes with possibly decreased allergenic potential, as well as the subsequent effects of lipid changes on allergy reactions, require additional exploration. Further research will be needed, including expanding the sampling distribution throughout the degranulation course, tracking the effects of curcumin and EGCG in vivo, and a combined analysis including transcriptomic, biochemical, and immunological results.

5. Conclusions

Two typical dietary polyphenols, curcumin and EGCG, were confirmed to show anti-allergic potential in the present study. Both curcumin and EGCG significantly suppressed the release of β-hexosaminidase, IL-4, and TNF-α from IgE/antigen-stimulated RBL-2H3 cells. As compared to the alteration pattern associated with IgE/antigen-stimulated degranulation (Veh vs. AG), the lipidome modulation by curcumin/EGCG could be identified (Cur vs. AG and EGCG vs. AG). Comprehensive lipidomics analysis revealed that the ability to disturb lipid metabolism was stronger with curcumin than EGCG, in accordance with the superior ability of curcumin to suppress the degranulation process. These key lipidome disturbances provide novel insights into the effects of curcumin/EGCG intervention underlying the progression of degranulation. Our findings open the possibility of preventing immediate allergic reactions via antigen-stimulated mast cells in vitro and will help guide future attempts to use dietary polyphenols.

Supplementary Materials: The following supporting information can be downloaded at: https://www.mdpi.com/article/10.3390/metabo13050628/s1, Supplementary data associated with this article can be found in the online version. Materials and Methods of Chemicals and Reagents, Lipidome Extraction, and Lipidomics Analysis by UPLC-Q-Exactive MS. Results of Analytical Performance of Lipid Profiling [19–21]; Figure S1: Characteristics of the experimental model; Figure S2: Cell morphology; Figure S3: Lipidome profiling; Figure S4: The distributions of %number for identified lipid species; Figure S5: Analytical Performance of Lipid Profiling; Figure S6: The comparison between AG and Veh groups; Figure S7: The comparison between AG and curcumin groups; Figure S8: The comparison between AG and EGCG groups; Figure S9: LPC-O changes associated with acyl chain composition; Figure S10: Identification of the potential biomarker LPC-O 22:0; Table S1: Statistical results of identification; Table S2: Information of identified lipids; Table S3: Statistical information of differential lipids changed by degranulation; Table S4: Changes in the content and composition of important lipids.

Author Contributions: Conceptualization, J.Z. and J.L.; methodology, J.Z. and J.H.; validation, Z.Y. and C.M.; investigation, J.H. and Z.Y.; data curation, J.H., L.G., Y.C. and G.L.; writing—original draft preparation, J.Z., J.H. and Z.Y.; writing—review and editing, J.Z. and J.L. All authors have read and agreed to the published version of the manuscript.

Funding: The study has been supported by National Key R&D Program of China (2019YFD0901704), Natural Science Foundation of Fujian Province of China (Grants No. 2022J01330) and Natural Science Foundation of Xiamen City of China (Grants No. 3502Z20227208).

Data Availability Statement: The datasets generated and/or analysed during the current study are available in the Metabolomics Workbench (https://www.metabolomicsworkbench.org/, accessed on 1 January 2023) repository (Study ID ST002384, DatatrackID: 3356 and Project DOI: http://dx.doi.org/10.21228/M8DH62, accessed on 1 January 2023).

Conflicts of Interest: The authors declare no conflict of interest.

References

1. Galli, S.J.; Tsai, M. IgE and mast cells in allergic disease. *Nat. Med.* **2012**, *18*, 693–704. [CrossRef] [PubMed]
2. Yang, H.; Qu, Y.Z.; Gao, Y.R.; Sun, S.Y.; Ding, R.X.; Cang, W.H.; Wu, R.N.; Wu, J.R. Role of the dietary components in food allergy: A comprehensive review. *Food Chem.* **2022**, *386*, 12. [CrossRef]
3. Bessa, C.; Francisco, T.; Dias, R.; Mateus, N.; de Freitas, V.; Perez-Gregorio, R. Use of Polyphenols as Modulators of Food Allergies. From Chemistry to Biological Implications. *Front. Sustain. Food Syst.* **2021**, *5*, 18. [CrossRef]
4. Ohishi, T.; Hayakawa, S.; Miyoshi, N. Involvement of microRNA modifications in anticancer effects of major polyphenols from green tea, coffee, wine, and curry. *Crit. Rev. Food Sci. Nutr.* **2022**, 1–32. [CrossRef]
5. Plundrich, N.J.; Bansode, R.R.; Foegeding, E.A.; Williams, L.L.; Lila, M.A. Protein-bound Vaccinium fruit polyphenols decrease IgE binding to peanut allergens and RBL-2H3 mast cell degranulation in vitro. *Food Funct.* **2017**, *8*, 1611–1621. [CrossRef] [PubMed]
6. Li, Y.C.; Mattison, C.P. Polyphenol-rich pomegranate juice reduces IgE binding to cashew nut allergens. *J. Sci. Food Agric.* **2018**, *98*, 1632–1638. [CrossRef]
7. Sun, S.F.; Jiang, T.Y.; Gu, Y.J.; Yao, L.; Du, H.; Luo, J.Z.; Che, H.L. Contribution of five major apple polyphenols in reducing peanut protein sensitization and alleviating allergencitiy of peanut by changing allergen structure. *Food Res. Int.* **2023**, *164*, 10. [CrossRef]
8. Yan, X.H.; Zeng, Z.L.; McClements, D.J.; Gong, X.F.; Yu, P.; Xia, J.H.; Gong, D.M. A review of the structure, function, and application of plant-based protein-phenolic conjugates and complexes. *Compr. Rev. Food Sci. Food Saf.* **2023**, *22*, 1312–1336. [CrossRef]
9. Hagemann, P.M.; Nsiah-Dosu, S.; Hundt, J.E.; Hartmann, K.; Orinska, Z. Modulation of Mast Cell Reactivity by Lipids: The Neglected Side of Allergic Diseases. *Front. Immunol.* **2019**, *10*, 1174. [CrossRef]
10. Gilfillan, A.M.; Tkaczyk, C. Integrated signalling pathways for mast-cell activation. *Nat. Rev. Immunol.* **2006**, *6*, 218–230. [CrossRef]
11. Bankova, L.G.; Lai, J.Y.; Yoshimoto, E.; Boyce, J.A.; Austen, K.F.; Kanaoka, Y.; Barrett, N.A. Leukotriene E-4 elicits respiratory epithelial cell mucin release through the G-protein-coupled receptor, GPR99. *Proc. Natl. Acad. Sci. USA* **2016**, *113*, 6242–6247. [CrossRef] [PubMed]
12. Schauberger, E.; Peinhaupt, M.; Cazares, T.; Lindsley, A.W. Lipid Mediators of Allergic Disease: Pathways, Treatments, and Emerging Therapeutic Targets. *Curr. Allergy Asthma Rep.* **2016**, *16*, 48. [CrossRef]
13. Han, X.L.; Gross, R.W. Global analyses of cellular lipidomes directly from crude extracts of biological samples by ESI mass spectrometry: A bridge to lipidomics. *J. Lipid Res.* **2003**, *44*, 1071–1079. [CrossRef]
14. Abu-Hijleh, H.M.; Al-Zoubi, R.M.; Zarour, A.; Ansari, A.A.; Bawadi, H. The Therapeutic Role of Curcumin in Inflammation and Post-Surgical Outcomes. *Food Rev. Int.* **2023**, *16*. [CrossRef]
15. Mokra, D.; Joskova, M.; Mokry, J. Therapeutic Effects of Green Tea Polyphenol (-)-Epigallocatechin-3-Gallate (EGCG) in Relation to Molecular Pathways Controlling Inflammation, Oxidative Stress, and Apoptosis. *Int. J. Mol. Sci.* **2023**, *24*, 340. [CrossRef]
16. Matsui, T.; Ito, C.; Itoigawa, M.; Shibata, T. Three phlorotannins from Sargassum carpophyllum are effective against the secretion of allergic mediators from antigen-stimulated rat basophilic leukemia cells. *Food Chem.* **2022**, *377*, 131992. [CrossRef]
17. Tang, F.; Chen, F.; Ling, X.; Huang, Y.; Zheng, X.; Tang, Q.; Tan, X. Inhibitory effect of methyleugenol on IgE-mediated allergic inflammation in RBL-2H3 cells. *Mediat. Inflamm.* **2015**, *2015*, 463530. [CrossRef]
18. Nagai, H.; Abe, T.; Yamaguchi, I.; Mito, K.; Tsunematsu, M.; Kimata, M.; Inagaki, N. Role of mast cells in the onset of IgE-mediated late-phase cutaneous response in mice. *J. Allergy Clin. Immunol.* **2000**, *106*, S91–S98. [CrossRef]
19. Zeng, J.; Li, J.L.; Liu, S.S.Y.; Yang, Z.Q.; Zhong, Y.; Chen, X.M.; Li, G.L.; Li, J. Lipidome disturbances in preadipocyte differentiation associated with bisphenol A and replacement bisphenol S exposure. *Sci. Total Environ.* **2021**, *753*, 10. [CrossRef]
20. Zeng, J.; Liu, S.; Cai, W.; Jiang, H.; Lu, X.; Li, G.; Li, J.; Liu, J. Emerging lipidome patterns associated with marine Emiliania huxleyi-virus model system. *Sci. Total Environ.* **2019**, *688*, 521–528. [CrossRef]

21. Li, J.; Yuan, H.B.; Rong, Y.T.; Qian, M.C.; Liu, F.Q.; Hua, J.J.; Zhou, Q.H.; Deng, Y.L.; Zeng, J.; Jiang, Y.W. Lipid metabolic characteristics and marker compounds of ripened Pu-erh tea during pile fermentation revealed by LC-MS-based lipidomics. *Food Chem.* **2023**, *404*, 12. [CrossRef] [PubMed]
22. Matsui, T.; Ito, C.; Masubuchi, S.; Itoigawa, M. Licarin A is a candidate compound for the treatment of immediate hypersensitivity via inhibition of rat mast cell line RBL-2H3 cells. *J. Pharm. Pharmacol.* **2015**, *67*, 1723–1732. [CrossRef]
23. Singh, B.K.; Kambayashi, T. The Immunomodulatory Functions of Diacylglycerol Kinase zeta. *Front. Cell Dev. Biol.* **2016**, *4*, 96. [CrossRef] [PubMed]
24. Sakuma, M.; Shirai, Y.; Ueyama, T.; Saito, N. Diacylglycerol kinase gamma regulates antigen-induced mast cell degranulation by mediating Ca(2+) influxes. *Biochem. Biophys. Res. Commun.* **2014**, *445*, 340–345. [CrossRef]
25. Arita, M. Eosinophil polyunsaturated fatty acid metabolism and its potential control of inflammation and allergy. *Allergol. Int.* **2016**, *65*, S2–S5. [CrossRef]
26. Wang, X.; Kulka, M. n-3 Polyunsaturated fatty acids and mast cell activation. *J. Leukoc. Biol.* **2015**, *97*, 859–871. [CrossRef]
27. Sugiyama, H.; Nonaka, T.; Kishimoto, T.; Komoriya, K.; Tsuji, K.; Nakahata, T. Peroxisome proliferator-activated receptors are expressed in human cultured mast cells: A possible role of these receptors in negative regulation of mast cell activation. *Eur. J. Immunol.* **2000**, *30*, 3363–3370. [CrossRef]
28. Tachibana, M.; Wada, K.; Katayama, K.; Kamisaki, Y.; Maeyama, K.; Kadowaki, T.; Blumberg, R.S.; Nakajima, A. Activation of peroxisome proliferator-activated receptor gamma suppresses mast cell maturation involved in allergic diseases. *Allergy* **2008**, *63*, 1136–1147. [CrossRef]
29. Field, K.A.; Holowka, D.; Baird, B. Compartmentalized activation of the high affinity immunoglobulin E receptor within membrane domains. *J. Biol. Chem.* **1997**, *272*, 4276–4280. [CrossRef]
30. Simons, K.; Sampaio, J.L. Membrane Organization and Lipid Rafts. *Cold Spring Harb. Perspect. Biol.* **2011**, *3*, 17. [CrossRef]
31. Trinchese, G.; Paparo, L.; Aitoro, R.; Fierro, C.; Varchetta, M.; Nocerino, R.; Mollica, M.P.; Berni Canani, R. Hepatic Mitochondrial Dysfunction and Immune Response in a Murine Model of Peanut Allergy. *Nutrients* **2018**, *10*, 744. [CrossRef]
32. Showalter, M.R.; Berg, A.L.; Nagourney, A.; Heil, H.; Carraway, K.L., 3rd; Fiehn, O. The Emerging and Diverse Roles of Bis(monoacylglycero) Phosphate Lipids in Cellular Physiology and Disease. *Int. J. Mol. Sci.* **2020**, *21*, 8067. [CrossRef]
33. Akgoc, Z.; Sena-Esteves, M.; Martin, D.R.; Han, X.; d'Azzo, A.; Seyfried, T.N. Bis(monoacylglycero)phosphate: A secondary storage lipid in the gangliosidoses. *J. Lipid Res.* **2015**, *56*, 1006–1013. [CrossRef]
34. Gallala, H.D.; Sandhoff, K. Biological function of the cellular lipid BMP-BMP as a key activator for cholesterol sorting and membrane digestion. *Neurochem. Res.* **2011**, *36*, 1594–1600. [CrossRef]
35. Grabner, G.F.; Fawzy, N.; Pribasnig, M.A.; Trieb, M.; Taschler, U.; Holzer, M.; Schweiger, M.; Wolinski, H.; Kolb, D.; Horvath, A.; et al. Metabolic disease and ABHD6 alter the circulating bis(monoacylglycero)phosphate profile in mice and humans. *J. Lipid Res.* **2019**, *60*, 1020–1031. [CrossRef]
36. Groot Kormelink, T.; Arkesteijn, G.J.; van de Lest, C.H.; Geerts, W.J.; Goerdayal, S.S.; Altelaar, M.A.; Redegeld, F.A.; Nolte-'t Hoen, E.N.; Wauben, M.H. Mast Cell Degranulation Is Accompanied by the Release of a Selective Subset of Extracellular Vesicles That Contain Mast Cell-Specific Proteases. *J. Immunol.* **2016**, *197*, 3382–3392. [CrossRef]
37. Valadi, H.; Ekstrom, K.; Bossios, A.; Sjostrand, M.; Lee, J.J.; Lotvall, J.O. Exosome-mediated transfer of mRNAs and microRNAs is a novel mechanism of genetic exchange between cells. *Nat. Cell Biol.* **2007**, *9*, U654–U672. [CrossRef]
38. Skokos, D.; Le Panse, S.; Villa, I.; Rousselle, J.C.; Peronet, R.; David, B.; Namane, A.; Mecheri, S. Mast cell-dependent B and T lymphocyte activation is mediated by the secretion of immunologically active exosomes. *J. Immunol.* **2001**, *166*, 868–876. [CrossRef]
39. Skokos, D.; Botros, H.G.; Demeure, C.; Morin, J.; Peronet, R.; Birkenmeier, G.; Boudaly, S.; Mecheri, S. Mast cell-derived exosomes induce phenotypic and functional maturation of dendritic cells and elicit specific immune responses in vivo. *J. Immunol.* **2003**, *170*, 3037–3045. [CrossRef]
40. Snyder, F. Platelet-Activating-Factor—The Biosynthetic and Catabolic Enzymes. *Biochem. J.* **1995**, *305*, 689–705. [CrossRef]

Disclaimer/Publisher's Note: The statements, opinions and data contained in all publications are solely those of the individual author(s) and contributor(s) and not of MDPI and/or the editor(s). MDPI and/or the editor(s) disclaim responsibility for any injury to people or property resulting from any ideas, methods, instructions or products referred to in the content.

Article

A Strategy for Uncovering the Serum Metabolome by Direct-Infusion High-Resolution Mass Spectrometry

Xiaoshan Sun [1,2,3], Zhen Jia [1,3,4], Yuqing Zhang [1,3,5], Xinjie Zhao [1,2,3], Chunxia Zhao [1,2,3], Xin Lu [1,2,3,*] and Guowang Xu [1,2,3,*]

[1] CAS Key Laboratory of Separation Science for Analytical Chemistry, Dalian Institute of Chemical Physics, Chinese Academy of Sciences, Dalian 116023, China
[2] University of Chinese Academy of Sciences, Beijing 100049, China
[3] Liaoning Province Key Laboratory of Metabolomics, Dalian 116023, China
[4] Department of Cell Biology, College of Life Sciences, China Medical University, Shenyang 110122, China
[5] Zhang Dayu School of Chemistry, Dalian University of Technology, Dalian 116024, China
* Correspondence: luxin001@dicp.ac.cn (X.L.); xugw@dicp.ac.cn (G.X.)

Abstract: Direct infusion nanoelectrospray high-resolution mass spectrometry (DI-nESI-HRMS) is a promising tool for high-throughput metabolomics analysis. However, metabolite assignment is limited by the inadequate mass accuracy and chemical space of the metabolome database. Here, a serum metabolome characterization method was proposed to make full use of the potential of DI-nESI-HRMS. Different from the widely used database search approach, unambiguous formula assignments were achieved by a reaction network combined with mass accuracy and isotopic patterns filter. To provide enough initial known nodes, an initial network was directly constructed by known metabolite formulas. Then experimental formula candidates were screened by the predefined reaction with the network. The effects of sources and scales of networks on assignment performance were investigated. Further, a scoring rule for filtering unambiguous formula candidates was proposed. The developed approach was validated by a pooled serum sample spiked with reference standards. The coverage and accuracy rates for the spiked standards were 98.9% and 93.6%, respectively. A total of 1958 monoisotopic features were assigned with unique formula candidates for the pooled serum, which is twice more than the database search. Finally, a case study of serum metabolomics in diabetes was carried out using the developed method.

Keywords: metabolomics; direct-infusion; high-resolution mass spectrometry; formula assignment; reaction network

1. Introduction

Direct infusion combined with high-resolution mass spectrometry (DI-HRMS) is an attractive alternative for high-throughput metabolomics analysis due to its simplicity, high resolution, high mass accuracy, and time-saving [1,2]. In metabolomics analysis, nanoelectrospray ionization (nESI) DI-HRMS provides similar discrimination capabilities to conventional liquid chromatography-mass spectrometry (LC-MS) while greatly reducing the total analysis time [3,4]. DI-HRMS approach has been widely applied in various metabolomics studies [5–7], such as urinary metabolic profiling in human epidemiological investigations [8], metabolite analysis for the diagnosis of inborn errors of metabolism [9–11], and serum metabolomics analysis in diabetes [12].

Metabolite annotation is widely recognized as a critical part of metabolomics analysis. Accurate molecular formula assignment is a prerequisite for identifying unknown metabolites [13]. Typically, metabolite annotation relies on metabolome database searching using accurate mass, where the extent of database coverage determines the annotation performance [14–16]. Sarvin et al. [17] recently analyzed the m/z value distribution of ions and identified optimal scan ranges, resulting in a ~50% increase in feature detection compared to conventional spectral-stitching FI-MS [18]. However, this study employed publicly available databases, such as the LIPID MAPS Structure Database (LMSD) for lipidomics data [19] and the Human Metabolome Database (HMDB) for metabolomics data [20,21], and only 564 out of 3233 features in metabolomics data and 401 out of 3339 features in lipidomics data obtained putative annotations with a 5 ppm mass tolerance, respectively. Alternatively, the stochastic molecular generator method, which incorporates chemical constraints (e.g., high mass accuracy, limited element number/species, and high-fidelity isotopic pattern), allows reliable formula assignment without database limitations [22–24]. But this approach is not suitable for high-resolution mass spectrometry analysis, especially for low-abundance and high-mass MS signals, for which multiple formula candidates may match [25–27].

Metabolites serve as substrates and products of vital biological reactions [28,29], so it is very promising to exploit the reaction relationships among them for metabolome annotation. The approach of utilizing mass differences has been applied for molecular formula assignment in ultra-high-resolution mass spectrometry [30–32]. In this method, spectral features are represented as nodes, while pre-defined metabolic reactions, as reflected by mass differences, serve as edges connecting these nodes [33,34]. High mass accuracy (<1 ppm) is crucial for minimizing the number of candidate molecular formulas and possible reaction types. Nevertheless, it is difficult to achieve in high-resolution mass spectrometry methods [35]. To address this challenge, Chen et al. [36] proposed a global network optimization approach, NetID, which annotates formulas for as many MS features as possible, including adducts, fragments and isotopes of metabolites in untargeted LC-HRMS data. In NetID seed nodes were limited to the detected features and assigned by matching peaks to the HMDB with a mass tolerance of 10 ppm, it is especially important to use highly reliable metabolite formulas as seed nodes [37].

In this work, a serum metabolome characterization method was proposed to fully explore the potential of DI-nESI-HRMS. To enhance the coverage and accuracy of unique formula assignment, a reaction network was constructed and endogenous metabolites from the HMDB served as initial seed nodes. A scoring system was established to further screen reliable molecular formulas from the possible candidates based on the topological relationship in the reaction network, mass accuracy, and isotopic fine structure. Finally, the developed method was applied to a high-throughput study of serum metabolomics in diabetes.

2. Materials and Methods

2.1. Chemicals

LC-MS grade formic acid was purchased from J&K Scientific Ltd. (Beijing, China). HPLC grade acetonitrile, methanol, and chloroform were purchased from Merck (Darmstadt, Germany). Ultrapure water was prepared by a Milli-Q Ultrapure water system (Millipore, Billerica, MA, USA). Seventy-eight metabolites used for evaluating annotation accuracy were supplied by Sigma-Aldrich (St. Louis, MO, USA) (Table S1). Nine stable isotope labeled internal standards (ISs) were used to normalize MS features in DI-nESI-HRMS. (Table S2). Carnitine C2:0-d_3, carnitine C12:0-d_3, and carnitine C16:0-d_3 were purchased from International Laboratory (South San Francisco, CA, USA). Choline-d_4, phenylalanine-d_5 (Phe-d5), tryptophan-(indole-d_5) (Trp-d5), leucine-d_3 (Leu-d3), glutamine-d_5, cholic acid-d_4 (CA-d_4), chenodeoxycholic acid-d_4 (CDCA-d_4) were purchased from Sigma-Aldrich (St. Louis, MO, USA).

2.2. Sample Information and Preparation

The serum samples were collected from 38 healthy control individuals and 31 type 2 diabetes (T2D) patients at The Second Hospital of Dalian Medical University. All enrolled participants had written consents and the research protocol (No. 2019(124)) was approved by the Ethics and Human Subjects Committee of the Second Hospital of Dalian Medical University. Detailed clinical data of the samples are summarized in Table S3.

A pooled sample was prepared by combining 10 µL aliquots from each serum sample as a quality control (QC) sample. The working solution of IS was mixed in methanol for subsequent protein precipitation of serum samples. Method linearity was assessed with the mixture of 10 metabolite standards at 6 different concentrate levels (Table S4).

For DI-nESI HRMS analysis, routine protein precipitation was first performed. Then, 200 µL of ice-cold methanol containing IS was added into a 1.5 mL Eppendorf tube containing 50 µL of a serum sample, followed by thorough vortexing for 1 min, and centrifuged for 10 min at 4 °C and $15,000 \times g$. 220 µL of the supernatant was vacuum concentrated at 4 °C by CentriVap Centrifugal Vacuum Concentrators (Labconco, Kansas City, MO, USA). Then, the freeze-dried residue was diluted 20 times with methanol/water 2:1 (v/v) containing 0.1% formic acid. After vortexing for 1 min, the sample solution was centrifuged at 4 °C and $15,000 \times g$ for 10 min. Finally, the supernatant was transferred into 96-well plates for DI-nESI HRMS analysis.

2.3. DI-nESI HRMS Analysis

In DI-nESI HRMS analysis, TriVersa Nanomate chip electrospray system (Advion BioSciences, Ithaca, NY, USA) was coupled to a Q Exactive HF (Thermo Fisher Scientific, Rockford, IL, USA). For the NanoMate system, the voltage and gas pressure were set as 1.7 kV and 0.6 psi, respectively. For mass spectrometry acquisition, the capillary temperature was set at 270 °C and S-lens level was 50. The data of the full scan based on the spectral-stitching acquisition were acquired in positive ionization mode and the mass resolution at m/z 200 was 240,000, the scan windows were set as follows: m/z 65–235, m/z 225–315, m/z 305–355, m/z 345–395, m/z 385–435, m/z 425–475, m/z 465–515, m/z 505–562, m/z 552–609. Maximum IT was 200 ms and Microscan was set as 3. The total analysis time was 0.6 min per sample. The spectrum data type above all acquisition methods was set as centroid mode.

2.4. Data Processing

For DI-nESI-Orbitrap MS analysis, Xcalibur software (Thermo Fisher Scientific, Rockford, IL, USA) was used to visualize and process rawdata files. Theoretical formulas for MS features were generated according to the rules of elemental type (C, H, O, N, P, S), hydrogen/carbon ratios (0.4~5.1) and isotopic fine structure, and further filtration restricting elemental composition and reaction network analysis was performed using in-house Python scripts. The subsequent steps were processed including peak lists export, noise filtration (S/N > 10), peaks alignment (5 ppm), and blank reduction.

The MS feature intensities were normalized by ISs and only those features with a relative standard deviation (RSD) less than 30% in QC were retained for further statistical analysis. Finally, the peak table was used for statistical analyses, and the partial least squares-discriminant analysis (PLS-DA) was performed. Significantly differential metabolites between health and T2D diabetes were screened out by nonparametric tests ($p < 0.05$) and VIP > 1.

An initial reaction network was constructed using unique molecular formulas (seed nodes) extracted from the Human Metabolome Database (HMDB, https://hmdb.ca/metabolites) (accessed on 16 December 2022). Three filtering criteria were employed for the selection of seed metabolites. Criterion 1 included metabolites filtered by the biospecimen of "Blood" and the origin of "Endogenous", criterion 2 included metabolites filtered by the origin of "Endogenous", and criterion 3 included all metabolites in the HMDB without any filtering. After restricting the elemental composition to C, H, O, N, P, and S, non-redundant formulas within a mass range of 50~800 were retained. To establish edges between nodes, the differences between any two formulas were obtained. If they matched the pre-defined reactions listed in Table S5, edges between nodes were established. Next, all possible formula candidates of experimental features were generated based on their accurate mass. The formula differences between the candidates and the nodes in the initial reaction network were calculated using custom Python scripts. If the formula differences matched the pre-defined reactions, edges were connected between the possible formula candidates and the nodes in the initial reaction network. The topological parameters were obtained by network analysis. The molecular network was visualized using Cytoscape (version 3.8.0).

3. Results

3.1. Workflow of the Developed Method

The entire workflow of the developed comprehensive characterization of serum metabolome is displayed in Figure 1. First, untargeted serum metabolome analysis was performed using spectral-stitching acquisition based on DI-nESI HRMS. Possible formula candidates for all the detected monoisotopic MS features were generated according to the rules outlined in the Data processing section with a mass accuracy of 2 ppm. A reaction network approach was then used to screen reliable formula candidates, using 76 predefined biological reactions as edge species (Table S5). The initial reaction network was constructed using unique molecular formula records in the metabolome database as seed nodes. The nodes were connected when their formula difference met with the predefined reactions. Then, the formula difference between all possible candidates and nodes in the initial reaction network was calculated. If the formula difference satisfies the predefined reaction, a connection is established between the candidate formula and seed nodes. For the monoisotopic MS features with multiple formula candidates connected with the initial reaction network, a scoring system was applied to select the top-ranked candidate as the unique formula. These unique formula candidates were newly-seeded nodes to integrate with the initial seeds and reconstructed the reaction network. For the monoisotopic MS features without any formula candidates connected to the initial reaction network, the formula candidates were screened using the reconstructed network again. This filtering process was repeated until no additional features were assigned with unique formulas. Finally, all the unique formula assignments were summarized.

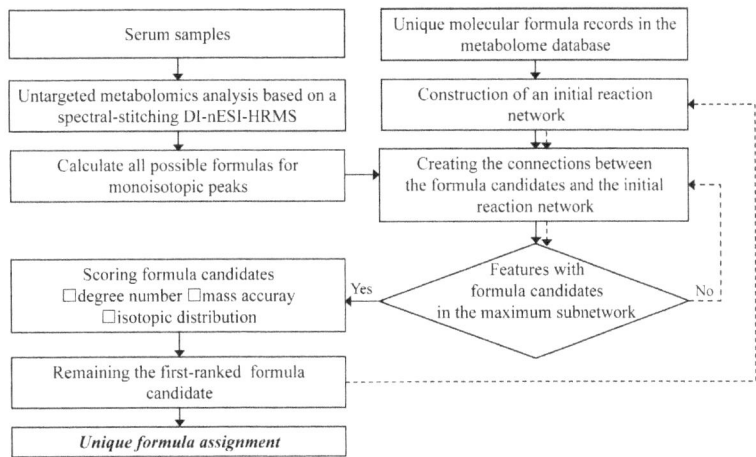

Figure 1. The workflow of a comprehensive serum metabolome analysis based on DI-nESI HRMS.

3.2. The Method Establishment

To evaluate the mass accuracy of spectral-stitching DI-nESI HRMS, 78 reference metabolites dissolved in pure solvent (standard mixture) were analyzed (Table S1). A total of 101 monoisotopic MS features related to the standard mixture, including three adduct types ([M+H]$^+$, [M+Na]$^+$, and [M+K]$^+$) were detected. As shown in Figure S1, the mass error distribution of the metabolites was mainly clustered within 2 ppm and near zero. Therefore, a mass error of 2 ppm was used in the following formula assignment.

Metabolism in a biological system is highly interconnected through metabolic reactions. Thus, the relevance between potential formula candidates and known endogenous metabolites in a reaction network can be leveraged to estimate the reliability of the formula assignment. The HMDB is a comprehensive database that contains known human metabolites [11]. To improve the accuracy and coverage of formula assignments, metabolites were selected from the HMDB and served as seed nodes for constructing an initial reaction network. To investigate the impact of seed nodes on formula assignment, three filtering criteria were employed for obtaining "endogenous metabolites from blood" (criterion 1), "endogenous metabolites" (criterion 2), and "all metabolites" (criterion 3) in the HMDB database. A total of 1650, 5620, and 12,351 non-redundant formulas were obtained for criteria 1, 2, and 3, respectively. Three initial metabolic reaction networks were constructed, in which 1457/4536, 5324/27,598, and 11,597/72,383 nodes/edges were obtained using criteria 1, 2, and 3 (Figure 2).

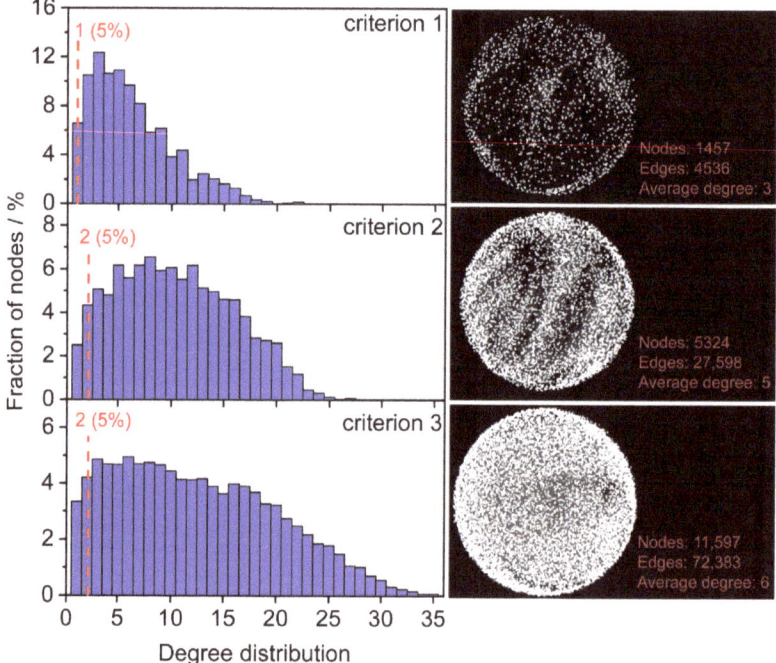

Figure 2. Degree distributions of seed nodes in the initial network. Criterion 1 represents endogenous metabolites from blood (1650), criterion 2 represents endogenous (5620) and criterion 3 represents all metabolites in the HMDB (12,351).

Degree distributions of the three initial metabolic reaction networks were evaluated (Figure 2). With the increase of non-redundant formulas from 1650 to 12,351, the nodes display closer connections of the average degree from 3 to 6. The fractions of nodes with a degree value less than three were 17.1% (criterion 1), 6.8% (criterion 2), and 7.5% (criterion 3), respectively. The initial network constructed by endogenous metabolites (criterion 2) had the lowest proportion of low-degree nodes. Considering formula candidates with a high degree may present higher confidence, the cut-off values of the degree to the three initial reaction networks were set 2, 3, and 3, respectively.

The above standard mixture was used to assess three initial reaction networks. The overlap of the molecular formula between the standard mixture and metabolites in criteria 1, 2, and 3 was 58, 72, and 77, respectively (Figure S2). Using a mass accuracy of 2 ppm, 457 possible formula candidates for 101 features in the standard mixture were generated. All the possible formula candidates were further connected with three initial reaction networks. Among these possible formula candidates, 140 possible formula candidates corresponding to 98 features (criterion 1), 166 formula candidates corresponding to 101 features (criterion 2), and 198 possible formula candidates corresponding to 101 features have a network connection. A total of 119/140, 127/166, and 171/198 possible formula candidates had one more (criterion 1) and two more neighbor nodes (criteria 2 and 3). Through the network filter, the number of MS features with multiple formula candidates increased from 19 (criterion 1) to 23 (criterion 2) and 32 (criterion 3), respectively.

After filtering by the reaction network, there were still many monoisotopic MS features with multiple formula candidates. A scoring rule was established to further rank formula candidates using three alternative scoring criteria, including mass accuracy, isotope pattern, and degree of formula candidates (Equation (1)).

$$\text{Score} = W_{\text{degree}} \text{Score}_{\text{degree}} - W_{m/z} \text{Score}_{m/z} + W_{\text{iso}} \text{Score}_{\text{iso}} \quad (1)$$

The W_{degree}, $W_{m/z}$ and W_{iso} represent the weight coefficients of each criterion. $\text{Score}_{m/z}$ represents the mass accuracy score between experimental m/z (m/z_E) and theoretical m/z (m/z_T) values (Equation (2)):

$$\text{Score}_{m/z} = \frac{|(m/z_E - m/z_T)|}{m/z_E \times 2} \times 10^6 \quad (2)$$

$\text{Score}_{\text{iso}}$ represents the similarity of isotopic distribution between experimental and theory isotopic patterns (Equation (3)):

$$\text{Score}_{\text{iso}} = 0.5 \times \text{Score}_{mz} + 0.5 \times \text{Score}_{\text{int}} \quad (3)$$

In which $\text{Score}_{mz} = 1 - \frac{|(m/z_E - m/z_T)|/(m/z_T) \times 10^6}{\text{Tolerance}_{m/z}}$ and $\text{Score}_{\text{int}} = 1 - \frac{|(\text{int}_E - \text{int}_T)|/(\text{int}_T) \times 10^6}{\text{Tolerance}_{\text{int}}}$. Where int_E and int_T are the experimental and theoretical relative intensity, respectively. $\text{Tolerance}_{m/z}$ and $\text{Tolerance}_{\text{int}}$ represent the mass tolerance of 2 ppm and the relative intensity tolerance of 500%, respectively [38].

The effects of the weight coefficients on formula assignments were evaluated using the above standard mixture (Figure S3). If only accurate mass filtering ($W_{\text{degree}} = 0$, $W_{m/z} = 1$, $W_{\text{iso}} = 0$) was used, the correct assignment rate for standards was 73.3~81.2%. If degree filtering was considered alone (with $W_{\text{degree}} = 1$, $W_{m/z} = 0$, $W_{\text{iso}} = 0$), the correct assignment rate for standards increased to 85.1~87.1%. The combination of all three scoring criteria resulted in a correct assignment rate of approximately 90%. The optimal weight coefficients were determined to be $W_{\text{degree}} = 0.5$, $W_{m/z} = 0.3$, and $W_{\text{iso}} = 0.2$.

The assignment performance of three initial reaction networks for the standard mixture was compared using the optimal weight coefficients (Figure 3). With the increase of non-redundant formulas from 1650 to 12,351, the correct assignment rate slightly increased from 89.1% to 93.1%. Using criteria 1 and 2, 4% of monoisotopic MS features were not retained any candidates because they lacked connections with the initial network. Furthermore, to assess the performance of the three initial networks, a target-decoy strategy was implemented. A set of 1000 known formulas presented in the HMDB database (blood matrix) and 1000 decoy formulas absent in the HMDB database (token from the ChEMBL database) were used. The line charts in Figure 3 demonstrate a similar filtering trend between the known/decoy formulas and the standard mixture. The percentage of formula assignment rates for known formulas increased slightly from 86.3%, 91.0% to 97.9% (solid line), while the percentage of formula assignment rates for decoy formulas significantly increased from 14.9%, 22.6% to 61.5% (dash line). The results implied that initial networks under criteria 1 and 2 exhibited more specific assignments for the biosample compared to criterion 3. As a result, the initial network consisting of 5620 endogenous metabolites (criterion 2) demonstrated superior assignment performance and was used for the subsequent characterization.

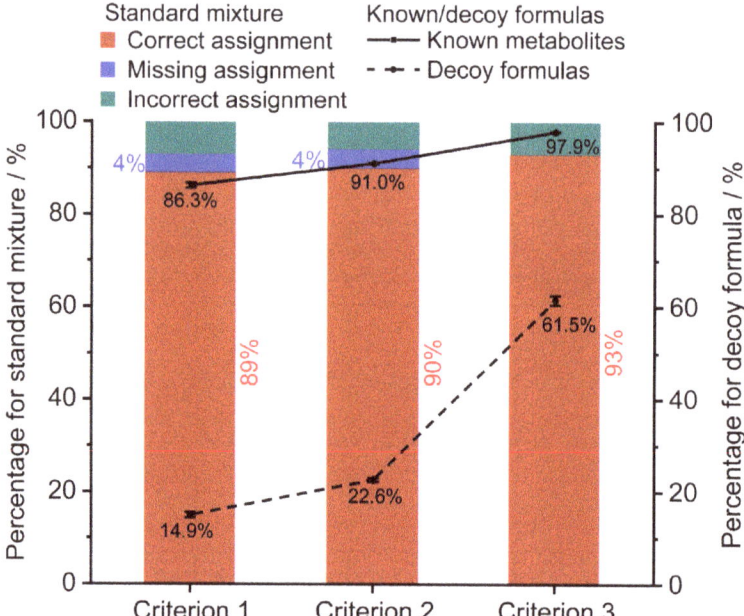

Figure 3. The effects of the initial networks on the assignment performance. The formula assignment results of the standard mixture (stacked column), 1000 blood metabolites (solid line), and 1000 decoy formulas (dash line). Criterion 1, 2, and 3 represent the initial network constructed by 1650 endogenous formulas from blood, 5620 endogenous formulas, and all the 12,351 formulas in the HMDB database, respectively. Error bars represent ±SE (n = 3).

3.3. Method Validation

A pooled serum sample spiked 78 reference standards was used for method validation. As shown in Figure 4A, 96 monoisotopic MS features were detected to be associated with 68 reference standards, which were assigned 353 potential formula candidates. Using the workflow described above, 95 out of the 96 monoisotopic MS features were successfully assigned with unique formula candidates through the two rounds of assignments (Figure 4A, stacked column). A total of 89 out of 95 assigned features were correctly allocated unique formulas. As a result, the coverage and accuracy rates were 98.9% (red solid line) and 93.6% (red dash line), respectively. For the spiked pooled serum sample, 3140 monoisotopic MS features were assigned with 19,474 possible formula candidates. Among these features, 1958 features were assigned with unique formula candidates after seven rounds of assignments (grey columns). The coverage rate was up to 62.3% (grey solid line). It is noteworthy that 18.1% of features were assigned between the second and seventh rounds.

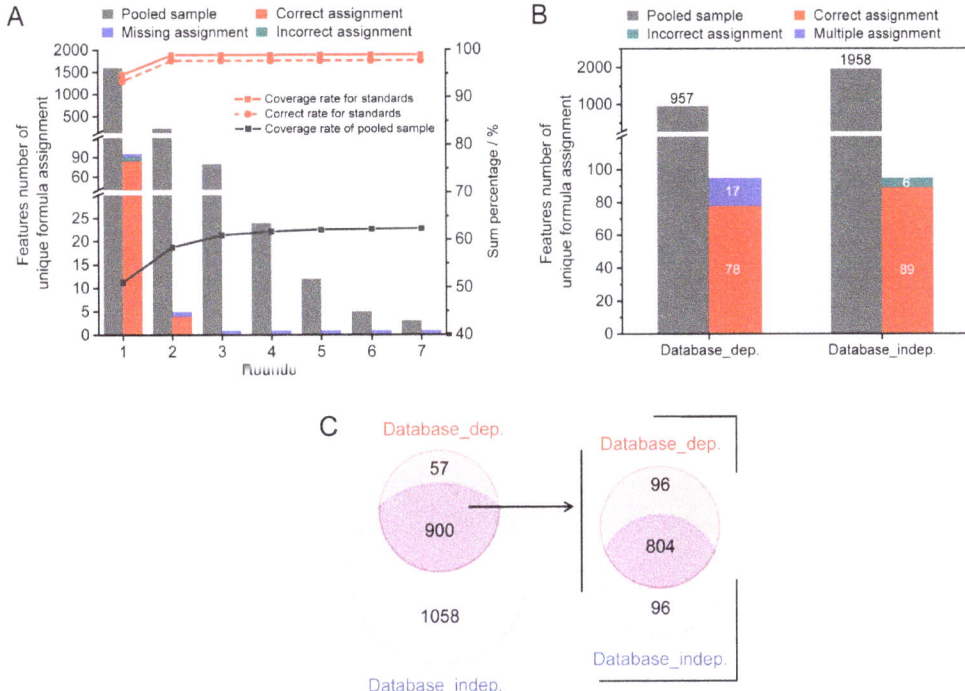

Figure 4. The method validation results using a spiked serum sample. (**A**) The coverage and accuracy rates of the developed method. (**B**) Comparison of assignment performance between the developed method and the database search. (**C**) The Venn diagram of assignment results of both methods for the spiked pooled serum sample.

The assignment performance of the developed method was compared with the conventional formula assignments method of searching m/z against the metabolome database (database-dependent method). The results are shown in Figure 4B. For the 68 spiked reference standards, the coverage rates of the two methods were similar (99%). However, the accuracy rates of unambiguous formula candidates were improved from 81.3% (78/96) to 92.7% (89/96) by the developed method. A total of 17.7 % (17/96) of MS features obtained multiple formula candidates by searching m/z against the HMDB (stacked column). For the spiked pooled serum sample, only 957 MS features were able to obtain unambiguous formula candidates by HMDB database search using a mass tolerance of 2 ppm, which yielded less than half the number of formulas compared to the developed method. The Venn diagram of the assigned monoisotopic MS features by two methods is shown in Figure 4C. It can be observed that the assigned monoisotopic MS features by the database search had good agreement with the developed method (900 out of 957). Among these 900 shared assigned monoisotopic MS features, about 89% (804/900) of MS features annotated consistent molecular formulas. The results indicated that the developed method showed high coverage and accuracy for formula assignment in the characterization of complex biosamples.

3.4. Application for Serum Metabolomic Analysis in Diabetes

To verify the practicability of the developed method, it was used to investigate the metabolic alterations in diabetes. Table S3 lists the detailed clinical information of participants corresponding to two groups of healthy control and diabetes patients. No significant differences except for blood glucose concentration were observed between the two groups.

The repeatability of the developed acquisition method was measured using QC samples. The RSD values were estimated (Figure S4A). It shows that 86% of the detected MS features had RSD less than 30%, accounting for 96% of the sum peak area. Then, only those assigned features with an RSD of less than 30% in QC were subjected to further statistical analysis. The method linearities were evaluated using 10 reference metabolites (Table S4). The method exhibited good linearities with linear correlation coefficients (R^2) of 0.9975–1, and linear ranges spanning two to four orders of magnitude. The results demonstrated that the current acquisition method was appropriate for serum metabolomics analysis.

Figure S4B represents a PLS-DA score plot of two groups. The separated clusters were observed between healthy control and patients. To further validate the classification models, the permutation test was performed (M = 200). The intercept values of R^2 and Q^2 were below 0.4 and −0.05, respectively, which indicated no overfitting of the PLS-DA model (Figure S4C).

A total of 57 significantly differential formulas were obtained between healthy control and T2D patients with VIP > 1 and $p < 0.05$ (Table S6). A heat map of the significantly changed metabolic features revealed good clustering between the control and T2D groups (Figure 5A). Four typical metabolites ($[C_6H_{12}O_6+Na]^+$, $[C_6H_{13}NO_2+H]^+$, $[C_6H_{11}NO_2+Na]^+$ and $[C_7H_{15}NO_3+H]^+$) are displayed in Figure 5B. Compared with the controls, significantly increased $[C_6H_{12}O_6+Na]^+$ (glucose), $[C_6H_{13}NO_2+H]^+$ (Leucine/Isoleucine) and decreased $[C_6H_{11}NO_2+Na]^+$ (pipecolinic acid), $[C_7H_{15}NO_3+H]^+$ (carnitine) in diabetes were observed. As shown in the clinical information (Table S3), blood glucose existed a significant difference between healthy and T2D groups, it was consistent with our result as $[C_6H_{12}O_6+Na]^+$. In the previous study, the abnormal concentrations of carnitines also appeared in the T2D groups owing to the fatty acid oxidation dysregulation [39,40], our result showed a decrease of $[C_7H_{15}NO_3+H]^+$ in T2D group. A positive association of branch-chain amino acids in diabetes had been concluded by modulating insulin secretion and leading to the pancreatic b-cell exhaustion [41,42]. Leucine/isoleucine ($C_6H_{13}NO_2$) recognized as branch chain amino acids had a robust association with the risk of T2D, a similar trend with $[C_6H_{13}NO_2+H]^+$ is shown in Figure 5B. In addition, Ouyang et al. observed that pipecolinic acid ($C_6H_{11}NO_2$) was decreased in T2D patients in a Chinese prospective cohort study [43].

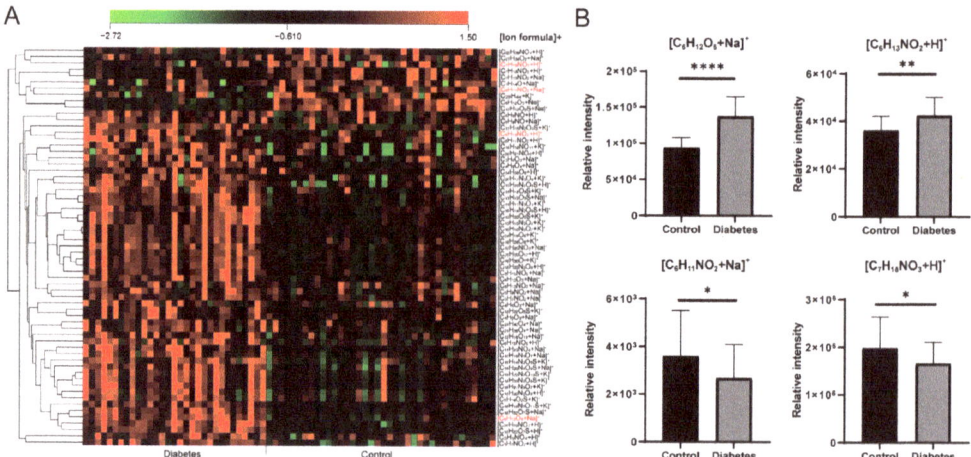

Figure 5. (**A**) The heat map of significantly changed ion formulas between control and diabetes groups, (**B**) Changes of $[C_6H_{12}O_6+Na]^+$, $[C_6H_{13}NO_2+H]^+$, $[C_6H_{11}NO_2+Na]^+$ and $[C_7H_{15}NO_3+H]^+$ between control and diabetes groups, respectively. **** $p < 0.0001$, ** $p < 0.01$, * $p < 0.05$.

4. Discussion

DI-nESI HRMS is a promising high-throughput analytical method for large-scale cohort metabolomic study [44,45]. High coverage and reliable assignment of MS features is a critical prerequisite for comprehensive metabolomics analysis [46]. In the present study, we aimed to develop a comprehensive metabolomics analysis method in direct-infusion high-resolution mass spectrometry. Our result revealed that based on a reaction network combing with mass accuracy and isotopic distribution, more MS peaks could be assigned with a confident unique formula.

For untargeted metabolomics analysis based on DI-HRMS, the most widely used approach for formula assignments was searching m/z against metabolome databases. According to the performance of mass resolution (Rs) and mass accuracy in the high-resolution mass spectrometer, appropriate mass tolerance parameter was obtained, such as 10~20 ppm for time-of-flight (Rs on the order of 35,000), 5~15 ppm for orbitrap (Rs on the order of 200,000 at m/z 200) and 0.5 ppm for Fourier transform ion cyclotron mass spectrometers (Rs 900,000 at m/z 200) [47,48]. However, even with tremendous advances in database size and scape, the identification of metabolites still is a primary challenge in the field [49,50].

The relationship between metabolites attracted wide attention for formula identification in untargeted DI ultrahigh resolution MS. Reaction network approach based on mass difference had been used for MS feature assignments, in which experimental features as nodes and chemical reactions as edges [19,22]. However, reliable and enough experimental features need to be first assigned as initial seed nodes. It is difficult to achieve for the untargeted metabolomics analysis using HRMS. Instead of only experimental features used for the reaction network, the formula records from the metabolome database were used to construct the initial reaction network in this study. The connections between formula candidates and the initial reaction network were established for filtering multiple candidates. The assignment performance was improved due to sufficient and accurate known formulas from the initial network. Furthermore, using assigned unambiguous formula candidates as newly added seeds to reconstruct networks, multiple rounds of annotation were designed to further improve annotation efficiency (1.22-fold increase after 7 rounds in this study). Integration of mass accuracy and isotopic pattern filter resulted in a further improvement in assignment accuracy.

Although the developed method demonstrated more than twice the number of unique assigned formulas for a pooled serum sample compared to the database search, still about 38% of MS features could not obtain unambiguous formula candidates. The false negatives are possibly caused by the initial network filter. To avoid false positives, the initial network in this study was constructed using only the most common metabolic reactions and endogenous formulas from the HMDB database. Furthermore, the excellent assignment performance for serum untargeted metabolic features is partly attributed to the comprehensive HMDB database for the construction of the initial reaction network. Nevertheless, for the less studied objects, the method performance may be suboptimal because the initial network can't be established or the initial network size is too small. Furthermore, it is essential to accurately confirm the structures of differential metabolites and then determine key differential metabolites from them in subsequent studies.

5. Conclusions

A comprehensive serum metabolome characterization method in DI-nESI HRMS was proposed in this study. High coverage, high confidence, and database-independent formula assignments were achieved by reaction network combined with mass accuracy and isotopic pattern filter. A total of 1958 monoisotopic features were assigned the unique formula in a pooled serum sample, while only 957 were annotated by searching m/z against HMDB. This method has the advantages of comprehensive formula assignment and high-speed acquisition, it has great application potential in large-scale cohort metabolomic studies.

Supplementary Materials: The following supporting information can be downloaded at: https://www.mdpi.com/article/10.3390/metabo13030460/s1, Figure S1: Mass error distribution of metabolites standards; Figure S2: The overlap between the standards formulas and seed nodes in three initial reaction networks; Figure S3: Weight coefficients optimization for the degree, mass error and isotopic distribution; Figure S4: (A) Estimation of detection repeatability. (B) PLS-DA scores plot of healthy and diabetes groups. (C) Cross-validation plots of the PLS-DA model; Table S1: Information of metabolites standards; Table S2: Information of internal standards; Table S3: Clinical information of participants; Table S4: Information of 10 standards for linearity; Table S5. Reactions types; Table S6. Significantly changed features between healthy and T2D groups.

Author Contributions: Conceptualization, X.S., C.Z., X.Z., X.L. and G.X.; methodology, X.S. and Z.J.; software, Z.J.; validation, X.S.; formal analysis, X.S.; investigation, X.S.; resources, Y.Z.; data curation, X.S. and Z.J.; writing—original draft preparation, X.S.; writing—review and editing, C.Z., X.Z., X.L. and G.X.; supervision, X.Z., X.L. and G.X.; funding acquisition, X.L. and G.X. All authors have read and agreed to the published version of the manuscript.

Funding: This research was funded by the National Natural Science Foundation of China (21934006, 22274153, 22174141, 22274151, and 22074145), the Innovation Program (DICP I202004) of Science and Research from Dalian Institute of Chemical Physics, Chinese Academy of Sciences, and the AI S&T Program (Grant No. DNL-YL A202202) from Yulin Branch, Dalian National Laboratory for Clean Energy, CAS, China.

Institutional Review Board Statement: The study was approved by the Ethics Committee of the Second Hospital of Dalian Medical University (2019(124) and October 2019).

Informed Consent Statement: Informed consent was obtained from all subjects involved in the study.

Data Availability Statement: Data will be made available upon request from the corresponding author. The data are not publicly available due to privacy.

Conflicts of Interest: The authors declare no conflict of interest.

References

1. Southam, A.D.; Weber, R.J.M.; Engel, J.; Jones, M.R.; Viant, M.R. A complete workflow for high-resolution spectral-stitching nanoelectrospray direct-infusion mass-spectrometry-based metabolomics and lipidomics. *Nat. Protoc.* **2016**, *12*, 310–328. [CrossRef] [PubMed]
2. Habchi, B.; Alves, S.; Jouan-Rimbaud Bouveresse, D.; Moslah, B.; Paris, A.; Lécluse, Y.; Gauduchon, P.; Lebailly, P.; Rutledge, D.N.; Rathahao-Paris, E. An innovative chemometric method for processing direct introduction high resolution mass spectrometry metabolomic data: Independent component–discriminant analysis (IC–DA). *Metabolomics* **2017**, *13*, 45. [CrossRef]
3. Chekmeneva, E.; Dos Santos Correia, G.; Gómez-Romero, M.; Stamler, J.; Chan, Q.; Elliott, P.; Nicholson, J.K.; Holmes, E. Ultra-Performance Liquid Chromatography–High-Resolution Mass Spectrometry and Direct Infusion–High-Resolution Mass Spectrometry for Combined Exploratory and Targeted Metabolic Profiling of Human Urine. *J. Proteome Res.* **2018**, *17*, 3492–3502. [CrossRef]
4. Pöhö, P.; Lipponen, K.; Bespalov, M.M.; Sikanen, T.; Kotiaho, T.; Kostiainen, R. Comparison of liquid chromatography-mass spectrometry and direct infusion microchip electrospray ionization mass spectrometry in global metabolomics of cell samples. *Eur. J. Pharm. Sci.* **2019**, *138*, 104991. [CrossRef] [PubMed]
5. Malinowska, J.M.; Palosaari, T.; Sund, J.; Carpi, D.; Bouhifd, M.; Weber, R.J.M.; Whelan, M.; Viant, M.R. Integrating in vitro metabolomics with a 96-well high-throughput screening platform. *Metabolomics* **2022**, *18*, 11. [CrossRef] [PubMed]
6. Bowen, T.J.; Hall, A.R.; Lloyd, G.R.; Weber, R.J.M.; Wilson, A.; Pointon, A.; Viant, M.R. An Extensive Metabolomics Workflow to Discover Cardiotoxin-Induced Molecular Perturbations in Microtissues. *Metabolites* **2021**, *11*, 644. [CrossRef]
7. González-Domínguez, R.; Sayago, A.; Fernández-Recamales, Á. Direct infusion mass spectrometry for metabolomic phenotyping of diseases. *Bioanalysis* **2017**, *9*, 131–148. [CrossRef]
8. Chekmeneva, E.; Dos Santos Correia, G.; Chan, Q.; Wijeyesekera, A.; Tin, A.; Young, J.H.; Elliott, P.; Nicholson, J.K.; Holmes, E. Optimization and Application of Direct Infusion Nanoelectrospray HRMS Method for Large-Scale Urinary Metabolic Phenotyping in Molecular Epidemiology. *J. Proteome Res.* **2017**, *16*, 1646–1658. [CrossRef]
9. de Sain-van der Velden, M.G.M.; van der Ham, M.; Gerrits, J.; Prinsen, H.C.M.T.; Willemsen, M.; Pras-Raves, M.L.; Jans, J.J.; Verhoeven-Duif, N.M. Quantification of metabolites in dried blood spots by direct infusion high resolution mass spectrometry. *Anal. Chim. Acta* **2017**, *979*, 45–50. [CrossRef]
10. Haijes, H.A.; van der Ham, M.; Gerrits, J.; van Hasselt, P.M.; Prinsen, H.C.M.T.; de Sain-van der Velden, M.G.M.; Verhoeven-Duif, N.M.; Jans, J.J.M. Direct-infusion based metabolomics unveils biochemical profiles of inborn errors of metabolism in cerebrospinal fluid. *Mol. Genet. Metab.* **2019**, *127*, 51–57. [CrossRef] [PubMed]

11. Haijes, H.A.; Willemsen, M.; Van der Ham, M.; Gerrits, J.; Pras-Raves, M.L.; Prinsen, H.C.M.T.; Van Hasselt, P.M.; De Sain-van der Velden, M.G.M.; Verhoeven-Duif, N.M.; Jans, J.J.M. Direct Infusion Based Metabolomics Identifies Metabolic Disease in Patients' Dried Blood Spots and Plasma. *Metabolites* **2019**, *9*, 12. [CrossRef] [PubMed]
12. Wang, L.; Lv, W.; Sun, X.; Zheng, F.; Xu, T.; Liu, X.; Li, H.; Lu, X.; Peng, X.; Hu, C.; et al. Strategy for Nontargeted Metabolomic Annotation and Quantitation Using a High-Resolution Spectral-Stitching Nanoelectrospray Direct-Infusion Mass Spectrometry with Data-Independent Acquisition. *Anal. Chem.* **2021**, *93*, 10528–10537. [CrossRef] [PubMed]
13. Ludwig, M.; Nothias, L.-F.; Dührkop, K.; Koester, I.; Fleischauer, M.; Hoffmann, M.A.; Petras, D.; Vargas, F.; Morsy, M.; Aluwihare, L.; et al. Database-independent molecular formula annotation using Gibbs sampling through ZODIAC. *Nat. Mach. Intell.* **2020**, *2*, 629–641. [CrossRef]
14. Wolthuis, J.C.; Magnusdottir, S.; Pras-Raves, M.; Moshiri, M.; Jans, J.J.M.; Burgering, B.; van Mil, S.; de Ridder, J. MetaboShiny: Interactive analysis and metabolite annotation of mass spectrometry-based metabolomics data. *Metabolomics* **2020**, *16*, 99. [CrossRef]
15. Mitchell, J.M.; Flight, R.M.; Moseley, H.N.B. Small Molecule Isotope Resolved Formula Enumeration: A Methodology for Assigning Isotopologues and Metabolite Formulas in Fourier Transform Mass Spectra. *Anal. Chem.* **2019**, *91*, 8933–8940. [CrossRef] [PubMed]
16. Haijes, H.A.; Willemse, E.A.; Gerrits, J.; van der Flier, W.M.; Teunissen, C.E.; Verhoeven-Duif, N.M.; Jans, J.J.M. Assessing the Pre-Analytical Stability of Small-Molecule Metabolites in Cerebrospinal Fluid Using Direct-Infusion Metabolomics. *Metabolites* **2019**, *9*, 236. [CrossRef] [PubMed]
17. Sarvin, B.; Lagziel, S.; Sarvin, N.; Mukha, D.; Kumar, P.; Aizenshtein, E.; Shlomi, T. Fast and sensitive flow-injection mass spectrometry metabolomics by analyzing sample-specific ion distributions. *Nat. Commun.* **2020**, *11*, 3186. [CrossRef]
18. Southam, A.D.; Payne, T.G.; Cooper, H.J.; Arvanitis, T.N.; Viant, M.R. Dynamic Range and Mass Accuracy of Wide-Scan Direct Infusion Nanoelectrospray Fourier Transform Ion Cyclotron Resonance Mass Spectrometry-Based Metabolomics Increased by the Spectral Stitching Method. *Anal. Chem.* **2007**, *79*, 4595–4602. [CrossRef]
19. Sud, M.; Fahy, E.; Cotter, D.; Brown, A.; Dennis, E.A.; Glass, C.K.; Merrill, A.H.; Murphy, R.C.; Raetz, C.R.H.; Russell, D.; et al. LMSD: LIPID MAPS structure database. *Nucleic Acids Res.* **2007**, *35*, D527–D532. [CrossRef]
20. Wishart, D.S.; Feunang, Y.D.; Marcu, A.; Guo, A.C.; Liang, K.; Vázquez-Fresno, R.; Sajed, T.; Johnson, D.; Li, C.; Karu, N.; et al. HMDB 4.0: The human metabolome database for 2018. *Nucleic Acids Res.* **2018**, *46*, D608–D617. [CrossRef]
21. Wishart, D.S.; Guo, A.; Oler, E.; Wang, F.; Anjum, A.; Peters, H.; Dizon, R.; Sayeeda, Z.; Tian, S.; Lee, B.L.; et al. HMDB 5.0: The Human Metabolome Database for 2022. *Nucleic Acids Res.* **2022**, *50*, D622–D631. [CrossRef] [PubMed]
22. Kind, T.; Fiehn, O. Seven Golden Rules for heuristic filtering of molecular formulas obtained by accurate mass spectrometry. *BMC Bioinform.* **2007**, *8*, 105. [CrossRef] [PubMed]
23. Huang, D.; Bouza, M.; Gaul, D.A.; Leach, F.E.; Amster, I.J.; Schroeder, F.C.; Edison, A.S.; Fernández, F.M. Comparison of High-Resolution Fourier Transform Mass Spectrometry Platforms for Putative Metabolite Annotation. *Anal. Chem.* **2021**, *93*, 12374–12382. [CrossRef] [PubMed]
24. Zielinski, A.T.; Kourtchev, I.; Bortolini, C.; Fuller, S.J.; Giorio, C.; Popoola, O.A.M.; Bogialli, S.; Tapparo, A.; Jones, R.L.; Kalberer, M. A new processing scheme for ultra-high resolution direct infusion mass spectrometry data. *Atmospheric Environ.* **2018**, *178*, 129–139. [CrossRef]
25. Cao, D.; Hao, Z.; Hu, M.; Geng, F.; Rao, Z.; Niu, H.; Shi, Y.; Cai, Y.; Zhou, Y.; Liu, J.; et al. A feasible strategy to improve confident elemental composition determination of compounds in complex organic mixture such as natural organic matter by FTICR-MS without internal calibration. *Sci. Total. Environ.* **2021**, *751*, 142255. [CrossRef] [PubMed]
26. Fiehn, O.; Robertson, D.; Griffin, J.; van der Werf, M.; Nikolau, B.; Morrison, N.; Sumner, L.W.; Goodacre, R.; Hardy, N.W.; Taylor, C.; et al. The metabolomics standards initiative (MSI). *Metabolomics* **2007**, *3*, 175–178. [CrossRef]
27. Kind, T.; Fiehn, O. Metabolomic database annotations via query of elemental compositions: Mass accuracy is insufficient even at less than 1 ppm. *BMC Bioinform.* **2006**, *7*, 234. [CrossRef] [PubMed]
28. Alden, N.; Krishnan, S.; Porokhin, V.; Raju, R.; McElearney, K.; Gilbert, A.; Lee, K. Biologically Consistent Annotation of Metabolomics Data. *Anal. Chem.* **2017**, *89*, 13097–13104. [CrossRef] [PubMed]
29. Nielsen, J. Systems biology of metabolism. *Annu. Rev. Biochem.* **2017**, *86*, 245–275. [CrossRef]
30. Moritz, F.; Kaling, M.; Schnitzler, J.-P.; Schmitt-Kopplin, P. Characterization of poplar metabotypes via mass difference enrichment analysis. *Plant, Cell Environ.* **2017**, *40*, 1057–1073. [CrossRef]
31. Tziotis, D.; Hertkorn, N.; Schmitt-Kopplin, P. Kendrick-Analogous Network Visualisation of Ion Cyclotron Resonance Fourier Transform Mass Spectra: Improved Options for the Assignment of Elemental Compositions and the Classification of Organic Molecular Complexity. *Eur. J. Mass Spectrom.* **2011**, *17*, 415–421. [CrossRef]
32. Witting, M.; Lucio, M.; Tziotis, D.; Wägele, B.; Suhre, K.; Voulhoux, R.; Garvis, S.; Schmitt-Kopplin, P. DI-ICR-FT-MS-based high-throughput deep metabotyping: A case study of the Caenorhabditis elegans–Pseudomonas aeruginosa infection model. *Anal. Bioanal. Chem.* **2015**, *407*, 1059–1073. [CrossRef]
33. Amara, A.; Frainay, C.; Jourdan, F.; Naake, T.; Neumann, S.; Novoa-Del-Toro, E.M.; Salek, R.M.; Salzer, L.; Scharfenberg, S.; Witting, M. Networks and Graphs Discovery in Metabolomics Data Analysis and Interpretation. *Front. Mol. Biosci.* **2022**, *9*, 841373. [CrossRef] [PubMed]

34. Schmitt-Kopplin, P.; Hemmler, D.; Moritz, F.; Gougeon, R.D.; Lucio, M.; Meringer, M.; Müller, C.; Harir, M.; Hertkorn, N. Systems chemical analytics: Introduction to the challenges of chemical complexity analysis. *Faraday Discuss.* **2019**, *218*, 9–28. [CrossRef] [PubMed]
35. Traquete, F.; Luz, J.; Cordeiro, C.; Silva, M.S.; Ferreira, A.E.N. Graph Properties of Mass-Difference Networks for Profiling and Discrimination in Untargeted Metabolomics. *Front. Mol. Biosci.* **2022**, *9*, 917911. [CrossRef]
36. Chen, L.; Lu, W.; Wang, L.; Xing, X.; Chen, Z.; Teng, X.; Zeng, X.; Muscarella, A.D.; Shen, Y.; Cowan, A.; et al. Metabolite discovery through global annotation of untargeted metabolomics data. *Nat. Methods* **2021**, *18*, 1377–1385. [CrossRef] [PubMed]
37. Forcisi, S.; Moritz, F.; Lucio, M.; Lehmann, R.; Stefan, N.; Schmitt-Kopplin, P. Solutions for Low and High Accuracy Mass Spectrometric Data Matching: A Data-Driven Annotation Strategy in Nontargeted Metabolomics. *Anal. Chem.* **2015**, *87*, 8917–8924. [CrossRef] [PubMed]
38. Shen, X.; Wang, R.; Xiong, X.; Yin, Y.; Cai, Y.; Ma, Z.; Liu, N.; Zhu, Z.-J. Metabolic reaction network-based recursive metabolite annotation for untargeted metabolomics. *Nat. Commun.* **2019**, *10*, 1516. [CrossRef]
39. Sun, L.; Liang, L.; Gao, X.; Zhang, H.; Yao, P.; Hu, Y.; Ma, Y.; Wang, F.; Jin, Q.; Li, H.; et al. Early Prediction of Developing Type 2 Diabetes by Plasma Acyl-carnitines: A Population-Based Study. *Diabetes Care* **2016**, *39*, 1563–1570. [CrossRef] [PubMed]
40. Guasch-Ferré, M.; Ruiz-Canela, M.; Li, J.; Zheng, Y.; Bullo, M.; Wang, D.D.; Toledo, E.; Clish, C.; Corella, D.; Estruch, R.; et al. Plasma Acylcarnitines and Risk of Type 2 Diabetes in a Mediterranean Population at High Cardiovascular Risk. *J. Clin. Endocrinol. Metab.* **2019**, *104*, 1508–1519. [CrossRef]
41. Guasch-Ferré, M.; Hruby, A.; Toledo, E.; Clish, C.B.; Martínez-González, M.A.; Salas-Salvadó, J.; Hu, F.B. Metabolomics in Prediabetes and Diabetes: A Systematic Review and Meta-analysis. *Diabetes Care* **2016**, *39*, 833–846. [CrossRef] [PubMed]
42. Vanweert, F.; Schrauwen, P.; Phielix, E. Role of branched-chain amino acid metabolism in the pathogenesis of obesity and type 2 diabetes-related metabolic disturbances BCAA metabolism in type 2 diabetes. *Nutr. Diabetes* **2022**, *12*, 35. [CrossRef] [PubMed]
43. Ouyang, Y.; Qiu, G.; Zhao, X.; Su, B.; Feng, D.; Lv, W.; Xuan, Q.; Wang, L.; Yu, D.; Wang, Q.; et al. Metabolome-Genome-Wide Association Study (mGWAS) Reveals Novel Metabolites Associated with Future Type 2 Diabetes Risk and Susceptibility Loci in a Case-Control Study in a Chinese Prospective Cohort. *Glob. Chall.* **2021**, *5*, 2000088. [CrossRef]
44. Fuhrer, T.; Zamboni, N. High-throughput discovery metabolomics. *Curr. Opin. Biotechnol.* **2015**, *31*, 73–78. [CrossRef] [PubMed]
45. Kempa, E.E.; Hollywood, K.A.; Smith, C.A.; Barran, P.E. High throughput screening of complex biological samples with mass spectrometry–from bulk measurements to single cell analysis. *Analyst* **2019**, *144*, 872–891. [CrossRef] [PubMed]
46. Kozlova, A.; Shkrigunov, T.; Gusev, S.; Guseva, M.; Ponomarenko, E.; Lisitsa, A. An Open-Source Pipeline for Processing Direct Infusion Mass Spectrometry Data of the Human Plasma Metabolome. *Metabolites* **2022**, *12*, 768. [CrossRef] [PubMed]
47. Forsberg, E.M.; Huan, T.; Rinehart, D.; Benton, H.P.; Warth, B.; Hilmers, B.; Siuzdak, G. Data processing, multi-omic pathway mapping, and metabolite activity analysis using XCMS Online. *Nat. Protoc.* **2018**, *13*, 633–651. [CrossRef]
48. Thompson, C.J.; Witt, M.; Forcisi, S.; Moritz, F.; Kessler, N.; Laukien, F.H.; Schmitt-Kopplin, P. An Enhanced Isotopic Fine Structure Method for Exact Mass Analysis in Discovery Metabolomics: FIA-CASI-FTMS. *J. Am. Soc. Mass Spectrom.* **2020**, *31*, 2025–2034. [CrossRef] [PubMed]
49. Chaleckis, R.; Meister, I.; Zhang, P.; Wheelock, C.E. Challenges, progress and promises of metabolite annotation for LC–MS-based metabolomics. *Curr. Opin. Biotechnol.* **2019**, *55*, 44–50. [CrossRef]
50. Baygi, S.F.; Banerjee, S.K.; Chakraborty, P.; Kumar, Y.; Barupal, D.K. IDSL.UFA Assigns High-Confidence Molecular Formula Annotations for Untargeted LC/HRMS Data Sets in Metabolomics and Exposomics. *Anal. Chem.* **2022**, *94*, 13315–13322. [CrossRef] [PubMed]

Disclaimer/Publisher's Note: The statements, opinions and data contained in all publications are solely those of the individual author(s) and contributor(s) and not of MDPI and/or the editor(s). MDPI and/or the editor(s) disclaim responsibility for any injury to people or property resulting from any ideas, methods, instructions or products referred to in the content.

Review

Unlocking the Potential: Amino Acids' Role in Predicting and Exploring Therapeutic Avenues for Type 2 Diabetes Mellitus

Yilan Ding [1,2,†], Shuangyuan Wang [1,2,†] and Jieli Lu [1,2,*]

[1] Department of Endocrine and Metabolic Diseases, Shanghai Institute of Endocrine and Metabolic Diseases, Ruijin Hospital, Shanghai Jiao Tong University School of Medicine, Shanghai 200025, China; dingyilan@sjtu.edu.cn (Y.D.); wsy12060@rjh.com.cn (S.W.)

[2] Shanghai National Clinical Research Center for Endocrine and Metabolic Diseases, Key Laboratory for Endocrine and Metabolic Diseases of the National Health Commission of the PR China, Shanghai National Center for Translational Medicine, Ruijin Hospital, Shanghai Jiao Tong University School of Medicine, Shanghai 200025, China

* Correspondence: ljl11319@rjh.com.cn

† These authors contributed equally to this work.

Abstract: Diabetes mellitus, particularly type 2 diabetes mellitus (T2DM), imposes a significant global burden with adverse clinical outcomes and escalating healthcare expenditures. Early identification of biomarkers can facilitate better screening, earlier diagnosis, and the prevention of diabetes. However, current clinical predictors often fail to detect abnormalities during the prediabetic state. Emerging studies have identified specific amino acids as potential biomarkers for predicting the onset and progression of diabetes. Understanding the underlying pathophysiological mechanisms can offer valuable insights into disease prevention and therapeutic interventions. This review provides a comprehensive summary of evidence supporting the use of amino acids and metabolites as clinical biomarkers for insulin resistance and diabetes. We discuss promising combinations of amino acids, including branched-chain amino acids, aromatic amino acids, glycine, asparagine and aspartate, in the prediction of T2DM. Furthermore, we delve into the mechanisms involving various signaling pathways and the metabolism underlying the role of amino acids in disease development. Finally, we highlight the potential of targeting predictive amino acids for preventive and therapeutic interventions, aiming to inspire further clinical investigations and mitigate the progression of T2DM, particularly in the prediabetic stage.

Keywords: amino acids; type 2 diabetes mellitus; prediction; mechanism; intervention

1. Introduction

Diabetes mellitus is a prevalent chronic metabolic disorder characterized by disrupted glucose homeostasis resulting in hyperglycemia [1]. It can be attributed to the progressive impairment of pancreatic beta cell function or the development of insulin resistance, leading to inadequate insulin action [2,3]. The global prevalence of diabetes is estimated to affect over 643 million individuals by 2030, with approximately 6.7 million deaths attributed to diabetes or its complications [4]. Type 2 diabetes mellitus (T2DM) is the predominant subgroup of diabetes, accounting for at least 90% of cases worldwide [5]. T2DM significantly impacts quality of life and life expectancy, and is a major contributor to disability or mortality [6]. Chronic exposure to hyperglycemia gives rise to numerous detrimental clinical consequences, including microvascular and macrovascular complications such as nephropathy, retinopathy and neuropathy [7]. Hence, a comprehensive understanding of the metabolic disturbances in T2DM is crucial for effective management.

Amino acids serve as fundamental building blocks for proteins and peptides, possessing both carboxylic and amino functional groups. Beyond their role in protein synthesis, these molecules play critical roles in various cellular processes, including cell signaling, oxidative stress, nutrition metabolism or maintenance [8]. Recent cross-sectional or prospective studies have shed light on the important involvement of amino acids in the development of T2DM [9] and insulin resistance [10]. Amino acids are closely associated with glucose dysregulation, particularly in their role as a primary source for gluconeogenesis, a process wherein glucose is synthesized from non-carbohydrate sources. The over-activation of gluconeogenesis is a key factor in prediabetes, which precedes the onset of T2DM [11]. Disturbances of amino acid metabolism may break the balance of muscle breakdown, protein synthesis and gluconeogenesis in the liver and kidneys. Under specific circumstances, amino acids may enhance glucose-stimulated insulin secretion or modulate insulin sensitivity early in the pathogenesis of T2DM [12]. The response effects are complex and depend on the different types of amino acids [13]. In addition to from the impairment of insulin sensitivity, certain altered amino acids could also block insulin signaling and affect lipid metabolism and mitochondrial oxidation. These changes have initiating and causal roles before insulin resistance is established. As T2DM advances and insulin resistance worsens, the combination of impaired beta cell function and resistance may lead to accelerated muscle proteolysis and disturbances in metabolic signals, further contributing to amino acid dysmetabolism [14]. These findings reveal the interconnectedness of amino acid metabolism and prediabetes, underscoring the great potential of amino acids in the early detection of metabolic abnormalities and prediction of future T2DM.

This review provides a comprehensive description of amino acids as strong correlative factors and mechanistic implications for insulin resistance, prediabetes, and future incident T2DM. Furthermore, we explore the promising inventions targeting amino acids, aiming to inspire advances in clinical research and therapeutic strategies for T2DM.

2. The Interrelation between Amino Acids and T2DM

The roles and underlying mechanisms of various amino acids in relation to T2DM represent a prominent area of investigation within the field of glucose metabolism regulation. The dysregulation of amino acid metabolism is crucial in prediabetes and future diabetes onset risk. The following sections elaborate the interrelation between several amino acids and T2DM, including branched-chain amino acids, aromatic amino acids, tryptophan, glycine, asparagine and aspartate.

2.1. Branched-Chain Amino Acids

2.1.1. Branched-Chain Amino Acid Metabolism in Health and T2DM

Branched-chain amino acids (BCAAs), including leucine, isoleucine and valine, are often studied as a collective unit. They share structural characteristics with a branched-side chain and undergo the common initiation steps of catabolism. BCAAs cannot be synthesized in higher organisms, making them nutritionally essential amino acids derived from protein-containing foods [15,16]. Through a specialized signaling network, BCAAs significantly modulate and regulate various metabolic and physiological processes, such as glucose, lipid or energy homeostasis [17].

Consistent correlations between elevated plasma BCAAs and T2DM have been observed in both human [18,19] and rodent models [20,21]. In the early 1970s, elevated plasma BCAAs were first described in diabetic patients with impaired insulin signaling [22]. In 2011, a prospective cohort study conducted within the Framingham Offspring Study revealed highly significant associations between baseline plasma BCAAs and future diabetes risk among 2422 normoglycemic individuals followed for 12 years. These results were further validated in many independent, prospective cohorts [23,24]. We conducted a nested case-control study on 3414 normoglycemic Chinese populations from a nation-wide prospective cohort of the China Cardiometabolic Disease and Cancer Cohort (4C) to explore the link between amino acids and incident diabetes. The results showed that higher levels of BCAAs were significantly associated with an increased risk of type 2 diabetes mellitus (T2DM) after accounting for various factors. Additionally, we found that triglycerides (TG) and waist-to-hip ratio (WHR) partially mediated the association between BCAAs and incident T2DM, providing further insights into the underlying mechanisms [23].A case-cohort study, including a random sample of 694 participants from PREDIMED trial, revealed that increases in the BCAA score at 1 year were correlated with a higher T2DM risk in the control group, with hazard ratio (HR) per SD = 1.61 (95% CI 1.02, 2.54) [25,26]. However, when participants were treated with a Mediterranean diet rich in extra-virgin olive oil, they experienced decreased BCAA levels, which attenuated the positive association between BCAAs and T2DM incidence [25]. Recent large prospective and cross-sectional cohort studies have also revealed close associations between elevated blood levels of BCAAs and insulin resistance, homeostasis model assessment (HOMA) insulin sensitivity and HbA1c level [12,27–29]. Higher plasma BCAA levels were found to be inversely correlated with insulin sensitivity but positively associated with fasting insulin levels [30]. These consistent results have led to speculation about a potential causative role for BCAAs [31].

Despite an overall consistency, varied effect sizes are observed based on age, gender or ethnicity [32]. Generally, BCAAs exhibit higher levels and a closer relationship with T2DM in male participants due to the catabolic differences in the liver [33–35]. Fewer studies have investigated the correlations between BCAA concentrations and adverse metabolic outcomes in children and adolescents. Nevertheless, in view of growth hormone secretion and protein turnover during pubertal growth, statistically significant associations have been identified between BCAA levels and future insulin resistance in a pediatric population [36]. Furthermore, individual BCAAs exhibit varying predictive capabilities for future T2DM among different populations [37]. In comparison, valine often stands out in the Chinese population compared to participants of South Asian descent [38]. Presumably, this can be attributed to specific genetic loci, along with earlier beta-cell dysfunction in diabetic patients in China [39]. BCAAs also showed significant differences in the associations with glycemic index and insulin resistance among ethnic groups [40]. African Americans generally exhibit greater insulin resistant, low muscle mass, and higher central obesity, which may explain their higher levels of valine and leucine compared with Hispanics [41].

2.1.2. Mechanisms Underlying Branched-Chain Amino Acids in T2DM

Several mechanisms indicate a direct link between BCAAs and, as leucine and isoleucine exhibit insulinotropic effects, while valine and isoleucine are gluconeogenic. Leucine can affect insulin receptor function via the activation of the mammalian target of rapamycin complex 1 (mTORC1), and insulin mediates the branched-chain α-ketoacid dehydrogenase complex (BCKDH). Despite emerging evidence supporting the predictive capability of increased BCAAs levels in T2DM, researchers still contemplate whether BCAAs are true causative factors in insulin resistance and T2DM or merely passive biomarkers of impaired insulin action [27]. Several acknowledged hypothesized mechanisms explaining how BCAAs might contribute to insulin resistance and T2DM are depicted in Figure 1.

Role of mTORC1

One speculated mechanism focuses on the leucine-mediated activation of mTORC1, which leads to the uncoupling of insulin signaling at an early stage. The mammalian target of rapamycin (mTOR) is a serine/threonine kinase that belong to the phosphoinositide 3-kinase (PI3K)-related kinase family and interacts with a series of proteins to form two distinct complexes named mTORC1 and mTORC2 [42]. mTORC1 promotes cell growth, such as protein synthesis, and drives cell cycle progression in response to various stimuli, including growth factors, stress, energy status, oxygen and amino acids. Amino acids, particularly leucine, regulate Rag guanine nucleotide binding, managing the interaction between Rag GTPases and mTORC1 [43]. Various growth factors and signaling molecules regulate the nucleotide state of the small GTPase Rheb (GDP- versus GTP-bound), activating mTORC1-dependent phosphorylation [44]. mTORC1 directly phosphorylates the translational regulators' eukaryotic translation initiation factor 4E (eIF4E)-binding protein 1 (4E-BP1) and S6 kinase 1 (S6K1), accelerating protein synthesis [45]. Several pathways converge on the tuberous sclerosis complex (TSC) protein, the major upstream negative mediator, to regulate mTORC1 in response to growth factors or cellular stress signaling, notably the prominent upstream growth factor/PI3K/AKT signaling [46]. Apart from the input of growth factor signaling, the 5′ AMP-activated protein kinase (AMPK) signaling is activated and suppresses mTORC1 through activating the phosphorylation of TSC and enhancing its activity during states of energy deficiency [47].

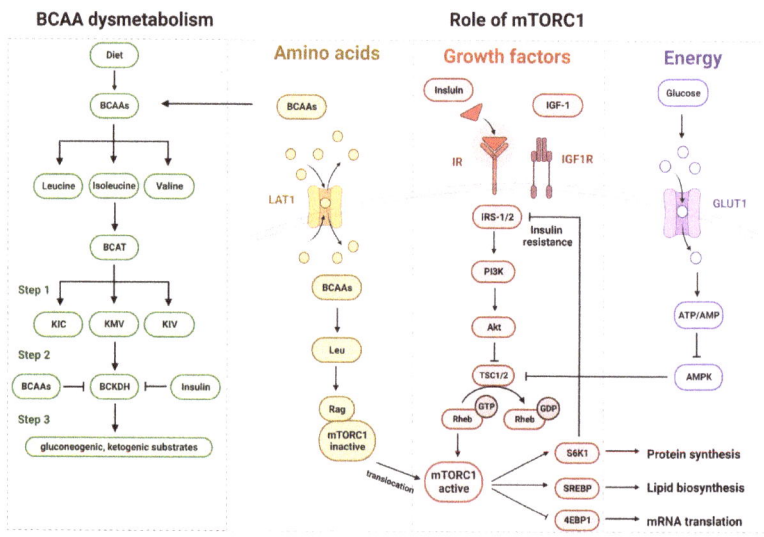

Figure 1. Mechanisms underlying branched-chain amino acids in T2DM. The figure depicts several acknowledged hypothesized mechanisms explaining how BCAAs might contribute to insulin resistance and T2DM. This figure was created with BioRender.com.

mTOR is a crucial regulator of cellular metabolism and catabolism, while the deregulation of mTOR signaling can induce many human diseases, including diabetes, degenerative disorders and cancer [48]. Persistent nutrient signaling results in insulin resistance by BCAA activation of the mTORC1 signaling pathway. The mTORC1-S6K-mediated negative feedback loops have a deleterious effect on the regulation of insulin signaling, maintaining beta cell function and survival [49]. Continuous stimulation of the serine kinases S6K1 and mTORC1 induces insulin resistance through the recruitment and phosphorylation of insulin receptor substrate (IRS)-1 and IRS-2 at multiple tyrosine residues [50]. These sites function as docking motifs for PI3K and the subsequent phosphorylation of Akt, which disrupts its interaction with insulin signaling. This negative feedback loop attenuates insulin responses, resulting in a reduction in glucose utilization. Under a chronic diabetic milieu, the overwhelming demand for insulin invites impaired insulin action and the potentiation of beta cell dysfunction, resulting in T2DM eventually becoming evident [27].

BCAA Dysmetabolism

The second hypothetical mechanism analyses how BCAA dysmetabolism generates insulin resistance and T2DM. This hypothesis derives from studies of maple syrup urine disease (MSUD) and organic acidurias, which are inborn errors in metabolism caused by defects in BCKDH, leading to the elevation of BCAAs in plasma and α-ketoacids in urine [51].

BCAAs are imported into cells by L-type amino acid transporters (LATs). BCAA catabolism involves three main steps. Firstly, intracellular BCAAs are converted to branched-chain α-keto acids (BCKAs), including α-ketoisocaproate (KIC), α-keto-β-methylvalerate (KMV), and α-ketoisovalerate (KIV). BCAA transaminase (BCAT) catalyzes the transamination of corresponding BCKAs. Then, BCKAs undergo decarboxylation and dehydrogenation to yield respective ketoacids via BCKDH. These first two steps are shared by all three BCAAs, while the latter process is rate-controlling. Numerous metabolic factors altered in insulin resistance and T2DM impair BCAA catabolism by coordinating the downregulation of multiple enzymes, including BCAT and BCKDH. Insulin can directly inhibit BCKDH under insulin-resistant states, giving rise to elevated BCAA and BCKAs levels that are widely believed to be the toxic factors in the disease. Furthermore, elevated plasma BCAA levels can also arise from protein breakdown and degradation in the body [52,53]. The administration of BCKDH kinase (BDK) inhibitors might impair glucose tolerance and reduce plasma BCAA concentrations in rats, independent of the action of insulin [54]. When fed a BCAA-supplemented diet, spontaneous type 2 diabetes Otsuka Long-Evans Tokushima Fatty (OLETF) rats showed improved glucose tolerance upon repeated administration [55]. Animal models discussed the effects of administration of BCAAs, but data in humans are lacking. However, based on observational study, elevated BCAA levels can be perceived long before the occurrence of insulin resistance and seemingly contribute to insulin resistance and T2DM. It is reasonable that BCAAs give rise to an altered insulin-regulated metabolism in the early stages, while insulin, in turn, incurs the accumulation of BCAAs in the later stages of insulin resistance [56]. An impaired BCAAs metabolism induces higher levels of BCAAs and the accumulation of toxic metabolites, leading to mitochondrial bioenergetic dysfunction and subsequent apoptosis of beta cells [27]. It has been confirmed that individuals or animal models with impaired or incomplete BCAA metabolism could be more susceptible to insulin resistance or T2DM [57].

Moreover, BCAA dysmetabolism is involved in the regulation of macrophage activity in the onset of chronic low-grade inflammation under diabetic state. BCAA oxidative defects may promote inflammatory response and organ damage in T2DM conditions by inducing macrophage activation [58]. BCKAs can significantly increase the production of detrimental mediators such as ROS, cytokines, and chemokines in primary macrophages. It is demonstrated that BCKA stimulation could alter the expression of a key glucose transporter (Glut1) and enhance the utilization of glucose for ROS overproduction in macrophages via glut1-mediated glucose metabolism. Elevated glycolysis induced ROS-driven proinflammatory phenotype in macrophages, accelerating the promotion of insulin resistance [59]. Additionally, BCKAs also enhance cytokine release by incurring mitochondrial oxidative stress in macrophages. These findings reveal the possible mechanisms by which BCAA dysmetabolism plays an integral role in insulin resistance and T2DM.

2.2. Aromatic Amino Acids
2.2.1. Phenylalanine and Tyrosine Metabolism

Aromatic amino acids (AAAs) are precursors of many significant biological compounds and are necessary for the normal functioning of the human organism. Two kinds of AAAs, phenylalanine and tyrosine, have been observed to be connected with a tendency of increased risk of T2DM [41]. Mice fed with diets rich in phenylalanine could develop insulin resistance and T2DM symptoms [59]. Furthermore, changes in phenylalanine and tyrosine levels closely parallel the changes in fasting blood glucose (FPG) and 2 h postprandial blood glucose (2hPG) levels of individuals [60]. These two AAAs were also demonstrated to be elevated in subjects who developed T2DM from normal glucose tolerance. The levels of phenylalanine and tyrosine increased in hyperglycemia and particularly affected non-diabetic participants, who later developed T2DM over a 5-year follow-up period [41]. Using a nested case-control design from the China Cardiometabolic Disease and Cancer Cohort (4C) Study, our team also revealed that per SD increments in two AAAs had strong associations with the onset of T2DM [23]. Phenylalanine and tyrosine were positively correlated with T2DM, with a 23% increased risk of incident diabetes in the fully adjusted model including all the confounding factors, including diet score, liver enzymes, 2hPG, and HOMA-IR. These results align with previous findings, confirming the predictive value of two AAAs for risk of diabetes in normoglycemic Chinese individuals. AAAs have important implications for the pathogenesis of diabetes. It has been reported that phenylalanine modified insulin receptor beta (IRβ) and inhibited insulin signaling and glucose uptake [61]. Using phenylalanine and aspartame to mimic extremes for serum phenylalanine elevation in humans, phenylalanyl-tRNA synthetase (FARS) sensed phenylalanine concentrations and converted them into the phenylalanine signal by modifying proteins. For IRβ, the phenylalanine signal led to impairments in the components of the insulin signaling cascade and hindered glucose uptake by cells. This disruption of insulin signaling transmission by modifying IRβ has adverse effects on insulin sensitivity and accelerates T2DM progression. Tyrosine is involved in gluconeogenesis and glucose transport [62]. Superfluous tyrosine can be rapidly catabolized, weakening the clearance of blood glucose and enhancing gluconeogenesis. Free tyrosine may combine free radicals forming 3-nitrotyrosine, a more cytotoxic mediator that injuries pancreatic islet beta cells. In the case of accelerated rates of oxygen radical and nitric oxide generation in beta cells, insulin could be a potential target [63]. Interaction with insulin affects the receptor binding and hypoglycemic capacities. This research sheds light on the dysmetabolism of two AAAs and the disturbed insulin signaling pathway, leading to a higher risk of T2DM.

2.2.2. Tryptophan Metabolism
Tryptophan Metabolism in Health and T2DM

Tryptophan is an indispensable and essential amino acid that can only be obtained through the diet. Systemic and cellular concentrations of tryptophan mainly depend on the balance between biological conversion and degradation pathways. The tryptophan metabolism generally involves three metabolic pathways: the kynurenine (KYN) pathway, the 5-hydroxytryptamine (HT) pathway, and the indole pathway [64]. Most blood tryptophan is bound to albumin and not present in its free form. The majority of free tryptophan in humans is metabolized through the tryptophan–kynurenine pathway, which is involved in extensive physiological functions such as immune activation, growth and feed intake, and alterations in peripheral tissue conditions during aging, obesity and diabetes [65]. The first and rate-limiting step of the tryptophan–kynurenine pathway is catalyzed by the enzyme indoleamine 2,3-dioxygenase (IDO1) and tryptophan-2,3-dioxygenase (TDO). IDO1 exists in most cells, such as macrophages or central nervous cells, while TDO is almost exclusively expressed in the liver and mainly controls tryptophan concentrations in the blood [66]. Chronic stress or inflammatory factors activate enzymes of the upstream steps of tryptophan metabolism and convert tryptophan into KYN and KYN into kynurenic acid (KYNA), 3-hydroxykynurenine (3-HK) and anthranilic acid (AA). The further conversion of 3-HK to 3-hydroxyanthranilic acid (3-HAA) and alanine is catalyzed by kynureninase (KYNU), whereas xanthurenic acid (XA) is another conversion of 3-HK. Subsequently, 3-HAA transforms into the neurotoxic quinolinic acid (QA) and is also crucial in the production of the coenzyme NAD$^+$, contributing to energy metabolism and mitochondrial functions. Three pathways of tryptophan metabolism including the kynurenine, serotonin, and indole are depicted in Figure 2.

The gut microbial function in the tryptophan metabolism has emerged as a vital driving force [67]. The gut microbiome can mediate three pathways of tryptophan metabolism and produce correlative metabolites [68]. Tryptophan and its metabolites serve as critical communication regulators between the host and gut microorganisms, maintaining metabolic homeostasis. In germ-free mice, the KYN pathway was inhibited, leading to decreased tryptophan levels. However, after supplementation with intestinal flora, this KYN pathway was normalized [69]. In addition, several metabolites produced by gut microbes play a significant role in adjusting the tryptophan–kynurenine pathway, including the inhibition of IDO transcription [70].

Recent metabolomics screens have confirmed tryptophan metabolites as potential biological mediators in the onset of T2DM [71]. A ten-year longitudinal Shanghai Diabetes Study (SHDS) with 213 participants claimed that serum tryptophan level was significantly higher in individuals who developed future T2DM and was positively and independently related to diabetes onset risk [72]. Higher tryptophan concentration could adversely contribute to a higher degree of insulin resistance and secretion, triglyceride level and blood pressure.

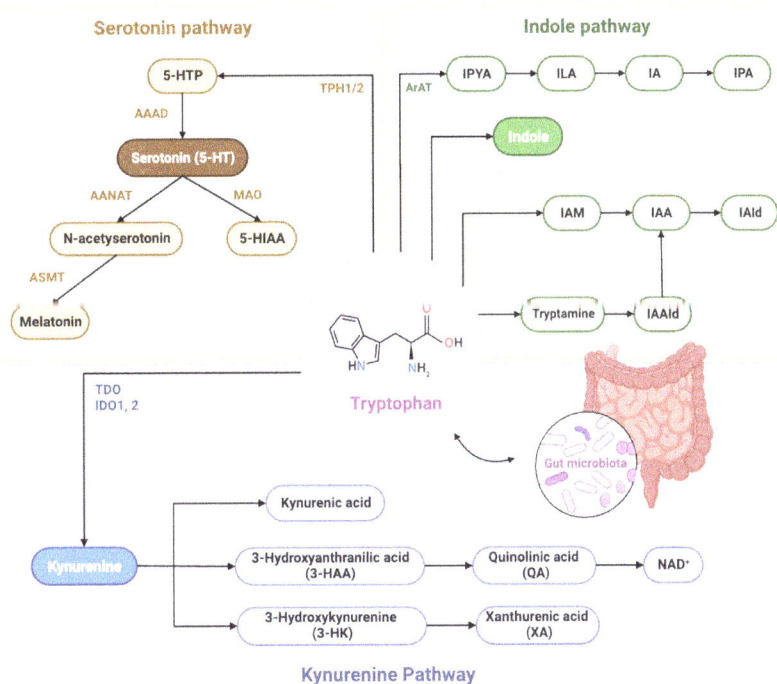

Figure 2. Overview of tryptophan metabolism via the kynurenine, serotonin, and indole pathways. The figure depicts three pathways of tryptophan metabolism including the kynurenine, serotonin, and indole. The gut microbiome can mediate three pathways of tryptophan metabolism and produce correlative metabolites. This figure was created with BioRender.com.

Generally, diabetics often show elevated tryptophan metabolism with decreased tryptophan and increased concentrations of downstream metabolites along the tryptophan–kynurenine pathway [73]. Cross-sectional studies have confirmed increased plasma levels of KYN and KYNA in subjects with insulin resistance prior to the evident manifestation of hyperglycemia and lower levels of tryptophan in nondiabetics [74]. In a metabolomics study including 5181 participants from the cross-sectional Metabolic Syndrome in Men study, the levels of KYNA and other downstream metabolites weakened insulin secretion and insulin sensitivity but enhanced susceptibility to T2DM [75]. In our prospective cohort study, in the fully adjusted model with all confounding factors, serum N-acetyltryptophan, but not tryptophan or kynurenine, was associated with increased risk of diabetes [23]. On this basis, another study involving 2519 individuals with coronary artery disease (CAD) but without T2DM focused on tryptophan and its downstream metabolite kynurenine for a median of 7.6 years [76]. The plasma and urine kynurenine-to-tryptophan ratio (KTR) provided a more suitable measure of tryptophan catabolism than the absolute level of kynurenine or tryptophan. It was observed that KTR in urine, but not in plasma, had a strong positive relationship with incident T2DM during 7 years of follow-up in this large cohort. In two cohorts comprising 856 individuals with T2DM, the serum KTR was associated with and improved the prediction of all-cause mortality among patients [77].

Mechanisms Underlying Tryptophan in T2DM

Various observational clinical studies analyzing clinical parameters in the tryptophan–kynurenine pathway reveal the existence of disturbed tryptophan metabolism in prediabetes or diabetes. However, these studies have not definitively determined whether the changes are causative or secondary to the disorder [78]. The diabetogenic tryptophan–kynurenine pathway is affected by many factors, including genetic factors, metabolic status, the degree of pancreas destruction and insulin resistance. These findings underline the potential value of tryptophan and downstream metabolites in identifying high-risk individuals before the occurrence of T2DM, even before remarkable alterations in metabolic markers are observed. However, the mechanisms behind the effects of the tryptophan pathways on T2DM are diverse. In cases of inflammation or stress, the KYN and KYN-NAD metabolic pathways particularly rely on pyridoxal-5-phosphate (P5P), an active form of vitamin B6, as a cofactor. The relative absence of P5P shifts the KYN-NAD metabolism from the common production of NAD^+ to the excessive formation of XA [79]. The accumulation of XA or other KYN metabolites have diabetogenic effects and can impair the biological function of insulin, promoting the progression of T2DM from prediabetes [80]. Additionally, systemic low-grade inflammation triggered by the dysregulation of tryptophan metabolism can lead to insulin resistance [79]. The severity of insulin resistance varies with central and peripheral concentrations of tryptophan and downstream metabolites. These metabolites can form less active chelate complexes with insulin and interfere with the glucose regulatory network at the prediabetic stage [74]. Serum tryptophan levels may initially increase during prediabetes and then gradually diminish with the advent of a full diabetic state. Nonetheless, monitoring of the tryptophan–kynurenine pathway, especially the KTR, is beneficial in recognizing individuals at risk for T2DM.

2.3. Glycine Metabolism

2.3.1. Glycine Metabolism in Health and T2DM

Glycine is a kind of nonessential amino acid in humans or other mammals. It is associated with various metabolic pathways and involved in numerous human physiological processes. Generally, the amount of glycine synthesized in vivo is insufficient to satisfy metabolic demands. A brief period of glycine shortage may not be a great hazard to health status while chronic depletion can affect growth, immune responses or health metabolism. Glycine usually functions as a precursor for various crucial metabolites of low molecular weight like glutathione synthesis or as a regulator of protein configuration and activity [81].

Decreased plasma glycine concentration is now regarded as a promising predictive factor for reduced glucose tolerance and T2DM. Prospective studies demonstrate that higher serum glycine level indicates a reduced risk of incident T2DM and hypoglycemia at baseline can suggest an inclination for T2DM. Patients with obesity or diabetes observed a relative lower plasma glycine concentration compared with control subjects [82]. Particularly, this metabolic change occurs before obvious clinical manifestations of the disorder. Moreover, the level of plasma glycine correlates positively with insulin sensitivity but negatively with insulin resistance in view of the homeostasis model assessment for the beta cell function index [83]. A randomized trial also reveals that diabetic patients treated with insulin sensitizer therapy including pioglitazone and metformin have higher plasma glycine in comparison to placebo [84]. However, in the prospective cohort, a 4C study did not discover remarkable changes in the ORs of glycine before onset of glucose dysregulation [23]. This calls for further studies to explore the potential usefulness of glycine as a clinical diagnostic tool for T2DM.

2.3.2. Mechanisms Underlying Glycine in T2DM

The pathophysiological mechanisms behind glycine insufficiency and the homeostasis of glycine and T2DM still need to be elaborated. A variety of hypotheses are widely accepted. First and foremost, glycine can directly adjust insulin secretion and has been identified as the strongest amino acid related to increased insulin sensitivity [85]. A positive feedback loop exists between human islet beta cells expressing glycine receptors and insulin depending on phosphoinositide 3-kinase [86]. In liver from sucrose-fed rats, glycine also diminishes the insulin-induced phosphorylation of insulin receptor substrate-1 in serine residue and enhances the phosphorylation of insulin receptor β-subunit in tyrosine residue, which elevates insulin sensitivity [87]. In T2DM with chronic low-grade inflammation, where pro-inflammatory markers are uplifted, glycine is certified as a new anti-inflammatory agent for increasing cytokine IL-10 in monocytes and decreasing TNF-alpha in monocytes and Kupffer cells [88]. These results provide clues of the glycine signaling mechanisms of significant metabolic benefits. Moreover, as one precursor amino acid of the antioxidant glutathione (GSH), glycine can directly affect the synthesis rate and availability of GSH. Animals fed with food short in GSH precursor amino acids were proved to suffer from GSH deficiency. In human cells, reduced GSH is one of the most abundant and common substances resisting damage caused by oxidative stress. However, uncontrolled blood glucose levels lead to oxidative stress and reactive oxygen species (ROS) formation, which is far beyond the capacity of GSH-driven antioxidant defense systems [89]. Deficiency of glycine gives rise to the insufficient synthesis of GSH, which fails to combat subsequent diabetic tissue damage. Accordingly, dietary supplementation with enough glycine may reverse the shortage of GSH synthesis and tackle oxidative stress.

2.4. Asparagine and Aspartate

2.4.1. Asparagine Metabolism in Health and T2DM

Asparagine is a kind of glucogenic amino acid whose byproduct, oxaloacetate, can be used in the TCA cycle to synthesize glucose. With metabolism fluctuation, asparagine and aspartate are readily converted to each other by corresponding enzymes and can undergo transamination to form glutamate. Amid the asparagine metabolism, asparagine synthetase (ASNS) catalyzes asparagine synthesis via aspartate, ATP and ammonia as substrates [90]. ASNS is ubiquitous in its organ distribution and highly associated with cellular nutritional imbalances like glucose deficiency and amino acid disturbances [91]. Asparagine and other biologically active molecules have a vital role in cell catabolism and signaling, host anti-oxidative ability, and immunity under physiological and pathological conditions [92]. Controversy has been arisen regarding the relationship between asparagine, aspartate and T2DM. Circulating concentrations of asparagine are correlated with incidence of T2DM. Only rarely has the literature provided connections between plasma asparagine levels and a lower risk for future T2DM. Animal experimental results revealed lower concentrations of asparagine in diabetic rats [93]. Prospective observational studies demonstrated that baseline plasma asparagine was a protective biomarker of diabetes [94]. According to the Framingham Heart Study, asparagine was negatively related to fasting insulin while aspartate was inversely associated with fasting glucose [95]. Additionally, asparagine was proved to be the sole protective and predictable metabolite for T2DM in a subset of 2939 Atherosclerosis Risk in Communities (ARIC) study participants with metabolomics data and without prevalent diabetes [96]. However, in our previous study, serum asparagine was shown to be associated with an increased risk of diabetes in the multivariable-adjusted model in addition to the adjustment of diet score, and in the fully adjusted model [23]. Particularly, a ratio of asparagine to aspartate of >1.5 contributed to the increased risk of T2DM, which could be further elevated by female gender and being >50 years of age [97]. Thus, whether asparagine is protective or diabetogenic is still a subject of debate.

2.4.2. Mechanisms Underlying Asparagine and Aspartate in T2DM

Findings about the correlation of asparagine and aspartate with the risk of T2DM are inconclusive, and the underlying mechanisms are also conflicting. On the one hand, asparagine can easily turn into aspartate and then undergo a transamination for glutamate, which is also a constituent of the tripeptide glutathione against oxidative stress and chronic diseases [98]. This assumption explains some of the inverse associations with diabetes risk. On the other hand, some results about asparagine are opposite to the protective correlations with T2DM. The discrepancy between different studies may derive from the sampling stage. Before the occurrence of diabetes, asparagine is likely inadequate while persistent adverse stimulation may upregulate ASNS and induce an excess of asparagine and aspartate deficiency in later periods. Thus, the increasing conversion of asparagine to aspartate contributes to lower concentrations of circulating asparagine and onset of hyperglycemia [99]. This might point toward a potential causal association between low plasma asparagine levels and prediabetes. Beyond that, asparagine can also hinder the phosphorylation of AMPK and upregulate mTORC1, causing increased insulin resistance and decreased beta cell reserve [100]. Uncommon asparagine and aspartate homeostasis with a higher risk of T2DM can be greatly amplified by the specific effect of older age and female gender [96]. Females at later stages of adult life are proven to suffer from a lack of estrogen, which is an important regulator of metabolic status [101]. An absence of estrogen can similarly generate insulin resistance, impaired insulin function and beta cell apoptosis [102]. The cooccurrence of abnormal asparagine and aspartate homeostasis and estrogen insufficiency promotes the process of insulin resistance and accelerates the development of T2DM by the AMPK-mTORC1 pathway. Further investigations into the latent molecular mechanisms are warranted for a better understanding of the cause of T2DM and asparagine and aspartate homeostasis.

2.5. Serine Metabolism

2.5.1. Serine Metabolism in Health and T2DM

As a type of nutritionally non-essential amino acid (NEAA), serine can be derived from the diet, or synthesized from 3-phosphoglycerate (3-PG) and glycine. Serine metabolism makes profound contributions to many cellular functions, particularly in the turnover of proteins and phospholipids as necessary building blocks in cellular membranes [103]. In addition, L-serine is required to promote the growth and differentiation of neurons. Deficiency in L-serine is highly correlated with the abnormal synthesis of phospholipids like phosphatidylserine (PS) and sphingolipids (SL), which impairs regular functions of the nervous system. The induction of systemic L-serine deficiency is linked to the risk of driving future T2DM and diabetic peripheral neuropathy [104,105]. Evidence that alterations in serine concentration play a role in T2DM is growing [106,107]. Systemic serine deficiency has been demonstrated to coexist with severe obesity, insulin resistance and hyperglycemia in a mouse model of T2DM [104]. Oral serine supplementation can correct the underlying serine deficiency, thus hindering the development of diabetic peripheral neuropathy in diabetic animals. Moreover, prior studies have shown that, compared with non-diabetic controls, plasma serine concentrations are markedly reduced in the T2DM patients [106,108]. Similar research analyzing the levels of amino acids in diabetic patients has also confirmed a decrease in L-serine concentration in blood [107].

2.5.2. Mechanisms Underlying Serine in T2DM

Irregular serine metabolism is believed to contribute to the pathogenesis of T2DM and related complications, although there is no general consensus regarding the potential mechanisms. Serine has been shown to be capable of promoting insulin secretion, increasing insulin sensitivity, and enhancing glucose tolerance [29]. As deoxysphingolipids accumulate under conditions of decreased serine availability, it is hypothesized that cytotoxic deoxysphingolipids may directly compromise pancreatic beta cells. Diabetic patients have significantly higher plasma levels of deoxysphingolipid in comparison with the control group [109]. The increased concentrations of these lipids impairs normal glucose homeostasis and induces beta cell failure in response to chronic hyperglycemia [110]. Nevertheless, further studies are required, since the mechanisms underlying serine in T2DM remain largely unknown.

2.6. Amino Acid Combination

To further promote the predictive performance of plasma amino acids, a combination of three amino acids predicting future diabetes has been reported. The top combination of these three amino acids, namely isoleucine, phenylalanine, and tyrosine, has been identified based on robust statistical measures such as the likelihood ratio (LHR) statistic and c-statistic. When compared to the use of a single amino acid, this combination significantly improves the LHR statistic by from +6 to +9 points ($p < 0.05$). However, the incremental improvement is relatively modest when five additional amino acids are included in the combination. Clinical models incorporating this predictive three-amino acid combination have demonstrated a substantially higher risk, from 5- to 7-fold higher, of developing diabetes among individuals in the top quartile, in contrast with those in the lowest quartile (p for trend, from 0.007 to 0.0009) [24]. The results have been successfully replicated in an independent, prospective cohort, confirming the reliability and robustness of the predictive value of the three amino acid combination in identifying individuals at higher risk for future diabetes. Meanwhile, the elevated concentrations of this three-amino acid combination not only serve as predictors of future diabetes but also provide early signals for the subsequent development of cardiovascular disease and its functional consequences during long-term follow-up [111].

Moreover, in addition to the three BCAAs mentioned earlier, two aromatic amino acids (AAAs)—phenylalanine and tyrosine—are commonly grouped together to indicate the occurrence and development of T2DM [112]. Experimental and clinical data have suggested five specific amino acids that may serve as effectors in prediabetes status [113]. Previous research demonstrated associations between these five amino acids and incident diabetes, its precursor states and insulin resistance. Individuals with hyperinsulinemia tend to have greater concentrations of AAAs [114]. Prospective cohorts comprising over 3000 participants highlighted that elevated levels of these five elevated amino acids were associated with 60–100% increases in the relative risk of T2DM [24]. When considering factors such as age and gender, this five-amino acid combination was significantly related to HOMA-IR at baseline and for men at 6-year follow-up (odds ratio 2.09 [95% CI 1.38–3.17]; $p = 0.0005$), while, for women, only leucine, valine and phenylalanine predicted HOMA-IR at the 6-year follow-up ($p < 0.05$) [115]. The combined impact of branched-chain amino acids and aromatic amino acids on promoting insulin resistance and future T2DM is most evident in young normoglycemic adults, particularly in male individuals. In light of these findings, instead of screening for individual amino acids, exploring optimal combinations of amino acids that demonstrate a stronger correlation by including or excluding certain amino acids may be a more meaningful approach.

In addition, a series of amino acid combinations exhibit temporal characteristics in predicting diabetes. Previous research has observed that, at various timepoints, specific amino acids undergo changes in their concentrations and metabolic pathways that may be related to the onset and progression of diabetes. These timepoints include but are not limited to various stages such as pre-diabetes, early-stage diabetes, and disease progression. The elevated of BCAAs and AAAs prior to T2DM could potentially predict the onset of incident diabetes years before the clinical T2DM manifestation. When compared with individuals with NGR, participants with impaired fasting glucose (IFG) showed lower levels of leucine while diabetic patients exhibited increased levels of leucine, tyrosine, and asparagine in contrast to IFG patients [116]. The fluctuations in leucine levels might be characteristic of the transition from NGR to prediabetes or differences between individual patients, which still needs further study [116].

Overall, the characteristics of studies investigating associations between amino acids and type 2 diabetes are summarized in Table 1.

Table 1. Characteristics of studies investigating associations between amino acids and type 2 diabetes.

Reference	Study Population Location	N, Follow-Up Time	Study Design	Biological Sample	Methods/Tested Amino Acids	Key Findings
Wang TJ et al., 2011 Nat Med [24]	Discovery analyses: Framingham Offspring Study, U.S. Replication analyses: Malmö Diet and Cancer Study, Sweden	Discovery analyses: 189 cases who developed diabetes during a 12-year follow-up period, and 189 propensity-matched controls who did not develop diabetes. Replication analyses: 163 cases and 163 controls	Cohort Nested case-control study	Plasma	LC-MS isoleucine, leucine, valine, tyrosine, phenylalanine, tryptophan, arginine, lysine, histidine, aspartate, glutamic acid, asparagine, glutamine, methionine, serine, threonine, alanine, glycine, proline, cis/trans-hydroxyproline, taurine	Increased risk of T2DM (↑): isoleucine, leucine, valine, tyrosine, phenylalanine, sum of isoleucine, tyrosine and phenylalanine. Replication analyses: leucine, valine, tyrosine, and phenylalanine were significantly associated with increased risk of incident diabetes (↑).
Ruiz-Canela M et al., 2018 Diabetologia [25]	PREvención con DIeta MEDiterránea (PREDIMED) trial	251 T2DM, 694 controls (641 non-T2DM and 53 overlapping cases) 3.8 years	Cohort Nested case-control study	Plasma	LC-MS/MS leucine, isoleucine, valine, phenylalanine, tyrosine	Increased risk of T2DM (↑): baseline BCAA (sum of leucine, isoleucine and valine) and AAA (sum of phenylalanine and tyrosine) scores, BCAAs/AAAs, leucine, isoleucine, valine, phenylalanine, tyrosine.
Wang, S. et al., 2022 Cell Rep Med [23]	China Cardiometabolic Disease and Cancer Cohort (4C)	1707 matched case-control pairs with up to 5 years of follow-up	Cohort Nested case-control study	Serum	UPLC-MS/MS Branched-chain AAs, aromatic AAs, asparagine, alanine, glutamic acid, homoserine, 2-aminoadipic acid, histidine, methionine, asparagine, and proline	Increased risk of T2DM (↑): branched-chain AAs, aromatic AAs, asparagine, alanine, glutamic acid, homoserine, 2-aminoadipic acid, histidine, methionine, and proline.
Stančáková A et al., 2012 Diabetes [60]	METabolic Syndrome In Men (METSIM) study	3026 NGT, 4327 IFG, 312 IGT, 1058 IFG + IGT, 646 T2DM 4.7 years	Population-based cohort	Plasma	High-throughput serum nuclear magnetic resonance (NMR) platform alanine, phenylalanine, valine, leucine, isoleucine, tyrosine, histidine, glutamine	Increased risk of T2DM (↑): alanine, leucine, isoleucine, tyrosine, and glutamine predicted incident T2DM, and their effects were largely mediated by insulin resistance (except for glutamine).
Yamada, C. et al., 2015 J Diabetes Investig [40]	Volunteers Japan	94 non-diabetic Japanese men and women	Cross-sectional study	Plasma	LC/MC serine, asparagine, glutamic acid, glutamine, and other 39 amino acids in total	(↑) Positive correlations were observed between HOMA-IR and valine, isoleucine, leucine, tyrosine, phenylalanine and total BCAA concentration.
Floegel A et al., 2013 Diabetes [71]	European Prospective Investigation into Cancer and Nutrition (EPIC)-Potsdam	800 incident T2DM, 2282 controls 7 years	Cohort-nested case-control study	Serum	Targeted FIA-MS/MS 163 metabolites (including 14 amino acids)	Decreased risk of T2DM (↓): glycine; increased risk of T2DM (↑): Phenylalanine.

Table 1. Cont.

Reference	Study Population Location	N, Follow-Up Time	Study Design	Biological Sample	Methods/Tested Amino Acids	Key Findings
Vangipurapu, J. et al., 2020 Diabetes Care [75]	Metabolic Syndrome in Men (METSIM) study	5169 participants of METSIM having a follow-up of 7.4 years	Population-based cohort	Plasma	UHPLC 86 microbiome-based metabolites	Increased risk of T2DM (↑): xanthurenate, kynurenate as tryptophan-kynurenine downstream metabolites
Menni C et al., 2013 Diabetes [21]	Twins UK	2204 female (115 T2DM, 192 IFG, 1897 control)	Population-based cohort	Plasma	Nontargeted metabolomics provide Metabolon, Inc.42 metabolites	Increased risk of T2DM (↑): proline, 3-Methyl-2-oxovalerate, 4-Methyl-2-oxopentanoate, isoleucine, leucine, valine; decreased risk of T2DM (↓): N-acetylglycine, citrulline, dimethylarginine (SDMA + ADMA); increased risk of IFG (↑): 2-hydroxybutyrate (AHB), 3-methyl-2-oxobutyrate, 3-Methyl-2-oxovalerate, 4-methyl-2-oxopentanoate, isoleucine, leucine.
Tianlu Chen et al., 2016 Plos one [72]	Shanghai Diabetes Study (SHDS)	213 NGT 10 years	Cohort population-based cohort	Serum	UPLC-TQ/MS tryptophan	(↑) Serum tryptophan was positively associated with T2DM risk.
Rebholz, C.M. et al., 2018 Diabetologia [96]	Atherosclerosis Risk in Communities (ARIC) study	2939	Cohort population-based cohort	Serum	isoleucine, leucine, valine, asparagine, 3-(4-hydoxyphenyl) lactate	Increased risk of T2DM (↑): isoleucine, leucine, valine; decreased risk of T2DM (↓): asparagine

(↑), positive association (e.g., higher risk); (↓), inverse association (e.g., lower risk) with prediabetes traits or type 2 diabetes.

3. Targeting Predictive Amino Acids for Preventive and Therapeutic Interventions in T2DM

3.1. Lifestyle Interventions

In general, lifestyle interventions promoting a healthy diet with a limited intake of high sugar and saturated fat, combined with enhanced physical activity, are known to reduce the risk for T2DM, especially in high-risk individuals [117,118]. However, few studies have investigated how lifestyle interventions influence the relationship between amino acids and insulin resistance or T2DM. Although the altered levels of circulating amino acids can indicate a risk of imminent T2DM, the extent to which this correlation is independent of lifestyle factors needs to be further studied and established.

In experimental models, young and growing mice subjected to a specific decreased consumption of BCAAs exhibited beneficial effects on metabolic health and improved glucose tolerance [119]. This provides valuable insights into the potential of dietary interventions targeting reduced BCAA intake for the treatment of insulin resistance. Furthermore, the Finnish Diabetes Prevention Study (DPS) explored the association between BCAAs and T2DM in trajectory models via a lifestyle intervention setting. The intervention group underwent lifestyle changes that included supervised food intake and increased physical activity, leading to a decrease in BCAA levels and a diminished association with T2DM risk [120]. The intervention goals placed an emphasis on reducing total and saturated fat intake, increasing fiber density in the diet, and promoting moderate physical activity. Similarly, the above-mentioned PREDIMED trial using a Mediterranean diet supplemented with extra-virgin olive oil as the main intervention resulted in reductions in the plasma BCAA and AAA levels, thereby weakening the subsequent risk of T2DM [25,26]. These associations align with a previous study of BCAAs and CVD [121]. The effects of the MedDiet on BCAA concentrations persisted even after adjusting for changes in insulin or HOMA-IR. Further research is needed to gain a comprehensive understanding of these complex biological mechanisms. As the MedDiet does not specifically target the profile of amino acid intake, the most possible explanation for this intervention is that it alleviates the deleterious correlations between BCAAs and AAA on T2DM risk, potentially via downstream pathways or alternative protective mechanisms. Another lifestyle intervention implemented from 2016 to 2018 involving 5% weight reduction and a diet with increased consumption of whole grains, nuts, low-fat dairy, olive and rapeseed oils and decreased intake of snacks, fast foods, red and processed meat [122] led to a noticeable reduction in BCAA concentrations. This suggests that this approach holds promise in preventing or delaying the onset of T2DM. On the other hand, dietary supplementation with certain amino acids like glycine significantly increased insulin responses and led to remarkable decreases in systemic inflammation, highlighting their potential as protective biomarkers in blood glucose control [123]. In a nutshell, current research highlights the therapeutic potential of manipulating lifestyle interventions targeting amino acids for treating insulin resistance and preventing or delaying the future exacerbation of the T2DM pandemic [124]. Assessing the safe limits of amino acid intake may provide a useful metric to determine appropriate dietary amino acid recommendations, especially for individuals susceptible to T2DM or those already in the prediabetic state.

3.2. Pharmacologic Treatment Approaches

Compared with lifestyle interventions alone, different pharmacologic treatment approaches are also beneficial to prevent or delay various subgroups of T2DM. The establishment of appropriate pharmacological interventions could ameliorate the amino-acid-driven impairment of cell signaling and maladaptive phenotypes. Diverse amino acid metabolic pathways are likely to become potential targets of pharmacologic therapies. Common glucose-lowering medications including metformin, glipizide or empagliflozin might alter amino acid levels as a downstream consequence [125]. For instance, the effect of metformin dramatically alters the BCAA metabolism. Sustaining treatment with metformin could adjust circulating BCAA levels in a specific manner. Apart from activating AMPK and reducing hepatic gluconeogenesis and blood glucose, metformin also suppressed BCAT2 and BCKDHa mRNA expression, suggesting that metformin could function via downregulating BCAA catabolic enzyme expression or activity [126]. After the administration of glipizide and metformin, levels of branched-chain amino acids and aromatic amino acids experienced acute changes, which reflect an improvement in glycemic metabolism [112]. Low-dose metformin treatment could rectify glucose metabolic imbalance and integrative metabolomics analysis further investigated the elevation of amino acid levels including serine, glycine, glutamate, along with the decrease in aspartate [127,128]. Beyond this, accumulating evidence has demonstrated metabolic effects of the SGLT2 inhibition empagliflozin on increased concentrations of BCAA metabolites such as acylcarnitine [129]. The hypoglycemic agent DPP4 inhibitor, sitagliptin, induced glycemic improvements and led to a remarkable decrease in plasma valine levels [130]. Sitagliptin significantly changed the pattern of amino acids in both mice and T2DM patients [131]. The mechanisms behind these complex associations remain speculative and need further investigation for the development of novel effective pharmacologic T2DM therapies.

4. Conclusions and Future Perspectives

Based on flourishing metabolite profiling platforms, a panel of amino acids has been identified to predict the onset of future diabetes. Specific amino acids play crucial parts early in the pathogenesis of incident insulin resistance and future T2DM. As promising predictive metabolites, the better utilization of amino acids contributes to improving early diagnosis and clinical outcomes, allowing for precautionary measures to be taken to avoid complications, and delaying the onset and progression of T2DM. Enhanced insight into amino acid profiling has also increased interest in various inventions, as it holds therapeutic potential for the management of diabetes. To achieve a more personalized and precise control of T2DM, the translation of these insights to clinical application requires several additional steps. First and foremost, it is necessary to clarify whether these relevant amino acids are merely associated with impaired insulin function or can directly give rise to insulin resistance and subsequent T2DM. Furthermore, the mechanisms behind certain amino acids remain controversial and inconclusive, warranting further investigation to build comprehensive theories that elucidate the dominant signaling pathways and biological mechanisms linking amino acids with diabetes risk.

In conclusion, amino acids hold both predictive and therapeutic potential in future T2DM. A thorough understanding of amino acid dysmetabolism in T2DM is essential for the effective screening, diagnosis and prediction of future diabetic complications, allowing clinicians to make informed decisions and benefitting individuals at risk. Once this diagnosis approach passes to the clinical level, it is expected to achieve considerably high detection accuracy and offer more specific therapeutic possibilities for high-risk patients.

Author Contributions: Writing—original draft, Y.D. and S.W.; writing—review and editing, S.W. and J.L.; supervision—J.L. All authors have read and agreed to the published version of the manuscript.

Funding: This research was funded by the National Natural Science Foundation of China (grant nos. 81970691 and 82372347), Innovative research team of high-level local universities in Shanghai, Science and Technology Committee of Shanghai (grant nos. 20Y11905100 and 19MC1910100), and Clinical Research Project of Shanghai Municipal Health Commission (grant nos. 20224Y0087).

Acknowledgments: The collaboration of all the co-authors of our joint articles is greatly appreciated.

Conflicts of Interest: The authors declare no conflict of interest.

Abbreviations

Type 2 diabetes mellitus (T2DM); branched-chain amino acids (BCAAs); homeostasis model assessment (HOMA); mammalian target of rapamycin complex 1 (mTORC1); branched-chain α-ketoacid dehydrogenase complex (BCKDH); phosphoinositide 3-kinase (PI3K); mammalian target of rapamycin (mTOR); eukaryotic translation initiation factor

4E (eIF4E); 4E binding protein 1 (4E-BP1); S6 kinase 1 (S6K1); tuberous sclerosis tumor suppressor complex (TSC); AMP-activated protein kinase (AMPK); insulin receptor substrate (IRS); maple syrup urine disease (MSUD); L-type amino acid transporters (LATs); α-ketoisocaproate (KIC); α-keto-β-methylvalerate (KMV); α-ketoisovalerate (KIV); BCAA transaminase (BCAT); glucose transporter (Glut1); Aromatic amino acids (AAAs); China Cardiometabolic Disease and Cancer Cohort (4C); normal glucose regulation (NGR); insulin receptor beta (IRβ); phenylalanyl-tRNA synthetase (FARS); 5-hydroxytryptamine (HT); kynurenine (KYN); indoleamine 2,3-dioxygenase (IDO1); tryptophan-2,3-dioxygenase (TDO); kynurenic acid (KYNA); 3-hydroxykynurenine (3-HK); anthranilic acid (AA); 3-hydroxyanthranilic acid (3-HAA); kynureninase (KYNU); xanthurenic acid (XA); quinolinic acid (QA); Shanghai Diabetes Study (SHDS); coronary artery disease (CAD); kynurenine-to-tryptophan ratio (KTR); pyridoxal-5-phosphate (P5P); glutathione (GSH); reactive oxygen species (ROS); asparagine synthetase (ASNS); Atherosclerosis Risk in Communities (ARIC); likelihood ratio (LHR); Diabetes Prevention Study (DPS).

References

1. Sharma, A.; Chavali, S.; Mahajan, A.; Tabassum, R.; Banerjee, V.; Tandon, N.; Bharadwaj, D. Genetic association, post-translational modification, and protein-protein interactions in Type 2 diabetes mellitus. *Mol. Cell Proteom.* 2005, 4, 1029–1037. [CrossRef]
2. Qin, J.; Li, Y.; Cai, Z.; Li, S.; Zhu, J.; Zhang, F.; Liang, S.; Zhang, W.; Guan, Y.; Shen, D.; et al. A metagenome-wide association study of gut microbiota in type 2 diabetes. *Nature* 2012, 490, 55–60. [CrossRef] [PubMed]
3. Zhang, A.H.; Qiu, S.; Xu, H.Y.; Sun, H.; Wang, X.J. Metabolomics in diabetes. *Clin. Chim. Acta* 2014, 429, 106–110. [CrossRef] [PubMed]
4. Magliano, D.J.; Boyko, E.J.; IDF Diabetes Atlas 10th edition Scientific Committee. *IDF Diabetes Atlas*; International Diabetes Federation: Brussels, Belgium, 2021.
5. Zheng, Y.; Ley, S.H.; Hu, F.B. Global aetiology and epidemiology of type 2 diabetes mellitus and its complications. *Nat. Rev. Endocrinol.* 2018, 14, 88–98. [CrossRef] [PubMed]
6. Cole, J.B.; Florez, J.C. Genetics of diabetes mellitus and diabetes complications. *Nat. Rev. Nephrol.* 2020, 16, 377–390. [CrossRef] [PubMed]
7. Faselis, C.; Katsimardou, A.; Imprialos, K.; Deligkaris, P.; Kallistratos, M.; Dimitriadis, K. Microvascular Complications of Type 2 Diabetes Mellitus. *Curr. Vasc. Pharmacol.* 2020, 18, 117–124. [CrossRef] [PubMed]
8. Wu, G. Amino acids: Metabolism, functions, and nutrition. *Amino Acids* 2009, 37, 1–17. [CrossRef]
9. Yang, Q.; Vijayakumar, A.; Kahn, B.B. Metabolites as regulators of insulin sensitivity and metabolism. *Nat. Rev. Mol. Cell Biol.* 2018, 19, 654–672. [CrossRef]
10. Newgard, C.B. Interplay between Lipids and Branched-Chain Amino Acids in Development of Insulin Resistance. *Cell Metab.* 2012, 15, 606–614. [CrossRef]
11. Chung, S.T.; Hsia, D.S.; Chacko, S.K.; Rodriguez, L.M.; Haymond, M.W. Increased gluconeogenesis in youth with newly diagnosed type 2 diabetes. *Diabetologia* 2015, 58, 596–603. [CrossRef]
12. Newgard, C.B.; An, J.; Bain, J.R.; Muehlbauer, M.J.; Stevens, R.D.; Lien, L.F.; Haqq, A.M.; Shah, S.H.; Arlotto, M.; Slentz, C.A.; et al. A branched-chain amino acid-related metabolic signature that differentiates obese and lean humans and contributes to insulin resistance. *Cell Metab.* 2009, 9, 311–326. [CrossRef] [PubMed]
13. Würtz, P.; Tiainen, M.; Mäkinen, V.P.; Kangas, A.J.; Soininen, P.; Saltevo, J.; Keinänen-Kiukaanniemi, S.; Mäntyselkä, P.; Lehtimäki, T.; Laakso, M.; et al. Circulating metabolite predictors of glycemia in middle-aged men and women. *Diabetes Care* 2012, 35, 1749–1756. [CrossRef] [PubMed]
14. Lee, S.G.; Yim, Y.S.; Lee, Y.H.; Lee, B.W.; Kim, H.S.; Kim, K.S.; Lee, Y.W.; Kim, J.H. Fasting serum amino acids concentration is associated with insulin resistance and pro-inflammatory cytokines. *Diabetes Res. Clin. Pract.* 2018, 140, 107–117. [CrossRef] [PubMed]
15. Neinast, M.; Murashige, D.; Arany, Z. Branched Chain Amino Acids. *Annu. Rev. Physiol.* 2019, 81, 139–164. [CrossRef] [PubMed]
16. Le Couteur, D.G.; Solon-Biet, S.M.; Cogger, V.C.; Ribeiro, R.; de Cabo, R.; Raubenheimer, D.; Cooney, G.J.; Simpson, S.J. Branched chain amino acids, aging and age-related health. *Ageing Res. Rev.* 2020, 64, 101198. [CrossRef]
17. Nie, C.; He, T.; Zhang, W.; Zhang, G.; Ma, X. Branched Chain Amino Acids: Beyond Nutrition Metabolism. *Int. J. Mol. Sci.* 2018, 19, 954. [CrossRef] [PubMed]
18. Vanweert, F.; Neinast, M.; Tapia, E.E.; van de Weijer, T.; Hoeks, J.; Schrauwen-Hinderling, V.B.; Blair, M.C.; Bornstein, M.R.; Hesselink, M.K.C.; Schrauwen, P.; et al. A randomized placebo-controlled clinical trial for pharmacological activation of BCAA catabolism in patients with type 2 diabetes. *Nat. Commun.* 2022, 13, 3508. [CrossRef] [PubMed]
19. Bloomgarden, Z. Diabetes and branched-chain amino acids: What is the link? *J. Diabetes* 2018, 10, 350–352. [CrossRef]
20. Lu, J.; Xie, G.; Jia, W.; Jia, W. Insulin resistance and the metabolism of branched-chain amino acids. *Front. Med.* 2013, 7, 53–59. [CrossRef]

21. Menni, C.; Fauman, E.; Erte, I.; Perry, J.R.; Kastenmüller, G.; Shin, S.Y.; Petersen, A.K.; Hyde, C.; Psatha, M.; Ward, K.J.; et al. Biomarkers for type 2 diabetes and impaired fasting glucose using a nontargeted metabolomics approach. *Diabetes* **2013**, *62*, 4270–4276. [CrossRef]
22. Felig, P.; Marliss, E.; Ohman, J.L.; Cahill, C.F., Jr. Plasma amino acid levels in diabetic ketoacidosis. *Diabetes* **1970**, *19*, 727–728. [CrossRef] [PubMed]
23. Wang, S.; Li, M.; Lin, H.; Wang, G.; Xu, Y.; Zhao, X.; Hu, C.; Zhang, Y.; Zheng, R.; Hu, R.; et al. Amino acids, microbiota-related metabolites, and the risk of incident diabetes among normoglycemic Chinese adults: Findings from the 4C study. *Cell Rep. Med.* **2022**, *3*, 100727. [CrossRef] [PubMed]
24. Wang, T.J.; Larson, M.G.; Vasan, R.S.; Cheng, S.; Rhee, E.P.; McCabe, E.; Lewis, G.D.; Fox, C.S.; Jacques, P.F.; Fernandez, C.; et al. Metabolite profiles and the risk of developing diabetes. *Nat. Med.* **2011**, *17*, 448–453. [CrossRef] [PubMed]
25. Ruiz-Canela, M.; Guasch-Ferré, M.; Toledo, E.; Clish, C.B.; Razquin, C.; Liang, L.; Wang, D.D.; Corella, D.; Estruch, R.; Hernáez, Á.; et al. Plasma branched chain/aromatic amino acids, enriched Mediterranean diet and risk of type 2 diabetes: Case-cohort study within the PREDIMED Trial. *Diabetologia* **2018**, *61*, 1560–1571. [CrossRef]
26. Yu, E.; Papandreou, C.; Ruiz-Canela, M.; Guasch-Ferre, M.; Clish, C.B.; Dennis, C.; Liang, L.; Corella, D.; Fitó, M.; Razquin, C.; et al. Association of Tryptophan Metabolites with Incident Type 2 Diabetes in the PREDIMED Trial: A Case-Cohort Study. *Clin. Chem.* **2018**, *64*, 1211–1220. [CrossRef]
27. Lynch, C.J.; Adams, S.H. Branched-chain amino acids in metabolic signalling and insulin resistance. *Nat. Rev. Endocrinol.* **2014**, *10*, 723–736. [CrossRef] [PubMed]
28. Hamaya, R.; Mora, S.; Lawler, P.R.; Cook, N.R.; Ridker, P.M.; Buring, J.E.; Lee, I.M.; Manson, J.E.; Tobias, D.K. Association of Plasma Branched-Chain Amino Acid With Biomarkers of Inflammation and Lipid Metabolism in Women. *Circ. Genom. Precis. Med.* **2021**, *14*, e003330. [CrossRef]
29. Vangipurapu, J.; Stancáková, A.; Smith, U.; Kuusisto, J.; Laakso, M. Nine Amino Acids Are Associated With Decreased Insulin Secretion and Elevated Glucose Levels in a 7.4-Year Follow-up Study of 5,181 Finnish Men. *Diabetes* **2019**, *68*, 1353–1358. [CrossRef]
30. Lee, C.C.; Watkins, S.M.; Lorenzo, C.; Wagenknecht, L.E.; Il'yasova, D.; Chen, Y.D.; Haffner, S.M.; Hanley, A.J. Branched-Chain Amino Acids and Insulin Metabolism: The Insulin Resistance Atherosclerosis Study (IRAS). *Diabetes Care* **2016**, *39*, 582–588. [CrossRef] [PubMed]
31. White, P.J.; McGarrah, R.W.; Herman, M.A.; Bain, J.R.; Shah, S.H.; Newgard, C.B. Insulin action, type 2 diabetes, and branched-chain amino acids: A two-way street. *Mol. Metab.* **2021**, *52*, 101261. [CrossRef]
32. Giesbertz, P.; Daniel, H. Branched-chain amino acids as biomarkers in diabetes. *Curr. Opin. Clin. Nutr. Metab. Care* **2016**, *19*, 48–54. [CrossRef] [PubMed]
33. Hosseinkhani, S.; Arjmand, B.; Dilmaghani-Marand, A.; Mohammadi Fateh, S.; Dehghanbanadaki, H.; Najjar, N.; Alavi-Moghadam, S.; Ghodssi-Ghassemabadi, R.; Nasli-Esfahani, E.; Farzadfar, F.; et al. Targeted metabolomics analysis of amino acids and acylcarnitines as risk markers for diabetes by LC-MS/MS technique. *Sci. Rep.* **2022**, *12*, 8418. [CrossRef] [PubMed]
34. Newbern, D.; Gumus Balikcioglu, P.; Balikcioglu, M.; Bain, J.; Muehlbauer, M.; Stevens, R.; Ilkayeva, O.; Dolinsky, D.; Armstrong, S.; Irizarry, K.; et al. Sex differences in biomarkers associated with insulin resistance in obese adolescents: Metabolomic profiling and principal components analysis. *J. Clin. Endocrinol. Metab.* **2014**, *99*, 4730–4739. [CrossRef] [PubMed]
35. Shin, A.C.; Fasshauer, M.; Filatova, N.; Grundell, L.A.; Zielinski, E.; Zhou, J.Y.; Scherer, T.; Lindtner, C.; White, P.J.; Lapworth, A.L.; et al. Brain insulin lowers circulating BCAA levels by inducing hepatic BCAA catabolism. *Cell Metab.* **2014**, *20*, 898–909. [CrossRef]
36. McCormack, S.E.; Shaham, O.; McCarthy, M.A.; Deik, A.A.; Wang, T.J.; Gerszten, R.E.; Clish, C.B.; Mootha, V.K.; Grinspoon, S.K.; Fleischman, A. Circulating branched-chain amino acid concentrations are associated with obesity and future insulin resistance in children and adolescents. *Pediatr. Obes.* **2013**, *8*, 52–61. [CrossRef]
37. Yang, P.; Hu, W.; Fu, Z.; Sun, L.; Zhou, Y.; Gong, Y.; Yang, T.; Zhou, H. The positive association of branched-chain amino acids and metabolic dyslipidemia in Chinese Han population. *Lipids Health Dis.* **2016**, *15*, 120. [CrossRef]
38. Chen, T.; Ni, Y.; Ma, X.; Bao, Y.; Liu, J.; Huang, F.; Hu, C.; Xie, G.; Zhao, A.; Jia, W.; et al. Branched-chain and aromatic amino acid profiles and diabetes risk in Chinese populations. *Sci. Rep.* **2016**, *6*, 20594. [CrossRef]
39. Cho, Y.S.; Chen, C.H.; Hu, C.; Long, J.; Ong, R.T.; Sim, X.; Takeuchi, F.; Wu, Y.; Go, M.J.; Yamauchi, T.; et al. Meta-analysis of genome-wide association studies identifies eight new loci for type 2 diabetes in east Asians. *Nat. Genet.* **2011**, *44*, 67–72. [CrossRef]
40. Yamada, C.; Kondo, M.; Kishimoto, N.; Shibata, T.; Nagai, Y.; Imanishi, T.; Oroguchi, T.; Ishii, N.; Nishizaki, Y. Association between insulin resistance and plasma amino acid profile in non-diabetic Japanese subjects. *J. Diabetes Investig.* **2015**, *6*, 408–415. [CrossRef]
41. Palmer, N.D.; Stevens, R.D.; Antinozzi, P.A.; Anderson, A.; Bergman, R.N.; Wagenknecht, L.E.; Newgard, C.B.; Bowden, D.W. Metabolomic profile associated with insulin resistance and conversion to diabetes in the Insulin Resistance Atherosclerosis Study. *J. Clin. Endocrinol. Metab.* **2015**, *100*, E463–E468. [CrossRef]
42. Laplante, M.; Sabatini, D.M. mTOR signaling in growth control and disease. *Cell* **2012**, *149*, 274–293. [CrossRef] [PubMed]
43. Kim, E.; Goraksha-Hicks, P.; Li, L.; Neufeld, T.P.; Guan, K.L. Regulation of TORC1 by Rag GTPases in nutrient response. *Nat. Cell Biol.* **2008**, *10*, 935–945. [CrossRef] [PubMed]

44. Long, X.; Lin, Y.; Ortiz-Vega, S.; Yonezawa, K.; Avruch, J. Rheb binds and regulates the mTOR kinase. *Curr. Biol. CB* **2005**, *15*, 702–713. [CrossRef]
45. Ma, X.M.; Blenis, J. Molecular mechanisms of mTOR-mediated translational control. *Nat. Rev. Mol. Cell Biol.* **2009**, *10*, 307–318. [CrossRef] [PubMed]
46. Condon, K.J.; Sabatini, D.M. Nutrient regulation of mTORC1 at a glance. *J. Cell Sci.* **2019**, *132*, jcs222570. [CrossRef] [PubMed]
47. Inoki, K.; Zhu, T.; Guan, K.L. TSC2 mediates cellular energy response to control cell growth and survival. *Cell* **2003**, *115*, 577–590. [CrossRef]
48. Kim, Y.C.; Guan, K.L. mTOR: A pharmacologic target for autophagy regulation. *J. Clin. Investig.* **2015**, *125*, 25–32. [CrossRef]
49. Ardestani, A.; Lupse, B.; Kido, Y.; Leibowitz, G.; Maedler, K. mTORC1 Signaling: A Double-Edged Sword in Diabetic β Cells. *Cell Metab.* **2018**, *27*, 314–331. [CrossRef]
50. White, M.F. The IRS-signalling system: A network of docking proteins that mediate insulin action. *Mol. Cell Biochem.* **1998**, *182*, 3–11. [CrossRef]
51. Blackburn, P.R.; Gass, J.M.; Vairo, F.P.E.; Farnham, K.M.; Atwal, H.K.; Macklin, S.; Klee, E.W.; Atwal, P.S. Maple syrup urine disease: Mechanisms and management. *Appl. Clin. Genet.* **2017**, *10*, 57–66. [CrossRef]
52. She, P.; Olson, K.C.; Kadota, Y.; Inukai, A.; Shimomura, Y.; Hoppel, C.L.; Adams, S.H.; Kawamata, Y.; Matsumoto, H.; Sakai, R.; et al. Leucine and protein metabolism in obese Zucker rats. *PLoS ONE* **2013**, *8*, e59443. [CrossRef]
53. Sun, L.; Li, H.; Lin, X. Linking of metabolomic biomarkers with cardiometabolic health in Chinese population. *J. Diabetes* **2019**, *11*, 280–291. [CrossRef] [PubMed]
54. Kadota, Y.; Kazama, S.; Bajotto, G.; Kitaura, Y.; Shimomura, Y. Clofibrate-induced reduction of plasma branched-chain amino acid concentrations impairs glucose tolerance in rats. *JPEN J. Parenter. Enter. Nutr.* **2012**, *36*, 337–343. [CrossRef] [PubMed]
55. Kuzuya, T.; Katano, Y.; Nakano, I.; Hirooka, Y.; Itoh, A.; Ishigami, M.; Hayashi, K.; Honda, T.; Goto, H.; Fujita, Y.; et al. Regulation of branched-chain amino acid catabolism in rat models for spontaneous type 2 diabetes mellitus. *Biochem. Biophys. Res. Commun.* **2008**, *373*, 94–98. [CrossRef] [PubMed]
56. Arany, Z.; Neinast, M. Branched Chain Amino Acids in Metabolic Disease. *Curr. Diabetes Rep.* **2018**, *18*, 76. [CrossRef] [PubMed]
57. Olson, K.C.; Chen, G.; Xu, Y.; Hajnal, A.; Lynch, C.J. Alloisoleucine differentiates the branched-chain aminoacidemia of Zucker and dietary obese rats. *Obesity* **2014**, *22*, 1212–1215. [CrossRef]
58. Papathanassiu, A.E.; Ko, J.H.; Imprialou, M.; Bagnati, M.; Srivastava, P.K.; Vu, H.A.; Cucchi, D.; McAdoo, S.P.; Ananieva, E.A.; Mauro, C.; et al. BCAT1 controls metabolic reprogramming in activated human macrophages and is associated with inflammatory diseases. *Nat. Commun.* **2017**, *8*, 16040. [CrossRef]
59. Liu, S.; Li, L.; Lou, P.; Zhao, M.; Wang, Y.; Tang, M.; Gong, M.; Liao, G.; Yuan, Y.; Li, L.; et al. Elevated branched-chain α-keto acids exacerbate macrophage oxidative stress and chronic inflammatory damage in type 2 diabetes mellitus. *Free Radic. Biol. Med.* **2021**, *175*, 141–154. [CrossRef]
60. Stancáková, A.; Civelek, M.; Saleem, N.K.; Soininen, P.; Kangas, A.J.; Cederberg, H.; Paananen, J.; Pihlajamäki, J.; Bonnycastle, L.L.; Morken, M.A.; et al. Hyperglycemia and a common variant of GCKR are associated with the levels of eight amino acids in 9,369 Finnish men. *Diabetes* **2012**, *61*, 1895–1902. [CrossRef]
61. Zhou, Q.; Sun, W.W.; Chen, J.C.; Zhang, H.L.; Liu, J.; Lin, Y.; Lin, P.C.; Wu, B.X.; An, Y.P.; Huang, L.; et al. Phenylalanine impairs insulin signaling and inhibits glucose uptake through modification of IRβ. *Nat. Commun.* **2022**, *13*, 4291. [CrossRef]
62. Li, J.; Cao, Y.-F.; Sun, X.-Y.; Han, L.; Li, S.-N.; Gu, W.-Q.; Song, M.; Jiang, C.-t.; Yang, X.; Fang, Z.-z. Plasma tyrosine and its interaction with low high-density lipoprotein cholesterol and the risk of type 2 diabetes mellitus in Chinese. *J. Diabetes Investig.* **2019**, *10*, 491–498. [CrossRef] [PubMed]
63. Chi, Q.; Wang, T.; Huang, K. Effect of insulin nitration by peroxynitrite on its biological activity. *Biochem. Biophys. Res. Commun.* **2005**, *330*, 791–796. [CrossRef] [PubMed]
64. Zhai, L.; Ladomersky, E.; Lenzen, A.; Nguyen, B.; Patel, R.; Lauing, K.L.; Wu, M.; Wainwright, D.A. IDO1 in cancer: A Gemini of immune checkpoints. *Cell Mol. Immunol.* **2018**, *15*, 447–457. [CrossRef] [PubMed]
65. Le Floc'h, N.; Otten, W.; Merlot, E. Tryptophan metabolism, from nutrition to potential therapeutic applications. *Amino Acids* **2011**, *41*, 1195–1205. [CrossRef]
66. O'Mahony, S.M.; Clarke, G.; Borre, Y.E.; Dinan, T.G.; Cryan, J.F. Serotonin, tryptophan metabolism and the brain-gut-microbiome axis. *Behav. Brain Res.* **2015**, *277*, 32–48. [CrossRef] [PubMed]
67. Qi, Q.; Li, J.; Yu, B.; Moon, J.Y.; Chai, J.C.; Merino, J.; Hu, J.; Ruiz-Canela, M.; Rebholz, C.; Wang, Z.; et al. Host and gut microbial tryptophan metabolism and type 2 diabetes: An integrative analysis of host genetics, diet, gut microbiome and circulating metabolites in cohort studies. *Gut* **2022**, *71*, 1095–1105. [CrossRef] [PubMed]
68. Xue, C.; Li, G.; Zheng, Q.; Gu, X.; Shi, Q.; Su, Y.; Chu, Q.; Yuan, X.; Bao, Z.; Lu, J.; et al. Tryptophan metabolism in health and disease. *Cell Metab.* **2023**, *35*, 1304–1326. [CrossRef]
69. Gao, K.; Mu, C.-l.; Farzi, A.; Zhu, W.-y. Tryptophan Metabolism: A Link Between the Gut Microbiota and Brain. *Adv. Nutr.* **2020**, *11*, 709–723. [CrossRef]
70. Vujkovic-Cvijin, I.; Dunham, R.M.; Iwai, S.; Maher, M.C.; Albright, R.G.; Broadhurst, M.J.; Hernandez, R.D.; Lederman, M.M.; Huang, Y.; Somsouk, M.; et al. Dysbiosis of the gut microbiota is associated with HIV disease progression and tryptophan catabolism. *Sci. Transl. Med.* **2013**, *5*, 193ra191. [CrossRef]

71. Floegel, A.; Stefan, N.; Yu, Z.; Mühlenbruch, K.; Drogan, D.; Joost, H.G.; Fritsche, A.; Häring, H.U.; Hrabě de Angelis, M.; Peters, A.; et al. Identification of serum metabolites associated with risk of type 2 diabetes using a targeted metabolomic approach. *Diabetes* **2013**, *62*, 639–648. [CrossRef]
72. Chen, T.; Zheng, X.; Ma, X.; Bao, Y.; Ni, Y.; Hu, C.; Rajani, C.; Huang, F.; Zhao, A.; Jia, W.; et al. Tryptophan Predicts the Risk for Future Type 2 Diabetes. *PLoS ONE* **2016**, *11*, e0162192. [CrossRef] [PubMed]
73. Oxenkrug, G.F. Increased Plasma Levels of Xanthurenic and Kynurenic Acids in Type 2 Diabetes. *Mol. Neurobiol.* **2015**, *52*, 805–810. [CrossRef] [PubMed]
74. Muzik, O.; Burghardt, P.; Yi, Z.; Kumar, A.; Seyoum, B. Successful metformin treatment of insulin resistance is associated with down-regulation of the kynurenine pathway. *Biochem. Biophys. Res. Commun.* **2017**, *488*, 29–32. [CrossRef]
75. Vangipurapu, J.; Fernandes Silva, L.; Kuulasmaa, T.; Smith, U.; Laakso, M. Microbiota-Related Metabolites and the Risk of Type 2 Diabetes. *Diabetes Care* **2020**, *43*, 1319–1325. [CrossRef] [PubMed]
76. Rebnord, E.W.; Strand, E.; Midttun, Ø.; Svingen, G.F.T.; Christensen, M.H.E.; Ueland, P.M.; Mellgren, G.; Njølstad, P.R.; Tell, G.S.; Nygård, O.K.; et al. The kynurenine:tryptophan ratio as a predictor of incident type 2 diabetes mellitus in individuals with coronary artery disease. *Diabetologia* **2017**, *60*, 1712–1721. [CrossRef] [PubMed]
77. Scarale, M.G.; Mastroianno, M.; Prehn, C.; Copetti, M.; Salvemini, L.; Adamski, J.; De Cosmo, S.; Trischitta, V.; Menzaghi, C. Circulating Metabolites Associate With and Improve the Prediction of All-Cause Mortality in Type 2 Diabetes. *Diabetes* **2022**, *71*, 1363–1370. [CrossRef]
78. Kozieł, K.; Urbanska, E.M. Kynurenine Pathway in Diabetes Mellitus-Novel Pharmacological Target? *Cells* **2023**, *12*, 460. [CrossRef]
79. Oxenkrug, G. Insulin resistance and dysregulation of tryptophan-kynurenine and kynurenine-nicotinamide adenine dinucleotide metabolic pathways. *Mol. Neurobiol.* **2013**, *48*, 294–301. [CrossRef]
80. Kotake, Y. Xanthurenic acid, an abnormal metabolite of tryptophan and the diabetic symptoms caused in albino rats by its production. *J. Vitaminol.* **1955**, *1*, 73–87. [CrossRef]
81. Durkin, P.J.; Friedberg, F. The synthesis of glutathione in torula utilis studied with 14C-carboxyl labeled glycine. *Biochim. Biophys. Acta* **1952**, *9*, 105–106. [CrossRef]
82. Adeva-Andany, M.; Souto-Adeva, G.; Ameneiros-Rodríguez, E.; Fernández-Fernández, C.; Donapetry-García, C.; Domínguez-Montero, A. Insulin resistance and glycine metabolism in humans. *Amino Acids* **2018**, *50*, 11–27. [CrossRef] [PubMed]
83. Takashina, C.; Tsujino, I.; Watanabe, T.; Sakaue, S.; Ikeda, D.; Yamada, A.; Sato, T.; Ohira, H.; Otsuka, Y.; Oyama-Manabe, N.; et al. Associations among the plasma amino acid profile, obesity, and glucose metabolism in Japanese adults with normal glucose tolerance. *Nutr. Metab.* **2016**, *13*, 5. [CrossRef] [PubMed]
84. Irving, B.A.; Carter, R.E.; Soop, M.; Weymiller, A.; Syed, H.; Karakelides, H.; Bhagra, S.; Short, K.R.; Tatpati, L.; Barazzoni, R.; et al. Effect of insulin sensitizer therapy on amino acids and their metabolites. *Metabolism* **2015**, *64*, 720–728. [CrossRef] [PubMed]
85. Xie, W.; Wood, A.R.; Lyssenko, V.; Weedon, M.N.; Knowles, J.W.; Alkayyali, S.; Assimes, T.L.; Quertermous, T.; Abbasi, F.; Paananen, J.; et al. Genetic variants associated with glycine metabolism and their role in insulin sensitivity and type 2 diabetes. *Diabetes* **2013**, *62*, 2141–2150. [CrossRef] [PubMed]
86. Yan-Do, R.; Duong, E.; Manning Fox, J.E.; Dai, X.; Suzuki, K.; Khan, S.; Bautista, A.; Ferdaoussi, M.; Lyon, J.; Wu, X.; et al. A Glycine-Insulin Autocrine Feedback Loop Enhances Insulin Secretion From Human β-Cells and Is Impaired in Type 2 Diabetes. *Diabetes* **2016**, *65*, 2311–2321. [CrossRef]
87. El-Hafidi, M.; Franco, M.; Ramírez, A.R.; Sosa, J.S.; Flores, J.A.P.; Acosta, O.L.; Salgado, M.C.; Cardoso-Saldaña, G. Glycine Increases Insulin Sensitivity and Glutathione Biosynthesis and Protects against Oxidative Stress in a Model of Sucrose-Induced Insulin Resistance. *Oxidative Med. Cell. Longev.* **2018**, *2018*, 2101562. [CrossRef]
88. Garcia-Macedo, R.; Sanchez-Muñoz, F.; Almanza-Perez, J.C.; Duran-Reyes, G.; Alarcon-Aguilar, F.; Cruz, M. Glycine increases mRNA adiponectin and diminishes pro-inflammatory adipokines expression in 3T3-L1 cells. *Eur. J. Pharmacol.* **2008**, *587*, 317–321. [CrossRef]
89. Sekhar, R.V.; McKay, S.V.; Patel, S.G.; Guthikonda, A.P.; Reddy, V.T.; Balasubramanyam, A.; Jahoor, F. Glutathione synthesis is diminished in patients with uncontrolled diabetes and restored by dietary supplementation with cysteine and glycine. *Diabetes Care* **2011**, *34*, 162–167. [CrossRef]
90. Blaise, M.; Fréchin, M.; Oliéric, V.; Charron, C.; Sauter, C.; Lorber, B.; Roy, H.; Kern, D. Crystal structure of the archaeal asparagine synthetase: Interrelation with aspartyl-tRNA and asparaginyl-tRNA synthetases. *J. Mol. Biol.* **2011**, *412*, 437–452. [CrossRef]
91. Balasubramanian, M.N.; Butterworth, E.A.; Kilberg, M.S. Asparagine synthetase: Regulation by cell stress and involvement in tumor biology. *Am. J. Physiol. Endocrinol. Metab.* **2013**, *304*, E789–E799. [CrossRef]
92. Wu, G. Functional amino acids in nutrition and health. *Amino Acids* **2013**, *45*, 407–411. [CrossRef] [PubMed]
93. Giesbertz, P.; Padberg, I.; Rein, D.; Ecker, J.; Höfle, A.S.; Spanier, B.; Daniel, H. Metabolite profiling in plasma and tissues of ob/ob and db/db mice identifies novel markers of obesity and type 2 diabetes. *Diabetologia* **2015**, *58*, 2133–2143. [CrossRef] [PubMed]
94. Ottosson, F.; Smith, E.; Melander, O.; Fernandez, C. Altered Asparagine and Glutamate Homeostasis Precede Coronary Artery Disease and Type 2 Diabetes. *J. Clin. Endocrinol. Metab.* **2018**, *103*, 3060–3069. [CrossRef] [PubMed]
95. Cheng, S.; Rhee, E.P.; Larson, M.G.; Lewis, G.D.; McCabe, E.L.; Shen, D.; Palma, M.J.; Roberts, L.D.; Dejam, A.; Souza, A.L.; et al. Metabolite profiling identifies pathways associated with metabolic risk in humans. *Circulation* **2012**, *125*, 2222–2231. [CrossRef]

96. Rebholz, C.M.; Yu, B.; Zheng, Z.; Chang, P.; Tin, A.; Köttgen, A.; Wagenknecht, L.E.; Coresh, J.; Boerwinkle, E.; Selvin, E. Serum metabolomic profile of incident diabetes. *Diabetologia* **2018**, *61*, 1046–1054. [CrossRef]
97. Luo, H.H.; Feng, X.F.; Yang, X.L.; Hou, R.Q.; Fang, Z.Z. Interactive effects of asparagine and aspartate homeostasis with sex and age for the risk of type 2 diabetes risk. *Biol. Sex. Differ.* **2020**, *11*, 58. [CrossRef]
98. Wu, G.; Fang, Y.Z.; Yang, S.; Lupton, J.R.; Turner, N.D. Glutathione metabolism and its implications for health. *J. Nutr.* **2004**, *134*, 489–492. [CrossRef]
99. Banerji, J. Asparaginase treatment side-effects may be due to genes with homopolymeric Asn codons (Review-Hypothesis). *Int. J. Mol. Med.* **2015**, *36*, 607–626. [CrossRef]
100. Krall, A.S.; Xu, S.; Graeber, T.G.; Braas, D.; Christofk, H.R. Asparagine promotes cancer cell proliferation through use as an amino acid exchange factor. *Nat. Commun.* **2016**, *7*, 11457. [CrossRef]
101. Brown, L.M.; Clegg, D.J. Central effects of estradiol in the regulation of food intake, body weight, and adiposity. *J. Steroid Biochem. Mol. Biol.* **2010**, *122*, 65–73. [CrossRef]
102. Meyer, M.R.; Clegg, D.J.; Prossnitz, E.R.; Barton, M. Obesity, insulin resistance and diabetes: Sex differences and role of oestrogen receptors. *Acta Physiol.* **2011**, *203*, 259–269. [CrossRef] [PubMed]
103. Holeček, M. Serine Metabolism in Health and Disease and as a Conditionally Essential Amino Acid. *Nutrients* **2022**, *14*, 1987. [CrossRef] [PubMed]
104. Handzlik, M.K.; Gengatharan, J.M.; Frizzi, K.E.; McGregor, G.H.; Martino, C.; Rahman, G.; Gonzalez, A.; Moreno, A.M.; Green, C.R.; Guernsey, L.S.; et al. Insulin-regulated serine and lipid metabolism drive peripheral neuropathy. *Nature* **2023**, *614*, 118–124. [CrossRef]
105. Starling, S. Serine slows diabetic neuropathy in mice. *Nat. Rev. Endocrinol.* **2023**, *19*, 187. [CrossRef] [PubMed]
106. Bertea, M.; Rütti, M.F.; Othman, A.; Marti-Jaun, J.; Hersberger, M.; von Eckardstein, A.; Hornemann, T. Deoxysphingoid bases as plasma markers in diabetes mellitus. *Lipids Health Dis.* **2010**, *9*, 84. [CrossRef]
107. Drábková, P.; Šanderová, J.; Kovařík, J.; Kand'ár, R. An Assay of Selected Serum Amino Acids in Patients with Type 2 Diabetes Mellitus. *Adv. Clin. Exp. Med.* **2015**, *24*, 447–451. [CrossRef]
108. Mihalik, S.J.; Michaliszyn, S.F.; de las Heras, J.; Bacha, F.; Lee, S.; Chace, D.H.; DeJesus, V.R.; Vockley, J.; Arslanian, S.A. Metabolomic profiling of fatty acid and amino acid metabolism in youth with obesity and type 2 diabetes: Evidence for enhanced mitochondrial oxidation. *Diabetes Care* **2012**, *35*, 605–611. [CrossRef]
109. Wei, N.; Pan, J.; Pop-Busui, R.; Othman, A.; Alecu, I.; Hornemann, T.; Eichler, F.S. Altered sphingoid base profiles in type 1 compared to type 2 diabetes. *Lipids Health Dis.* **2014**, *13*, 161. [CrossRef]
110. Zuellig, R.A.; Hornemann, T.; Othman, A.; Hehl, A.B.; Bode, H.; Güntert, T.; Ogunshola, O.O.; Saponara, E.; Grabliauskaite, K.; Jang, J.-H.; et al. Deoxysphingolipids, Novel Biomarkers for Type 2 Diabetes, Are Cytotoxic for Insulin-Producing Cells. *Diabetes* **2014**, *63*, 1326–1339. [CrossRef]
111. Magnusson, M.; Lewis, G.D.; Ericson, U.; Orho-Melander, M.; Hedblad, B.; Engström, G.; Ostling, G.; Clish, C.; Wang, T.J.; Gerszten, R.E.; et al. A diabetes-predictive amino acid score and future cardiovascular disease. *Eur. Heart J.* **2013**, *34*, 1982–1989. [CrossRef]
112. Walford, G.A.; Davis, J.; Warner, A.S.; Ackerman, R.J.; Billings, L.K.; Chamarthi, B.; Fanelli, R.R.; Hernandez, A.M.; Huang, C.; Khan, S.Q.; et al. Branched chain and aromatic amino acids change acutely following two medical therapies for type 2 diabetes mellitus. *Metabolism* **2013**, *62*, 1772–1778. [CrossRef] [PubMed]
113. Morze, J.; Wittenbecher, C.; Schwingshackl, L.; Danielewicz, A.; Rynkiewicz, A.; Hu, F.B.; Guasch-Ferré, M. Metabolomics and Type 2 Diabetes Risk: An Updated Systematic Review and Meta-analysis of Prospective Cohort Studies. *Diabetes Care* **2022**, *45*, 1013–1024. [CrossRef] [PubMed]
114. Felig, P.; Marliss, E.; Cahill, G.F., Jr. Plasma amino acid levels and insulin secretion in obesity. *N. Engl. J. Med.* **1969**, *281*, 811–816. [CrossRef]
115. Würtz, P.; Soininen, P.; Kangas, A.J.; Rönnemaa, T.; Lehtimäki, T.; Kähönen, M.; Viikari, J.S.; Raitakari, O.T.; Ala-Korpela, M. Branched-chain and aromatic amino acids are predictors of insulin resistance in young adults. *Diabetes Care* **2013**, *36*, 648–655. [CrossRef]
116. Li, X.; Li, Y.; Liang, Y.; Hu, R.; Xu, W.; Liu, Y. Plasma Targeted Metabolomics Analysis for Amino Acids and Acylcarnitines in Patients with Prediabetes, Type 2 Diabetes Mellitus, and Diabetic Vascular Complications. *Diabetes Metab. J.* **2021**, *45*, 195–208. [CrossRef] [PubMed]
117. Uusitupa, M.; Khan, T.A.; Viguiliouk, E.; Kahleova, H.; Rivellese, A.A.; Hermansen, K.; Pfeiffer, A.; Thanopoulou, A.; Salas-Salvadó, J.; Schwab, U.; et al. Prevention of Type 2 Diabetes by Lifestyle Changes: A Systematic Review and Meta-Analysis. *Nutrients* **2019**, *11*, 2611. [CrossRef] [PubMed]
118. Dunkley, A.J.; Bodicoat, D.H.; Greaves, C.J.; Russell, C.; Yates, T.; Davies, M.J.; Khunti, K. Diabetes prevention in the real world: Effectiveness of pragmatic lifestyle interventions for the prevention of type 2 diabetes and of the impact of adherence to guideline recommendations: A systematic review and meta-analysis. *Diabetes Care* **2014**, *37*, 922–933. [CrossRef]
119. Cummings, N.E.; Williams, E.M.; Kasza, I.; Konon, E.N.; Schaid, M.D.; Schmidt, B.A.; Poudel, C.; Sherman, D.S.; Yu, D.; Arriola Apelo, S.I.; et al. Restoration of metabolic health by decreased consumption of branched-chain amino acids. *J. Physiol.* **2018**, *596*, 623–645. [CrossRef]

120. Kivelä, J.; Meinilä, J.; Uusitupa, M.; Tuomilehto, J.; Lindström, J. Longitudinal Branched-Chain Amino Acids, Lifestyle Intervention, and Type 2 Diabetes in the Finnish Diabetes Prevention Study. *J. Clin. Endocrinol. Metab.* **2022**, *107*, 2844–2853. [CrossRef]
121. Ruiz-Canela, M.; Toledo, E.; Clish, C.B.; Hruby, A.; Liang, L.; Salas-Salvadó, J.; Razquin, C.; Corella, D.; Estruch, R.; Ros, E.; et al. Plasma Branched-Chain Amino Acids and Incident Cardiovascular Disease in the PREDIMED Trial. *Clin. Chem.* **2016**, *62*, 582–592. [CrossRef]
122. Lamiquiz-Moneo, I.; Bea, A.M.; Palacios-Pérez, C.; Miguel-Etayo, P.; González-Gil, E.M.; López-Ariño, C.; Civeira, F.; Moreno, L.A.; Mateo-Gallego, R. Effect of Lifestyle Intervention in the Concentration of Adipoquines and Branched Chain Amino Acids in Subjects with High Risk of Developing Type 2 Diabetes: Feel4Diabetes Study. *Cells* **2020**, *9*, 693. [CrossRef] [PubMed]
123. Thalacker-Mercer, A.E.; Ingram, K.H.; Guo, F.; Ilkayeva, O.; Newgard, C.B.; Garvey, W.T. BMI, RQ, diabetes, and sex affect the relationships between amino acids and clamp measures of insulin action in humans. *Diabetes* **2014**, *63*, 791–800. [CrossRef] [PubMed]
124. Jachthuber Trub, C.; Balikcioglu, M.; Freemark, M.; Bain, J.; Muehlbauer, M.; Ilkayeva, O.; White, P.J.; Armstrong, S.; Østbye, T.; Grambow, S.; et al. Impact of lifestyle Intervention on branched-chain amino acid catabolism and insulin sensitivity in adolescents with obesity. *Endocrinol. Diabetes Metab.* **2021**, *4*, e00250. [CrossRef] [PubMed]
125. Preiss, D.; Rankin, N.; Welsh, P.; Holman, R.R.; Kangas, A.J.; Soininen, P.; Würtz, P.; Ala-Korpela, M.; Sattar, N. Effect of metformin therapy on circulating amino acids in a randomized trial: The CAMERA study. *Diabet. Med.* **2016**, *33*, 1569–1574. [CrossRef]
126. Rivera, M.E.; Lyon, E.S.; Vaughan, R.A. Effect of metformin on myotube BCAA catabolism. *J. Cell Biochem.* **2020**, *121*, 816–827. [CrossRef]
127. Yan, M.; Qi, H.; Xia, T.; Zhao, X.; Wang, W.; Wang, Z.; Lu, C.; Ning, Z.; Chen, H.; Li, T.; et al. Metabolomics profiling of metformin-mediated metabolic reprogramming bypassing AMPKα. *Metabolism* **2019**, *91*, 18–29. [CrossRef]
128. Lv, Y.; Tian, N.; Wang, J.; Yang, M.; Kong, L. Metabolic switching in the hypoglycemic and antitumor effects of metformin on high glucose induced HepG2 cells. *J. Pharm. Biomed. Anal.* **2018**, *156*, 153–162. [CrossRef]
129. Kappel, B.A.; Lehrke, M.; Schütt, K.; Artati, A.; Adamski, J.; Lebherz, C.; Marx, N. Effect of Empagliflozin on the Metabolic Signature of Patients With Type 2 Diabetes Mellitus and Cardiovascular Disease. *Circulation* **2017**, *136*, 969–972. [CrossRef]
130. Muscelli, E.; Frascerra, S.; Casolaro, A.; Baldi, S.; Mari, A.; Gall, W.; Cobb, J.; Ferrannini, E. The amino acid response to a mixed meal in patients with type 2 diabetes: Effect of sitagliptin treatment. *Diabetes Obes. Metab.* **2014**, *16*, 1140–1147. [CrossRef]
131. Liao, X.; Liu, B.; Qu, H.; Zhang, L.; Lu, Y.; Xu, Y.; Lyu, Z.; Zheng, H. A High Level of Circulating Valine Is a Biomarker for Type 2 Diabetes and Associated with the Hypoglycemic Effect of Sitagliptin. *Mediat. Inflamm.* **2019**, *2019*, 8247019. [CrossRef]

Disclaimer/Publisher's Note: The statements, opinions and data contained in all publications are solely those of the individual author(s) and contributor(s) and not of MDPI and/or the editor(s). MDPI and/or the editor(s) disclaim responsibility for any injury to people or property resulting from any ideas, methods, instructions or products referred to in the content.

Review

Recent Advances and Perspectives in Relation to the Metabolomics-Based Study of Diabetic Retinopathy

Shuling He, Lvyun Sun, Jiali Chen and Yang Ouyang *

Department of Health Inspection and Quarantine, School of Public Health, Fujian Medical University, Fuzhou 350122, China; heshuling@fjmu.edu.cn (S.H.)
* Correspondence: ouyangyang@fjmu.edu.cn; Tel.: +86-139-4086-5942

Abstract: Diabetic retinopathy (DR), a prevalent microvascular complication of diabetes, is a major cause of acquired blindness in adults. Currently, a clinical diagnosis of DR primarily relies on fundus fluorescein angiography, with a limited availability of effective biomarkers. Metabolomics, a discipline dedicated to scrutinizing the response of various metabolites within living organisms, has shown noteworthy advancements in uncovering metabolic disorders and identifying key metabolites associated with DR in recent years. Consequently, this review aims to present the latest advancements in metabolomics techniques and comprehensively discuss the principal metabolic outcomes derived from analyzing blood, vitreous humor, aqueous humor, urine, and fecal samples.

Keywords: diabetic retinopathy; metabolomics; biomarker; metabolic pathway

1. Introduction

Diabetic retinopathy, as one of the various microvascular complications of diabetes, represents a significant contributor to adult acquired blindness [1,2]. Extensive analysis carried out by Yau et al. encompassing 35 studies comprising over 20,000 participants worldwide revealed that approximately 35% of individuals with diabetes exhibited some form of retinopathy [3]. Type 1 diabetes patients demonstrated an even higher incidence rate, with up to 54% of cases being affected [4]. Factors such as prolonged diabetes duration and an inadequate control of blood glucose and blood pressure emerge as the primary etiological elements underlying DR [3]. In light of these findings, a concerning projection indicates a surge in the number of DR cases from 103 million in 2020 to an estimated 161 million by 2045, signifying an escalating global burden [5]. The disease progression of DR can typically be categorized into two stages: non-proliferative DR (NPDR) and proliferative DR (PDR), based on the presence or absence of neovascularization within the retina. NPDR generally precedes the development of PDR [6]. This systematic classification provides valuable insights into the temporal sequence and severity of DR manifestations. Kornbla et al. conducted a comprehensive literature review on the adverse reactions associated with fluorescein angiography by utilizing the PubMed database. The analysis encompassed 78 relevant publications spanning from 1961 to 2017, revealing a diverse range of adverse reaction incidences, ranging from 0.083% to 21.69%. Among these cases, mild adverse reactions were reported to occur at rates of 1.24% to 17.65%, while moderate reactions were observed in 0.2% to 6% of instances. Severe adverse reactions were relatively infrequent, with reported occurrences ranging from 0.04% to 0.59% [7]. Deaths caused by fundus fluorescein angiography have also been reported [8,9]. The treatment options available for retinopathy management, such as laser photocoagulation, vitreoretinal surgery, and the intravitreal injection of corticosteroids or anti-vascular endothelial growth factor, are employed exclusively when the patient's visual function is threatened. Nonetheless, these treatments are associated with potential side effects [10–12]. Although significant advancements have been made in retinal imaging technology in recent years, enabling enhanced retina structural visualization without the need for fluorescein administration [13], mild

Citation: He, S.; Sun, L.; Chen, J.; Ouyang, Y. Recent Advances and Perspectives in Relation to the Metabolomics-Based Study of Diabetic Retinopathy. *Metabolites* 2023, 13, 1007. https://doi.org/10.3390/metabo13091007

Academic Editor: Victor Gault

Received: 13 August 2023
Revised: 3 September 2023
Accepted: 5 September 2023
Published: 12 September 2023

Copyright: © 2023 by the authors. Licensee MDPI, Basel, Switzerland. This article is an open access article distributed under the terms and conditions of the Creative Commons Attribution (CC BY) license (https://creativecommons.org/licenses/by/4.0/).

DR cases often exhibit subtle or no discernible abnormalities, leading to challenges in early diagnosis and subsequent intervention [14]. Hence, a comprehensive exploration of the metabolic characteristics of DR holds promise in elucidating its underlying pathogenesis, discovering potential key metabolites, and facilitating the development of novel clinical diagnostic approaches and treatment strategies.

Metabolomics, a systematic discipline, investigates the dynamic response of multiple metabolites within living organisms when subjected to internal genetic mutations, pathophysiological alterations, or external environmental stimuli [15]. It is a high-throughput analysis technology that can qualitatively or quantitatively study metabolites with a relative molecular weight less than 1500. Currently, metabolomics emerged as a pivotal tool extensively employed in diverse domains such as plant biology [16], nutrition [17], medicine [18,19], and clinical [20,21] research. Remarkable progress has been made by metabolomics in the discovery of metabolic disorders and key metabolites associated with DR. Thus, this paper provides a comprehensive summary of major improvements in analytical platforms and recent advances in metabolomics research, and discusses the advantages and limitations of each approach. Subsequently, we review the utility of metabolomics in DR studies, specifically focusing on the major metabolic outcomes observed in clinical populations through the analysis of blood, vitreous fluid, aqueous humor, urine, and stool samples. By exploring the metabolomics landscape, we aim to shed light on the metabolic intricacies underlying DR pathogenesis, paving the way for potential diagnostic and therapeutic avenues.

2. Analytical Technologies for Metabolomics

Analytical methods constitute the fundamental components of metabolomics research. Nuclear magnetic resonance spectroscopy (NMR) and mass spectrometry (MS) are the two predominant analytical techniques employed in metabolomics investigations [22]. Notably, chromatography-MS coupling systems, including gas chromatography-MS (GC-MS) and liquid chromatography-MS (LC-MS), have gained substantial popularity and represent the most frequently utilized methodologies. These coupling systems provide robust analytical capabilities, empowering researchers to comprehensively profile and characterize metabolites in various biological samples. Their wide-scale adoption in metabolomics research underscores their efficacy and versatility in exploring the metabolic intricacies associated with DR.

2.1. Nuclear Magnetic Resonance Spectroscopy

NMR is a widely utilized tool for metabolite identification due to its distinctive characteristics. It offers simplicity in sample preparation, no damage to the structure and properties of the sample, good reproducibility, short analysis time, robust signal detection, and the capability for absolute quantification of metabolites [23]. ^{1}H-NMR is particularly extensively employed in metabolomics research as hydrogen atoms are prevalent in the majority of organic metabolites [24]. In the realm of biomolecular NMR, crucial nuclei like ^{13}C, ^{15}N, and ^{31}P play pivotal roles. For instance, ^{13}C NMR facilitates structure elucidation and molecular identification [25], while ^{31}P NMR offers a broad chemical shift range and sharp peaks. Nonetheless, the overlapping signals of many phosphorylated compounds pose challenges for ^{31}P NMR analysis [26]. It is worth noting that one-dimensional NMR techniques are relatively less sensitive, limiting their ability to detect metabolites present in low abundance. Two-dimensional NMR (2D NMR) has been developed to overcome the problem of overlapping resonance in one-dimensional NMR spectroscopy. It also provides enhanced sensitivity and the ability to detect and identify a broader range of metabolites. Within the field of metabolomics, techniques like heteronuclear single-quantum coherence spectroscopy (HSQC) and heteronuclear multiple-quantum correlation (HMQC) are predominantly employed. In addition, homonuclear experiments including correlated spectroscopy (COSY) [27] and total correlation spectroscopy (TOCSY) [28] play integral roles in metabolite characterization. Nonetheless, a significant drawback associated with

2D NMR techniques is the lengthy measurement time, which can extend to several hours for each sample, thus limiting their applicability to small sample sizes [29]. To address this limitation and expedite spectral acquisition without compromising sensitivity in high-throughput research, Ghosh et al. utilized the selective optimized flip-angle short-transient ^{1}H-^{13}C HMQC technique in combination with nonlinear sampling strategies. This approach allowed for the acquisition of urine and serum sample spectra with a significantly reduced experimental time, requiring only about one-seventh of the time compared to traditional ^{1}H-^{13}C HSQC experiments, while nearly retaining all molecular information [30]. Furthermore, recent years have witnessed remarkable advancements in NMR methods, such as hyperpolarized NMR [31] and cryogenic-probe-based Rheo-NMR [32], which have improved spectral resolution and metabolite identification ability. These developments hold promise for further enhancing the efficiency and accuracy of metabolomics studies.

2.2. Gas Chromatography-Mass Spectrometry

MS is a highly potent technique primarily employed for the identification of unknown compounds and the quantification of known molecules within samples. GC-MS possesses high sensitivity, good peak resolution, and extensive databases, rendering it suitable for the qualitative analysis of metabolites [33]. Leveraging the newly developed GC/quadrupole Orbitrap MS system for targeted metabolite analysis can enhance sensitivity and facilitate the utilization of quantitative strategies [34].

One inherent limitation of GC-MS is its restricted applicability to the separation and identification of low-molecular-weight (<650 Da) and volatile compounds. In order to detect polar, heat-resistant, non-volatile metabolites, chemical derivatization is necessitated prior to analysis [35]. The common metabolites typically analyzed using GC-MS analysis encompass amino acids, organic acids, and fatty acids [36], which predominantly pertain to biochemical processes such as the tricarboxylic acid (TCA) cycle, glycolysis, amino acid metabolism, and fatty acid metabolism. Cesare et al. have proposed an enhanced GC-MS methodology incorporating full-scan and multi-reaction monitoring acquisition mode, which is an effective tool for exploring intestinal microbial metabolism [37]. Furthermore, GC-MS finds extensive utility in plant metabolomics [38], microbiology [39], and clinical metabolomics [40].

2.3. Liquid Chromatography–Mass Spectrometry

LC-MS circumvents the complicated sample pretreatment in GC-MS, while ultra-performance liquid chromatography (UPLC) offers advantages such as high separation efficiency, rapid analysis speed, high detection sensitivity, and broad application scope. In comparison to GC, UPLC has been proven to be more suitable for the separation and analysis of compounds with elevated boiling points, macromolecules, and those with diminished thermal stability. LC-MS ionization sources include electrospray ionization (ESI), atmospheric pressure chemical ionization, atmospheric pressure photoionization, and fast atom bombardment. Presently, ESI represents a favored approach for LC-MS metabolomics investigations due to its "soft ionization" capability, which generates ions through charge exchange within the solution and typically results in intact molecular ions, aiding in initial recognition [29].

LC can be categorized into reverse-phase liquid chromatography (RPLC) and hydrophilic interaction liquid chromatography (HILIC). RPLC usually employs C18 columns to separate semi-polar compounds, including phenolic acids, flavonoids, alkaloids, and other glycosylated species. On the other hand, HILIC usually employs aminopropyl columns to separate polar compounds such as sugars, amino acids, carboxylic acids, and nucleotides [41]. Recently, two-dimensional and multidimensional liquid chromatography have gained prominence as potent platform technologies capable of enhancing peak capacity and resolution [42]. Nevertheless, there is currently no single chromatographic mode that can comprehensively analyze the entire metabolome with a single analysis. The integration of multiple analytical platforms facilitates improved metabolic coverage.

3. Metabolomics in Diabetic Retinopathy

Metabolomics, emerging as a prominent branch in the field of "omics" sciences subsequent to genomics, proteomics, and transcriptomics, amalgamates high-throughput analysis techniques with bioinformatics. It encompasses the quantitative and qualitative assessment of metabolites, which are important intermediates and final products of metabolism. The retina, an integral component of the central nervous system, exhibits a distinctive high metabolic activity akin to the brain, making it an immensely active tissue with substantial energy requirements [43]. Consequently, employing metabolomics in the context of DR can offer insights into the underlying mechanisms of the disease, facilitate diagnosis, and enable disease monitoring (Figure 1).

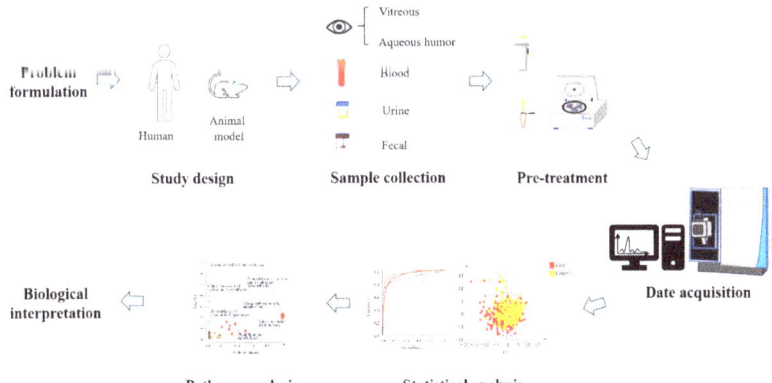

Figure 1. Schematic diagram of the research process design of metabolomics for DR.

Ever since the publication of "Metabolic fingerprints of proliferative diabetic retinopathy: an ^{1}H-NMR-based metabonomic approach using vitreous humor" in 2010 [44], there has been a growing body of research on the topic of metabolomics in the context of DR. This upsurge in studies, particularly observed in recent years (Figure 2), has predominantly focused on the analysis of vitreous and blood samples. However, several investigations have also explored alternative biological specimens including, but not limited to, aqueous humor, urine, and feces. Thus, in this review, we aim to summarize the key findings from metabolomics studies encompassing diverse sample matrices collected from individuals with DR.

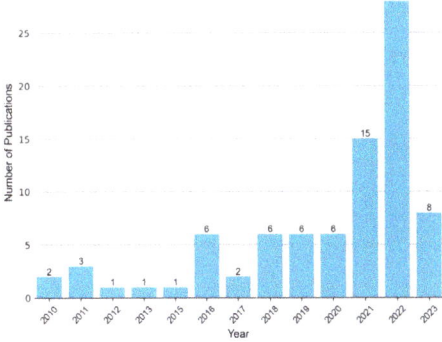

Figure 2. Annual publication trend of metabolomic studies in diabetic retinopathy. A total of 83 relevant articles were retrieved by searching the Web of Science and PubMed databases up until 5 June 2023. The search terms used were "diabetic retinopathy" AND ("metabolomics" or "metabonomics" or "metabolic profiling" or "metabolome").

3.1. Blood Metabolomics

Compared with other samples, a significant number of research studies have been undertaken in the field of blood-based metabolomics. This prevalence is primarily attributed to the relative ease of obtaining blood samples compared to other sample types. To further elucidate the subject, we compiled and summarized the key findings from previously published plasma metabolomics studies pertaining to DR in Table 1, as well as serum metabolomics studies in Table 2.

Table 1. Plasma metabolomics of DR clinical populations.

References	Subjects	Techniques	Statistical Methods	Differential Metabolites
Li X (2011) [45]	NPDR (n = 39) PDR (n = 25) DM (n = 25) Control (n = 30)	GC-TOFMS	PLS-DA	Pyruvic acids, L-aspartic acid, β-hydroxybutyric acid, methylmaleic acid, citric acid, glucose, stearic acid, transoleic acid, linoleic acid, and arachidonic acid
Xia JF (2011) [46]	DR (n = 38) DM (n = 37) Control (n = 41)	HPLC–UV/MS/MS	ROC	Cytosine, cytidine, uridine, thymine, thymidin, and 2′-deoxyuridine
Peng LY (2018) [47]	NPDR (n = 28) Control (n = 22)	LC-MS	OPLS-DA	Prostaglandin 2a
Rhee (2018) [48]	NPDR (n = 72) PDR (n = 52) Control (n = 59)	GC-TOF-MS, UPLC-Q-TOF-MS	OPLS-DA	Asparagine, aspartic acid, glutamine, glutamic acid, 1,5-anhydroglucitol, fructose, and *myo*-inositol
Sumarriva (2019) [49]	NPDR (n = 49) PDR (n = 34) Control (n = 90)	LC-MS/MS	PLS-DA	Arginine, citrulline, glutamic c-semialdehyde, acylcarnitine, and dehydroxycarnitine
Zhu XR (2019) [50]	PDR (n = 21) NDR (n = 21)	LC-MS	ROC	Fumaric acid, uridine, acetic acid, and cytidine
Sun Y (2021) [51]	DR (n = 42) Control (n = 32)	UHPLC-QE MS	OPLS-DA	Pseudouridine, N-acetyltryptophan, glutamate, leucylleucine, and HbA1c
Ding C (2022) [52]	PDR (n = 27) NPDR (n = 18) Control (n = 21)	UPLC-MS	OPLS-DA	Proline, threonine, glutamine, aspartate, glutamate, and tryptophan
Peters (2022) [53]	DM (n = 159) NPDR (n = 92) DR (n = 64)	LC-MS/MS	Wilcoxon Rank Sum test	Arginine, citrulline, asymmetric dimethylarginine, ornithine, proline, and argininosuccinic acid
Wang HY (2022) [54]	PDR (n = 88) Control (n = 51)	UPLC-MS/MS	OPLS-DA	Phenylacetyl glutamine, pantothenate, CoA, tyrosine, and phenylalanine
Wang ZY (2022) [55]	NPDR (n = 28) PDR (n = 28) DM (n = 27) Control (n = 27)	UHPLC-MS/MS	OPLS-DA	L-Citrulline, indoleacetic acid, 1-methylhistidine, phosphatidylcholines, hexanoylcarnitine, chenodeoxycholic acid, and eicosapentaenoic acid

Table 2. Serum metabolomics of DR clinical populations.

References	Subjects	Techniques	Statistical Methods	Differential Metabolites
Munipally (2011) [56]	NPDR (n = 22) PDR (n = 24) Control (n = 35)	HPLC	t-test	kynurenine, kynurenic acid, and 3-hydroxykynurenine
Curovic (2020) [57]	Mild PDR (n = 90) Moderate PDR (n = 186) PDR (n = 121) PDR with fibrosis (n = 107) Control (n = 141)	GC-TOFMS	Cox models	2,4-dihydroxybutyric acid, 3,4-dihydroxybutyric acid, ribonic acid, and ribitol
Xuan QH (2020) [58]	NDR (n = 111) NPDR (n = 99) MMPDR (n = 90) SNPDR (n = 85) PDR (n = 76)	GC-MS, LC-MS	PLS-DA	12-hydroxyeicosatetraenoic acid and 2-piperidone
Yun JH (2020) [59]	NDR (n = 143) NPDR (n = 123) PDR (n = 51)	LC-MS/MS	Stats	Total dimethylamine, tryptophan, kynurenine, carnitines, several amino acids, and phosphatidylcholines
Quek (2021) [60]	Moderate/above DR (n = 328) VTDR (n = 217) Control (n = 2211)	NMR	ROC	Tyrosine, 3-hydroxybutate, sphingomyelins, and creatinine
Zuo JJ (2021) [61]	DM (n = 46) DR (n = 46)	UPLC-ESI-MS/MS	OPLS-DA	Linoleic acid, nicotinuric acid, ornithine, and phenylacetylglutamine
Guo CG (2022) [62]	NPDR (n = 60) PDR (n = 9)	UPLC-MS/MS	PLS-DA	12-/15-HETE, PUFAs, thiamine triphosphate, L-cysteine, and glutamate
Li JS (2022) [63]	NDR (n = 112) DR (n = 83) Control (n = 755)	UPLC-ESI-MS/MS	ROC	Thiamine triphosphate and 2-pyrrolidone
Wang ZY (2022) [64]	NPDR (n = 15) PDR (n = 15) DM (n = 15) Control (n = 15)	UHPLC-MS/MS	PLS-DA	Aspartate, glutamine, N-acetyl-L-glutamate, N-acetyl-L-aspartate, pantothenate, dihomo-gamma-linolenate, docosahexaenoic acid, and icosapentaenoic acid
Yang J (2022) [65]	DR + DN (n = 20) Control (n = 20)	UPLC-MS/MS	OPLS-DA, PLS-DA	1-methylhistidine, coagulation factor, and fibrinogen
Shen YH (2023) [66]	NPDR (n = 105) PDR (n = 62)	LC-MS	ROC, PLS-DA	Methionine and taurine

3.1.1. Plasma Metabolomics

In the system of Western medicine, DR manifests in three intricate stages characterized by metabolic disorders: pre-clinical, NPDR, and PDR. In contrast, traditional Chinese medicine (TCM) categorizes DR into two syndrome types: non-Yang deficiency and Yang deficiency. The integration of TCM and Western medicine approaches has demonstrated promising results in alleviating fundus hemorrhage and diabetes-related symptoms. To discern and evaluate the similarities and differences between Western medicine staging and TCM syndrome-related biomarkers, plasma samples were collected and subjected to GC-TOFMS detection. Subsequent analysis identified that 10 metabolites exhibited the potential to discriminate between the various stages of Western medicine classification, while 4 metabolites could distinguish between the two TCM syndrome types. Notably, pyruvate and l-aspartic acid emerged as metabolites capable of differentiating both stages of DR according to Western medicine and TCM syndrome types. However, it is worth noting that the concentration of aspartic acid demonstrated an association with renal function, which is a potential confounding factor. In TCM theory, there is no significant correlation between aspartic acid concentration and renal function, and it is important to acknowledge the absence of pertinent renal-function-related information in this study [45]. Furthermore, another investigation indicated that the combination of glutamic acid and glutamine could improve the specificity of distinguishing DR from non-DR [48].

Furthermore, some studies have been conducted on the alterations in various types of amino acid metabolism through plasma metabolomics of DR patients. Sumarriva et al. identified 126 and 151 characteristic metabolites using LC-MS/MS to distinguish DR patients from diabetes, PDR, and NPDR patients, respectively. Among them, arginine, citrulline, dehydroxycarnitine, and glutamic c-semialdehyde can be used to effectively distinguish diabetes mellitus (DM) and DR patients, and carnitine served as a distinguishing factor between PDR and NPDR [49]. Arginine metabolic disorders may be related not only to urea cycle metabolites, but also to asymmetric dimethylarginine (ADMA) and nitric oxide. In order to gain further insight into their mechanistic roles, Peters et al. conducted a targeted metabolomics analysis on six arginine- and citrulline-related metabolites [53]. In comparison to the diabetic control group, plasma levels of arginine and citrulline were increased in DR patients, thus affirming the significance of arginine and citrulline metabolism in DR. Additionally, the study revealed elevated levels of plasma ADMA in PDR patients when compared to NPDR patients. However, this association lost significance after adjusting for creatinine values. Notably, a comparison between targeted and non-targeted metabolomics results unveiled the ability of citrulline and carnitine to differentiate the severity of DR using both approaches [55]. Combined with proteomics research, it has been discovered that tryptophan metabolism also plays a crucial role in the development of DR. Plasma tryptophan levels in PDR patients are lower than those in NPDR patients. The decrease in tryptophan will increase the content of vascular endothelial growth factor, which in turn promotes angiogenesis [52]. Moreover, using UHPLC-MS, Sun et al. identified 22 differentially expressed metabolites associated with different metabolic pathways and demonstrated the significance of 4 circulating plasma metabolites. Risk score analysis revealed a positive correlation between these metabolites and glycated hemoglobin levels [51]. Additionally, the combined use of alanine, histidine, leucine, pyruvate, tyrosine, and valine exhibited a superior correlation with type 2 diabetes and diabetic microangiopathy, which when combined may contribute to the subsequent triage of diabetic complications [67].

Numerous other types of metabolites have also been discovered as potential biomarkers for DR. Xia et al. employed a quantitative approach utilizing high-performance liquid chromatography coupled with ultraviolet and tandem MS to detect six pyrimidine-related metabolites. Significantly increased concentrations of cytosine, cytidine, and pyrimidine were observed in DR patients compared to the DM group. Receiver-operating characteristic (ROC) analysis revealed that cytidine exhibited superior performance as a potential marker, with an area under the curve (AUC) of 0.849 ± 0.048. The cytidine concentration of 0.076 mg/L was set as the cutoff point for distinguishing disease status, yielding a sensitivity of 73.7% and a specificity of 91.9% [46]. In another comparison between PDR and NPDR, cytidine displayed an AUC of 0.95. Moreover, fumaric acid, uridine, and acetic acid demonstrated potential as biomarkers for PDR, with AUCs of 0.96, 0.95, and 1.0, respectively. Notably, this study unveiled fumaric acid as a novel metabolite marker for DR, offering insights into potentially novel pathogenic pathways associated with DR [50]. Polyunsaturated fatty acids (PUFAs) and their metabolites have also demonstrated beneficial effects on various pathological processes, including diabetes. After screening and statistically analyzing the detected eicosane compounds, it was found that prostaglandin 2a (PGF2a) had a protective effect on NPDR patients. In vivo and in vitro experiments suggested that PGF2a may regulate the migration of retinal pericytes through FP receptors and reverse some retinal capillary damage, thus conferring a protective role [47]. In addition, the biosynthesis pathways of pantothenate and coenzyme A exhibited significant disruption and decreased plasma levels in DR patients, potentially attributed to impaired vitamin reabsorption in renal tubules, leading to reduced pantothenate conversion rates [54].

3.1.2. Serum Metabolomics

Serum metabolomics studies on DR have also shed light on alterations in amino acid metabolism. The changes of tryptophan metabolism in the pathophysiology of diabetes and its related complications have been confirmed. However, in the serum metabolomics study of NPDR and PDR patients, tryptophan metabolite levels were found to increase, but tryptophan itself did not exhibit significant changes [56]. Another high-throughput targeted metabolomics study identified 16 metabolites as common specific metabolites of NPDR and PDR. Among them, total dimethylamine, tryptophan, and total tryptophan were identified as potential factors contributing to the progression of DR in DM patients. Variations in the findings related to tryptophan could be influenced by factors such as sample size and different disease groups. The results further support the exploration of tryptophan and kynurenine in the treatment and understanding of DR [59]. A case–control study employing propensity score matching identified a set of multidimensional network biomarkers containing linoleic acid, nicotinic acid, ornithine, and phenylacetylglutamine with high specificity and sensitivity (96% and 78%, respectively) in distinguishing DR from patients with type 2 diabetes mellitus (T2DM). This multidimensional network biomarkers system may present the most effective means for identifying DR [61]. New DR-related metabolic changes, such as thiamine metabolic disorders, reduced trehalose, and increased choline and indole derivatives, were also revealed in another similar case–control study [62]. In order to gain further insights into the metabolic changes from T2DM to DR, Wang et al. conducted comparative analyses of metabolic profiles across various stages, including NPDR and T2DM patients, PDR and T2DM patients, as well as DR and non-DR patients. The study was the first to confirm the close association of serum phosphatidylcholine and 13-hydroperoxyoctadeca-9,11-dienoic acid levels with different stages of DR in T2DM in Asian populations. Furthermore, abnormalities in other pathways, such as arginine biosynthetic metabolism, linoleic acid metabolism, aspartic acid, and glutamic acid metabolism, were observed in DR patients of Asian population [64].

With the advancement of multi-platform and multi-omics analyses, novel potential metabolic biomarkers for DR have been discovered. The integration of lipidomics and metabolomics can characterize subtle disturbances in response to lipid and metabolic changes and provide new insights into diseases. Curovic et al., combining metabolomics and lipidomics approaches, discovered four metabolites that were associated with different stages of DR, and three triglycerides exhibiting a negative correlation with the DR stage. Among them, 3,4 dihydroxybutyric acid was identified as an independent marker for DR progression [57]. Retinopathy in DR encompasses various manifestations, including microaneurysms, hard exudates (HEs), soft exudates, fibrous hyperplasia, and neovascularization [68]. Persistent HEs will develop into subretinal fibrosis with irreversible visual loss. By studying the lipidomics and metabolic profiles of patients with different severity levels of HEs, 19 metabolites and 13 HE-related pathways were identified. The combination model containing 20 lipids, such as triglyceride, ceramide, and N-acylethanolamine, demonstrated the most effective discrimination ability for HEs, with an area under the curve of 0.804 [66]. Utilizing GC-MS metabolomics, LC-MS metabolomics, and LC-MS lipidomics, comprehensive insights on the metabolic pathways involved in DR development have been obtained, leading to the identification and verification of a novel biomarker panel. The panel, consisting of 12-hydroxyeicosatetraenoic acid and 2-piperidone biomarkers, showed high sensitivity and specificity in distinguishing NPDR and NDR (0.929 and 0.901), which was higher than that of HbA1c (0.611 and 0.686), underscoring its potential for early diagnosis [58]. DR represents a form of diabetic microangiopathy characterized by endothelial dysfunction associated with microvascular complications. By combining metabolomics and proteomics analyses of serum exosomes, it was found that the up-regulation of an alpha subunit of the coagulation factor fibrinogen (FIBA) and the down-regulation of

1-methylhistidine contribute to diabetic endothelial dysfunction, impacting both macrovascular and microvascular complications. However, further cohort studies are required to elucidate the specific role of FIBA and 1-methylhistidine in the development of DN and DR [65].

In addition to the abovementioned MS-based serum metabolomics analysis, there are various NMR-based metabolomics studies and studies combining metabolomics and algorithms to develop an optimal model for DR prediction. In a cross-sectional study involving three cohorts from China, Malay, and India, 16 serum metabolites associated with DR were identified using NMR technology. Among these metabolites, three were found to be significantly correlated with each stage of DR. Notably, elevated levels of tyrosine and cholesteryl ester to total lipid ratio demonstrated a protective effect against severe DR, whereas increased levels of creatinine were positively associated with all three DR outcomes [60]. Furthermore, Li et al. employed metabolomics techniques and machine learning algorithms, based on propensity score matching approach, to establish a nomogram model for DR prediction. This model incorporated factors including diabetes duration, systolic blood pressure, and thiamine triphosphate. Impressively, the developed nomogram model exhibited excellent classification performance, with AUCs (95% CI) of 0.99 (0.97–1.00) and 0.99 (0.95–1.00) in both the training and testing sets, respectively [63]. These findings provide valuable insights into the management and control of DR, thereby contributing to the advancement of this field.

3.2. Vitreous Metabolomics

The vitreous body serves as a water medium in direct contact with the retina, lens, and numerous cells. It contains valuable information regarding the etiology of eye and vitreoretinal diseases [69]. We summarized the main findings from previous studies on the vitreous metabolome in individuals with DR (Table 3). Barba et al. utilized ^{1}H-NMR to conduct a vitreous metabolomics analysis, aiming to explore the metabolic differences between macular hole surgery in non-diabetic patients and patients with type 1 diabetes accompanied by PDR. Partial least squares discriminant analysis was employed to develop a recognition model. The results demonstrated that, after removing the lactic acid peak, 19 out of 22 PDR patients and 18 out of 22 controls were accurately classified, yielding a sensitivity rate of 86% and a specificity rate of 81% [44]. Notably, the vitreous samples of PDR patients exhibited a significant depletion of ascorbic acid and galactose, along with elevated levels of lactic acid. However, the authors emphasized the study's limitations, considering that vitreous hemorrhage frequently occurs in PDR patients, which may render these identified metabolites unrelated to DR. In a separate investigation by Wang et al., a comparison between plasma and vitreous metabolic profiles was conducted. Interestingly, five metabolites were found to be overlapping. Specifically, phenylacetylglutamine exhibited a significant increase, whereas valeric acid displayed a significant decrease, contrary to previous studies. The authors hypothesized that these discrepancies may be attributed to racial variations [54].

Using UHPLC-MS non-targeted metabolomics analysis, it has been determined that purine metabolite xanthine serves as the primary biomarker for distinguishing individuals with DR from healthy controls. Moreover, proline and citrulline play essential roles in differentiating DR from control subjects well as those with rhegmatogenous retinal detachment (RD). Within the vitreous of individuals with DR, downstream glycolysis metabolites, such as glyceraldehyde 3-phosphate and 2/3-phosphoglycerate, as well as the ratio of lactic acid to pyruvic acid, demonstrate a significant decrease [70]. However, when comparing DR patients with non-diabetic patients with macular hole (MH), vitreous lactic acid levels are significantly higher in those with PDR [44]. Another study, utilizing GC-TOFMS, identified a group of metabolites (d-2,3-dihydroxypropanoic acid, isocitric acid, threonic acid, pyruvic acid, l-Lactic acid, pyroglutamic acid, fructose 6-phosphate, ornithine, l-threonine, l-glutamine, and l-alanine) that exhibited good discriminatory potential between PDR and control subjects [71]. More recently, untargeted UHPLC-MS/MS analysis identified creatine as a potential target for PDR, and the supplementation of creatine in mouse models demonstrated inhibitory effects on pathological retinal neovascularization [72].

Table 3. Vitreous humor metabolomics of DR clinical populations.

References	Subjects	Techniques	Statistical Methods	Differential Metabolites
Barba (2010) [44]	PDR (n = 22) Controls (n = 22)	^{1}H-NMR	PLS-DA	Lactate, acetate, galactitol, ascorbic acid, and ribose phosphate
Nathan R (2018) [70]	DR (n = 8) RD (n = 17) Controls (n = 9)	UHPLC-MS	PLS-DA, ROC	Xanthine, proline, citrulline, and long-chain acylcarnitines
Wang HY (2020) [71]	PDR (n = 28) MH (n = 22)	GC-TOFMS	OPLS-DA, ROC	Pyruvic acid, uric acid, ornithine, l-lysine, l-leucine, pyroglutamic acid, l-alanine, l-threonine, hydroxylamine, l valine, l-alloisoleucine, l-phenylalanine, creatinine, myoinositol, and l-glutamine
Tomita (2021) [72]	PDR (n = 35) Control (n = 19)	UHPLC-MS/MS	t test	Pyruvate, lactate, proline, glycine, citrulline, ornithine, allantoin, creatine, dimethylglycine, N-acetylserine, succinate, and α-ketoglutarate
Wang HY (2022) [54]	PDR (n = 51) Control (n = 23)	UPLC-MS/MS	OPLS-DA	Phenylacetyl glutamine, pantothenate, CoA, tyrosine, and phenylalanine

It is worth noting that the biochemical changes observed in the vitreous closely mirror alterations in retinal homeostasis. However, obtaining vitreous samples from healthy controls is considerably challenging, as it necessitates invasive surgical procedures. This limitation poses a hindrance to the advancement of vitreous metabolomics research.

3.3. Aqueous Humor Metabolomics

In the preclinical stage of DR, vitrectomy is not applicable, and obtaining vitreous samples for analysis is not feasible due to the inapplicability of vitrectomy. As an alternative, aqueous humor is a transparent liquid synthesized by ciliary epithelial cells. It circulates through the anterior and posterior chambers of the eye and eventually drains into the veins, providing nourishment to avascular ocular tissues and facilitating the removal of metabolic waste [73]. Aqueous humor has emerged as a viable substitute for vitreous samples in metabolomics investigations related to DR.

Studies have sought to explore the correlation between metabolite levels in aqueous humor and vitreous samples from individuals with DM and DR. Specifically, oxidized glutathione trisulfide, cystine, and cysteine persulfide levels were found to be correlated across aqueous and vitreous samples [74]. Moreover, a comparative analysis of metabolites in aqueous humor and vitreous revealed the presence of eight metabolites in aqueous humor samples, with three of these metabolites (citrulline, inositol, and d-glucose) also observed in vitreous samples [71]. These findings clearly illustrate the potential utility of aqueous humor as a suitable substitute for vitreous samples in metabolomics analyses pertaining to DR. Consequently, the use of aqueous humor in lieu of vitreous samples circumvents the challenges associated with obtaining vitreous samples in the preclinical stage of DR. This enables the investigation of metabolic alterations and biomarkers that help us understand the pathology of DR and may have implications in early diagnosis, therapeutic interventions, and improved clinical management strategies for this condition.

Limited research has focused on metabolomics studies involving human aqueous humor samples in the context of DR. One investigation utilized ^{1}H-NMR, employing 2D homonuclear TOCSY and 2D pulsed-field gradient COSY, to compare and analyze aqueous humor samples derived from patients with DM and cataracts, as well as DR and cataracts, alongside elderly individuals with cataracts. Notably, this study represents the first comprehensive metabolomics analysis study based on ^{1}H-NMR to explore the differential metabolic spectrum of aqueous humor in patients with DR. Following a comprehensive series of analyses, the study revealed several metabolites exhibiting the highest degree of variability. Notably, succinic acid, lactic acid, asparagine, histidine, glutamine, and threonine were identified as the most variable metabolites [75]. The identification of these distinct metabolic signatures within aqueous humor samples from DR patients adds valuable insights into the metabolic alterations associated with the disease. Nonetheless, additional studies involving larger sample sizes and rigorous validation are required to substantiate these findings and further enhance our understanding of the underlying metabolic perturbations in DR. These findings highlight the potential utility of aqueous humor metabolomics as a valuable approach to investigate the metabolic changes underlying DR. Moreover, this research may facilitate the identification of relevant biomarkers and inform the development of targeted therapeutic interventions. Continued investigations in this area hold significant promise for advancing our knowledge of DR pathogenesis and enhancing clinical management strategies for this sight-threatening condition.

3.4. Urine Metabolomics

Urine, being easily collectible and rich in metabolites, has emerged as a valuable source for non-invasive biomarker discovery in metabolomics studies [76]. Significantly, the field of urine metabolomics has gained prominence for its ability to reflect the imbalance of biochemical pathways in vivo.

In a study based on NMR technology used for the quantitative analysis of urine samples from two cohort populations in China and India, metabolites were examined for their correlation with different stages of DR, including any DR, moderate/severe DR, and vision-threatening DR. The analysis revealed that 10 metabolites (citrate, ethanolamine, formate, hypoxanthine, 3-hydroxyisovalerate, 3-hydroxyisobutyrate, alanine, glutamine, uracil, and glycolic acid) were associated with at least one of the DR stages. Among them, citrate, ethanolamine, formate, and hypoxanthine displayed a negative correlation with all three DR results [60]. In another investigation conducted by Wang et al., UPLC-MS was employed to study the metabolomics of DR in rat urine upon the administration of Bushen Huoxue prescription. In this study, nine potential biomarkers were identified, including cholic acid, p-cresol sulfuric acid, 5-l-glutamyl taurine, 3-methyldiglucoside, nephropathy and phenylacetylglycine, 3-methyldioxyindole, kynurenic acid, hippuric acid, indoxyl sulfate, cholic acid, p-cresol sulfate, p-cresol glucuronide, and 5-L-glutamyl-taurine. These biomarkers were found to be significantly correlated with tryptophan metabolism, lipid metabolism, and intestinal microbial metabolism [77]. Moreover, the urine metabolomics analysis of diabetic model rats demonstrated the impact of exogenous free N^ε-(carboxymethyl) lysine intake on various metabolic pathways, such as amino acid metabolism, TCA cycle, and carbohydrate metabolism [78]. In summary, urine metabolomics presents a promising avenue for non-invasive biomarker discovery in the context of DR. These studies demonstrate the potential of metabolite profiling in urine to identify specific biomarkers associated with different stages of DR and uncover metabolic perturbations. Further research in this field will contribute to an enhanced understanding, diagnosis, and management of DR.

3.5. Fecal Metabolomics

Recent studies have highlighted the significant influence of gut microbiota in the pathogenesis of diabetic complications, including the development of DR [79,80]. Fecal metabolomics provides a valuable approach for examining the metabolic interactions between the host, diet, and gut microbiota, enabling an in-depth exploration of their role in disease processes [81]. Moreover, emerging studies have begun to elucidate the connection between gut microbiota and ocular abnormalities, including uveitis, glaucoma, and age-related macular degeneration, leading to the proposition of the microbiota–gut–retina axis concept [80].

To explore the relationship between gut microbial metabolism and DR, fecal samples were subjected to metabolomic analysis using UHPLC-MS and LC-MS. In a study by Li et al., non-targeted metabolomic analysis was performed on stool samples obtained from DR and non-DR in type 2 diabetic patients. The findings revealed significant increases in *Acidaminococcus*, *Escherichia coli*, and *Enterobacteriaceae* in patients with DR, while bifidobacteria and lactic acid bacteria exhibited a significant decrease. Additionally, the proportion of Pasteurella was significantly reduced [82]. Moreover, the bacterial abundance and diversity of intestinal flora were found to be significantly lower in diabetic patients with PDR compared to those without DR. Furthermore, fecal metabolomics analysis demonstrated an elevation in arachidonic acid metabolites, including hydroxyeicosatetraenoic acid and leukotrienes, which are known mediators in the development of DR, in PDR patients [83]. In a comparison between DR patients and healthy individuals, DR patients exhibited an enrichment of fecal bacteria belonging to the Rochalimaea and Longevia genera, while Akermann bacteria are reduced. Furthermore, carnosine, succinic acid, niacin, and niacinamide levels in the DR group were significantly lower than those in the healthy control group. The KEGG annotation of metabolomic data revealed 17 pathways with substantial differences in metabolite composition between DR patients and healthy controls, while only 2 pathways exhibited significant differences between DR patients and DM patients [84].

3.6. Other Biological Samples: Metabolomics

In addition to the aforementioned types of samples, metabolomics studies on DR have also been conducted utilizing cerebrospinal fluid (CSF) and retinal samples. CSF, although a rare sample in DR metabolomics research, has been employed in certain investigations. For instance, the combination of alanine, histidine, leucine, pyruvate, tyrosine, and valine in CSF exhibited strong correlations with the presence of T2DM (AUC:0.951) and DR (AUC:0.858) [67]. Moreover, Wang et al. employed non-targeted UPLC-MS/MS and GC-MS techniques to explore the metabolic characteristics of retinas in streptozotocin-induced diabetic mice. Pathway enrichment analysis revealed alterations in 50 metabolic pathways. In conjunction with transcriptomics analysis, these findings suggest potential disturbances in the Wahlberg effect, amino acid metabolism, and retinol metabolism, shedding light on potential metabolic mechanisms and therapeutic targets for DR [85].

4. Metabolomics Studies in DR Models

At present, the main animal models used in DR research can be divided into two primary categories: induced animal models and genetic animal models [86,87], as summarized in Table 4. However, only a limited number of models have been utilized for metabolomics studies.

Among these models, the streptozotocin (STZ)-induced diabetic rat or mouse model has gained widespread use due to its similarity to the pathological changes observed in early DR [88]. Thus, this model has become the most commonly employed in DR research. Notably, a study investigating retinal metabolomics in STZ-induced diabetic mice revealed that the Warburg effect may play a pivotal role in the pathogenesis of DR. In the retina of diabetic mice, ornithine was significantly increased, proline was decreased, and arginine metabolism was changed, which were similar to the metabolic changes in the plasma and vitreous of DR patients [85]. The consistency between mouse retina and human samples indicates that STZ-induced diabetic mice can be used as valuable tools for the further exploration of metabolic mechanisms underlying DR. Additionally, comprehensive metabolomics, lipidomics, and RNA profiling studies conducted on the retinas of STZ-induced diabetic mice uncovered that activated microglia exhibit distinct metabolic characteristics and serve as the primary source of pro-inflammatory cytokines in early DR. Furthermore, the intracellular metabolic microenvironment, particularly glycolysis, appears to reprogram retinal inflammation. These findings pave the way for potential future studies aiming to reprogram the intracellular metabolism of retinal-specific microglia to mitigate local inflammation and prevent the progression of early DR [89].

Table 4. Animal models for studying DR.

Category		Animal Models
Induced model		STZ-induced models
		Alloxan-induced models [90]
		Diet-induced models
		Oxygen-induced models
Genetic models	Mice model	Ins2Akita [91]
		Non-obese diabetic [92]
		Leprdb [93]
		Kimba [94]
		Akimba [95]
	Rat model	Zucker diabetic fatty [96]
		Otsuka Long-Evans Tokushima Fatty [97]
		Biobreeding diabetes-prone [98]
		Wistar Bonn Kobori [99]
		Goto–Kakizaki [100]
		Spontaneously diabetic Torii [96]

However, it is important to note that STZ-induced models are unlikely to exhibit retinal neovascularization, which is a key pathological feature of PDR. Thus, researchers have turned to the oxygen-induced mouse retinopathy (OIR) model, which closely mimics the pathological manifestations observed in PDR patients, for investigating PDR [101]. In the vitreous of PDR patients, alterations in creatine and creatine-related pathways have been observed. Decreased levels of creatine have been found in the vitreous of PDR patients, as well as in the retina of the OIR model [72]. Moreover, creatine supplementation has been shown to reduce retinal neovascularization in OIR, suggesting its potential as a therapeutic target for PDR. However, it is important to recognize that while the OIR model simulates neovascularization in the retina, it does not encompass other aspects of diabetes that occur prior to neovascularization [72]. Consequently, further investigations are required to fully elucidate the underlying mechanism of creatine and its potential implications in DR.

The variations in metabolic changes observed among different induction or genetic models underscore the multifactorial nature of DR. In the future, animal-model-based metabolic studies will help to improve our understanding of the pathogenesis and progression of DR.

5. Dysregulation of Metabolic Pathways in DR

DR is a complex and chronic metabolic disease marked by the dysregulation of multiple metabolic pathways. Metabolomics studies have revealed that several metabolic pathways are disrupted in DR, including those associated with amino acid metabolism, energy metabolism (TCA cycle, glycolysis, and carnitine metabolism), pyrimidine metabolism, and lipid metabolism [49,58,71,102].

The retina, as a highly metabolically active tissue, has a significant energy demand. Glycolysis and the TCA cycle play crucial roles in DR. Glycolysis converts glucose into pyruvic acid, which can undergo anaerobic metabolism, leading to the production of lactic acid [67]. Moreover, pyruvic acid can be oxidized to generate acetyl-CoA, fueling the TCA cycle. Within the TCA cycle, succinic acid is an intermediate metabolite that plays a pivotal role in adenosine triphosphate production within the mitochondria. Therefore, succinic acid may hold potential as a biomarker for DR. Studies by Jin et al. demonstrated significant reductions in lactic acid and succinic acid levels in the aqueous humor of DR patients. However, in the vitreous humor, succinic acid levels were significantly increased and succinic acid may accumulate in the retina with an insufficient oxygen supply [75].

In the context of amino acid metabolism, the arginine metabolic pathway takes prominence and has been investigated in aqueous humor, plasma, serum, and vitreous metabolomics studies. Arginine can be metabolized through two different pathways: the arginase pathway, which involves the enzyme arginase II (Arg-II) and leads to the production of ornithine and urea, and the nitric oxide synthase (NOS) pathway, which produces citrulline and nitric oxide (NO) [64]. Ornithine has been linked to chronic inflammatory injury mediated by microglia and macrophages observed in type 2 diabetes. Elevated ornithine levels indicate an increase in Arg-II enzyme activity, potentially resulting in reduced NOS pathway activity, which predominantly synthesizes NO. NO is a crucial vasodilator that plays a vital role in maintaining vascular endothelial health [49]. Insufficient NO levels, combined with elevated levels of polyamines and proline, can lead to endothelial cell dysfunction and impaired vasodilation, and can also stimulate cell proliferation and fibrosis [55]. However, it should be noted that while arginine can stimulate insulin release in pancreatic β cells, poorly regulated arginine and citrulline levels can also contribute to retinal endothelial cell dysfunction [55,103]. Moreover, the metabolism of tryptophan, glutamic acid, and alanine has also been associated with the development of DR, although their precise mechanisms remain unclear [63,85].

Pyrimidine, a crucial component of DNA and RNA, performs various essential biological functions. Recent studies have revealed the association between pyrimidine metabolism and DR [46]. Pyrimidines participate in cellular processes as building blocks of genetic material and have implications for diverse cellular activities. Sphingolipids, on the other hand, have emerged as important components of lipids with pivotal roles in signal transduction, cell proliferation, apoptosis, and membrane structure. In recent years, their involvement in various cellular processes has been recognized. The sphingosine kinase 1/sphingosine 1 phosphate pathway, in particular, has been associated with cell fibrosis through its ability to stimulate mesangial cell proliferation and matrix formation. The plasma levels of sphingosine 1 phosphate were observed to be higher than those in individuals with DR compared to healthy individuals, and these levels demonstrated a positive correlation with HbA1c levels [102].

6. Conclusions and Future Perspectives

Metabolites are the final products of various changes in genome, transcriptome and proteome. Metabolomics offers a powerful approach to unravel the mechanisms underlying metabolic disorders and diseases, including DR. Metabolomics studies have provided insights into the metabolite changes that occur in various biological samples and have identified numerous potential biomarkers and therapeutic targets for DR. However, metabolomics is still in the early stage of development, and there are still many problems to be solved.

Firstly, the metabolomics database is still limited, and current findings may only represent a fraction of the overall etiology of DR. Moreover, differences in demographic data, such as race, region, and age, can affect outcomes and mask the direct effects of disease. Therefore, standardization of data analysis and reporting between institutions is essential for absolute quantification and reliable statistical results. Large-scale clinical metabolomics studies using standardized protocols and validated findings are necessary to reduce individual differences and improve the reliability of results. Additionally, advancements in metabolomics technology and analysis platforms are crucial for expanding metabolic coverage. Techniques such as hyperpolarized nuclear magnetic resonance and cryogenic-probe-based Rheo-NMR enhance spectral resolution and metabolite identification. Two-dimensional and multidimensional liquid chromatography techniques can offer enhanced peak capacity and improved resolution. The combination of multiple analysis platforms can help improve metabolic coverage, and simultaneous utilization of complementary analytical platforms can facilitate a more comprehensive understanding of the underlying biological processes.

Moreover, the integration of metabolomics with other omics approaches, such as transcriptomics and proteomics, holds great potential for unraveling the complex mechanisms driving the occurrence and progression of DR. This comprehensive approach can help to discover new biomarkers and effective therapeutic targets. In future studies, it is important to prioritize the development of non-invasive, rapid, and cost-effective DR biomarkers, which will greatly enhance our understanding of the complex pathogenesis of the disease.

In summary, metabolomics has demonstrated its power as a tool for characterizing metabolic alterations in DR and offers great potential for providing valuable insights into the disease. Standardization, large-scale studies, technological advancements, and interdisciplinary collaborations are key to advancing our understanding of DR and improving its clinical management.

Author Contributions: Writing—original draft preparation and visualization, S.H.; investigation, L.S. and J.C.; conceptualization, resources, writing—review and editing, supervision, and project administration, Y.O. All authors have read and agreed to the published version of the manuscript.

Funding: This study was funded by the Natural Science Foundation of Fujian Province (grant no. 2022J01236).

Acknowledgments: We sincerely thank all the participants in this study.

Conflicts of Interest: The authors declare no conflict of interest.

References

1. Simo-Servat, O.; Hernandez, C.; Simo, R. Usefulness of the vitreous fluid analysis in the translational research of diabetic retinopathy. *Mediators Inflamm.* **2012**, *2012*, 872978. [PubMed]
2. Cheung, N.; Mitchell, P.; Wong, T.Y. Diabetic retinopathy. *Lancet* **2010**, *376*, 124–136. [PubMed]
3. Yau, J.W.; Rogers, S.L.; Kawasaki, R.; Lamoureux, E.L.; Kowalski, J.W.; Bek, T.; Chen, S.J.; Dekker, J.M.; Fletcher, A.; Grauslund, J.; et al. Global prevalence and major risk factors of diabetic retinopathy. *Diabetes Care* **2012**, *35*, 556–564. [PubMed]
4. Li, J.Q.; Welchowski, T.; Schmid, M.; Letow, J.; Wolpers, C.; Pascual-Camps, I.; Holz, F.G.; Finger, R.P. Prevalence, incidence and future projection of diabetic eye disease in Europe: A systematic review and meta-analysis. *Eur. J. Epidemiol.* **2020**, *35*, 11–23.
5. Teo, Z.L.; Tham, Y.C.; Yu, M.; Chee, M.L.; Rim, T.H.; Cheung, N.; Bikbov, M.M.; Wang, Y.X.; Tang, Y.; Lu, Y.; et al. Global Prevalence of Diabetic Retinopathy and Projection of Burden through 2045: Systematic Review and Meta-analysis. *Ophthalmology* **2021**, *128*, 1580–1591. [CrossRef]
6. Lechner, J.; O'Leary, O.E.; Stitt, A.W. The pathology associated with diabetic retinopathy. *Vision Res.* **2017**, *139*, 7–14.
7. Kornblau, I.S.; El-Annan, J.F. Adverse reactions to fluorescein angiography: A comprehensive review of the literature. *Surv Ophthalmol* **2019**, *64*, 679–693.
8. Hitosugi, M.; Omura, K.; Yokoyama, T.; Kawato, H.; Motozawa, Y.; Nagai, T.; Tokudome, S. An autopsy case of fatal anaphylactic shock following fluorescein angiography: A case report. *Med. Sci. Law.* **2004**, *44*, 264–265. [CrossRef]
9. Ascaso, F.J.; Tiestos, M.T.; Navales, J.; Iturbe, F.; Palomar, A.; Ayala, J.I. Fatal acute myocardial infarction after intravenous fluorescein angiography. *Retina (Philadelphia, Pa.)* **1993**, *13*, 238–239. [CrossRef]
10. Mohamed, Q.; Gillies, M.C.; Wong, T.Y. Management of diabetic retinopathy: A systematic review. *JAMA.* **2007**, *298*, 902–916. [CrossRef]
11. Simo, R.; Hernandez, C. Intravitreous anti-VEGF for diabetic retinopathy: Hopes and fears for a new therapeutic strategy. *Diabetologia* **2008**, *51*, 1574–1580. [PubMed]
12. Simo, R.; Hernandez, C. Advances in the medical treatment of diabetic retinopathy. *Diabetes care* **2009**, *32*, 1556–1562. [PubMed]
13. Tan, T.E.; Wong, T.Y. Diabetic retinopathy: Looking forward to 2030. *Front. Endocrinol.* **2022**, *13*, 1077669.
14. Lind, M.; Pivodic, A.; Svensson, A.M.; Olafsdottir, A.F.; Wedel, H.; Ludvigsson, J. HbA$_{1c}$ level as a risk factor for retinopathy and nephropathy in children and adults with type 1 diabetes: Swedish population based cohort study. *BMJ* **2019**, *366*, l4894. [CrossRef]
15. Dunn, W.B.; Broadhurst, D.I.; Atherton, H.J.; Goodacre, R.; Griffin, J.L. Systems level studies of mammalian metabolomes: The roles of mass spectrometry and nuclear magnetic resonance spectroscopy. *Chem. Soc. Rev.* **2011**, *40*, 387–426.
16. Jorge, T.F.; Rodrigues, J.A.; Caldana, C.; Schmidt, R.; van Dongen, J.T.; Thomas-Oates, J.; Antonio, C. Mass spectrometry-based plant metabolomics: Metabolite responses to abiotic stress. *Mass Spectrom. Rev.* **2016**, *35*, 620–649.
17. Gonzalez-Pena, D.; Brennan, L. Recent Advances in the Application of Metabolomics for Nutrition and Health. *Annu. Rev. Food Sci. Technol.* **2019**, *10*, 479–519. [CrossRef]
18. Wishart, D.S. Emerging applications of metabolomics in drug discovery and precision medicine. *Nat. Rev. Drug Discov.* **2016**, *15*, 473–484.

19. Johnson, C.H.; Ivanisevic, J.; Siuzdak, G. Metabolomics: Beyond biomarkers and towards mechanisms. *Nat. Rev. Mol. Cell Biol.* 2016, *17*, 451–459.
20. Shao, Y.; Le, W. Recent advances and perspectives of metabolomics-based investigations in Parkinson's disease. *Mol. Neurodegener* 2019, *14*, 1–12.
21. Kaushik, A.K.; DeBerardinis, R.J. Applications of metabolomics to study cancer metabolism. *Biochim. Biophys. Acta Rev. Cancer* 2018, *1870*, 2–14. [PubMed]
22. Nagana Gowda, G.A.; Raftery, D. NMR-Based Metabolomics. *Adv. Exp. Med. Biol.* 2021, *1280*, 19–37. [PubMed]
23. Markley, J.L.; Bruschweiler, R.; Edison, A.S.; Eghbalnia, H.R.; Powers, R.; Raftery, D.; Wishart, D.S. The future of NMR-based metabolomics. *Curr. Opin. Biotechnol.* 2017, *43*, 34–40. [CrossRef] [PubMed]
24. Li, X.; Cai, S.; He, Z.; Reilly, J.; Zeng, Z.; Strang, N.; Shu, X. Metabolomics in Retinal Diseases: An Update. *Biology* 2021, *10*, 944. [CrossRef]
25. Tsujimoto, T.; Yoshitomi, T.; Maruyama, T.; Yamamoto, Y.; Hakamatsuka, T.; Uchiyama, N. (13)C-NMR-based metabolic fingerprinting of Citrus-type crude drugs. *J. Pharm. Biomed. Anal.* 2018, *161*, 305–312. [CrossRef]
26. Columbus, I.; Ghindes-Azaria, L.; Chen, R.; Yehezkel, L.; Redy-Keisar, O.; Fridkin, G.; Amir, D.; Marciano, D.; Drug, E.; Gershonov, E.; et al. Studying Lipophilicity Trends of Phosphorus Compounds by (31)P NMR Spectroscopy. A Powerful Tool for the Design of P-Containing Drugs. *J. Med. Chem.* 2022, *65*, 8511–8524. [CrossRef]
27. Goudar, C.; Biener, R.; Boisart, C.; Heidemann, R.; Piret, J.; de Graaf, A.; Konstantinov, K. Metabolic flux analysis of CHO cells in perfusion culture by metabolite balancing and 2D [13C, 1H] COSY NMR spectroscopy. *Metab. Eng.* 2010, *12*, 138–149. [CrossRef]
28. Bingol, K.; Zhang, F.; Bruschweiler-Li, L.; Brüschweiler, R. Quantitative analysis of metabolic mixtures by two-dimensional 13C constant-time TOCSY NMR spectroscopy. *Anal. Chem.* 2013, *85*, 6414–6420. [CrossRef]
29. Amberg, A.; Riefke, B.; Schlotterbeck, G.; Ross, A.; Senn, H.; Dieterle, F.; Keck, M. NMR and MS Methods for Metabolomics. *Methods Mol. Biol.* 2017, *1641*, 229–258.
30. Ghosh, S.; Sengupta, A.; Chandra, K. SOFAST-HMQC-an efficient tool for metabolomics. *Anal. Bioanal. Chem.* 2017, *409*, 6731–6738. [CrossRef]
31. Ribay, V.; Praud, C.; Letertre, M.P.M.; Dumez, J.N.; Giraudeau, P. Hyperpolarized NMR metabolomics. *Curr. Opin. Chem. Biol.* 2023, *74*, 102307. [CrossRef]
32. Morimoto, D.; Walinda, E.; Yamamoto, A.; Scheler, U.; Sugase, K. Rheo-NMR Spectroscopy for Cryogenic-Probe-Equipped NMR Instruments to Monitor Protein Aggregation. *Curr. Protoc.* 2022, *2*, e617. [CrossRef] [PubMed]
33. Spagou, K.; Theodoridis, G.; Wilson, I.; Raikos, N.; Greaves, P.; Edwards, R.; Nolan, B.; Klapa, M.I. A GC-MS metabolic profiling study of plasma samples from mice on low- and high-fat diets. *J. Chromatogr B.* 2011, *879*, 1467–1475. [CrossRef] [PubMed]
34. Peterson, A.C.; Balloon, A.J.; Westphall, M.S.; Coon, J.J. Development of a GC/Quadrupole-Orbitrap mass spectrometer, part II: New approaches for discovery metabolomics. *Anal. Chem.* 2014, *86*, 10044–10051. [CrossRef]
35. Fiehn, O. Metabolomics by Gas Chromatography-Mass Spectrometry: Combined Targeted and Untargeted Profiling. *Curr. Protoc. Mol. Biol.* 2016, *114*, 21.33.1–21.33.11. [CrossRef] [PubMed]
36. Beale, D.J.; Pinu, F.R.; Kouremenos, K.A.; Poojary, M.M.; Narayana, V.K.; Boughton, B.A.; Kanojia, K.; Dayalan, S.; Jones, O.A.H.; Dias, D.A. Review of recent developments in GC-MS approaches to metabolomics-based research. *Metabolomics* 2018, *14*, 1–31.
37. Lotti, C.; Rubert, J.; Fava, F.; Tuohy, K.; Mattivi, F.; Vrhovsek, U. Development of a fast and cost-effective gas chromatography-mass spectrometry method for the quantification of short-chain and medium-chain fatty acids in human biofluids. *Anal. Bioanal. Chem.* 2017, *409*, 5555–5567. [CrossRef]
38. Lima, V.F.; Erban, A.; Daubermann, A.G.; Freire, F.B.S.; Porto, N.P.; Cândido-Sobrinho, S.A.; Medeiros, D.B.; Schwarzländer, M.; Fernie, A.R.; Dos Anjos, L.; et al. Establishment of a GC-MS-based (13) C-positional isotopomer approach suitable for investigating metabolic fluxes in plant primary metabolism. *Plant J.* 2021, *108*, 1213–1233. [CrossRef]
39. Zhao, X.; Chen, M.; Zhao, Y.; Zha, L.; Yang, H.; Wu, Y. GC-MS-Based Nontargeted and Targeted Metabolic Profiling Identifies Changes in the Lentinula edodes Mycelial Metabolome under High-Temperature Stress. *Int. J. Mol. Sci.* 2019, *20*, 2330. [CrossRef]
40. Xie, H.; Wang, R.; Xie, L.; Wang, X.; Liu, C. Study on the pathogenesis and prevention strategies of kidney stones based on GC-MS combined with metabolic pathway analysis. *Rapid Commun. Mass Spectrom* 2022, *36*, e9387. [CrossRef]
41. Zhou, B.; Xiao, J.F.; Tuli, L.; Ressom, H.W. LC-MS-based metabolomics. *Mol. Biosyst.* 2012, *8*, 470–481. [CrossRef] [PubMed]
42. Stoll, D.R.; Harmes, D.C.; Staples, G.O.; Potter, O.G.; Dammann, C.T.; Guillarme, D.; Beck, A. Development of Comprehensive Online Two-Dimensional Liquid Chromatography/Mass Spectrometry Using Hydrophilic Interaction and Reversed-Phase Separations for Rapid and Deep Profiling of Therapeutic Antibodies. *Anal. Chem.* 2018, *90*, 5923–5929. [CrossRef] [PubMed]
43. Country, M.W. Retinal metabolism: A comparative look at energetics in the retina. *Brain Res.* 2017, *1672*, 50–57. [CrossRef]
44. Barba, I.; Garcia-Ramírez, M.; Hernández, C.; Alonso, M.A.; Masmiquel, L.; García-Dorado, D.; Simó, R. Metabolic fingerprints of proliferative diabetic retinopathy: An 1H-NMR-based metabonomic approach using vitreous humor. *Investig. Ophthalmol. Vis. Sci.* 2010, *51*, 4416–4421. [CrossRef] [PubMed]
45. Li, X.; Luo, X.; Lu, X.; Duan, J.; Xu, G. Metabolomics study of diabetic retinopathy using gas chromatography-mass spectrometry: A comparison of stages and subtypes diagnosed by Western and Chinese medicine. *Mol. Biosyst.* 2011, *7*, 2228–2237. [CrossRef]
46. Xia, J.F.; Wang, Z.H.; Liang, Q.L.; Wang, Y.M.; Li, P.; Luo, G.A. Correlations of six related pyrimidine metabolites and diabetic retinopathy in Chinese type 2 diabetic patients. *Clin. Chim. Acta* 2011, *412*, 940–945. [CrossRef]

47. Peng, L.Y.; Sun, B.; Liu, M.M.; Huang, J.; Liu, Y.J.; Xie, Z.P.; He, J.L.; Chen, L.M.; Wang, D.W.; Zhu, Y.; et al. Plasma metabolic profile reveals PGF2 alpha protecting against non-proliferative diabetic retinopathy in patients with type 2 diabetes. *Biochem. Biophys. Res. Commun.* **2018**, *496*, 1276–1283. [CrossRef]
48. Rhee, S.Y.; Jung, E.S.; Park, H.M.; Jeong, S.J.; Kim, K.; Chon, S.; Yu, S.Y.; Woo, J.T.; Lee, C.H. Plasma glutamine and glutamic acid are potential biomarkers for predicting diabetic retinopathy. *Metabolomics* **2018**, *14*, 89. [CrossRef]
49. Sumarriva, K.; Uppal, K.; Ma, C.; Herren, D.J.; Wang, Y.; Chocron, I.M.; Warden, C.; Mitchell, S.L.; Burgess, L.G.; Goodale, M.P.; et al. Arginine and Carnitine Metabolites Are Altered in Diabetic Retinopathy. *Investig. Ophthalmol. Vis. Sci.* **2019**, *60*, 3119–3126. [CrossRef]
50. Zhu, X.R.; Yang, F.Y.; Lu, J.; Zhang, H.R.; Sun, R.; Zhou, J.B.; Yang, J.K. Plasma metabolomic profiling of proliferative diabetic retinopathy. *Nutr. Metab.* **2019**, *16*, 37. [CrossRef]
51. Sun, Y.; Zou, H.; Li, X.; Xu, S.; Liu, C. Plasma Metabolomics Reveals Metabolic Profiling For Diabetic Retinopathy and Disease Progression. *Front. Endocrinol.* **2021**, *12*, 757088. [CrossRef] [PubMed]
52. Ding, C.; Wang, N.; Wang, Z.; Yue, W.; Li, B.; Zeng, J.; Yoshida, S.; Yang, Y.; Zhou, Y. Integrated Analysis of Metabolomics and Lipidomics in Plasma of T2DM Patients with Diabetic Retinopathy. *Pharmaceutics* **2022**, *14*, 2751. [CrossRef] [PubMed]
53. Peters, K.S.; Rivera, E.; Warden, C.; Harlow, P.A.; Mitchell, S.L.; Calcutt, M.W.; Samuels, D.C.; Brantley, M.A., Jr. Plasma Arginine and Citrulline are Elevated in Diabetic Retinopathy. *Am. J. Ophthalmol.* **2022**, *235*, 154–162. [CrossRef] [PubMed]
54. Wang, H.; Li, S.; Wang, C.; Wang, Y.; Fang, J.; Liu, K. Plasma and Vitreous Metabolomics Profiling of Proliferative Diabetic Retinopathy. *Investig. Ophthalmol. Vis. Sci.* **2022**, *63*, 17. [CrossRef] [PubMed]
55. Wang, Z.; Tang, J.; Jin, E.; Ren, C.; Li, S.; Zhang, L.; Zhong, Y.; Cao, Y.; Wang, J.; Zhou, W.; et al. Metabolomic comparison followed by cross-validation of enzyme-linked immunosorbent assay to reveal potential biomarkers of diabetic retinopathy in Chinese with type 2 diabetes. *Front. Endocrinol.* **2022**, *13*, 986303. [CrossRef]
56. Munipally, P.K.; Agraharm, S.G.; Valavala, V.K.; Gundae, S.; Turlapati, N.R. Evaluation of indoleamine 2,3-dioxygenase expression and kynurenine pathway metabolites levels in serum samples of diabetic retinopathy patients. *Arch. Physiol. Biochem.* **2011**, *117*, 254–258. [CrossRef]
57. Curovic, V.R.; Suvitaival, T.; Mattila, I.; Ahonen, L.; Trošt, K.; Theilade, S.; Hansen, T.W.; Legido-Quigley, C.; Rossing, P. Circulating Metabolites and Lipids Are Associated to Diabetic Retinopathy in Individuals With Type 1 Diabetes. *Diabetes* **2020**, *69*, 2217–2226. [CrossRef]
58. Xuan, Q.; Ouyang, Y.; Wang, Y.; Wu, L.; Li, H.; Luo, Y.; Zhao, X.; Feng, D.; Qin, W.; Hu, C.; et al. Multiplatform Metabolomics Reveals Novel Serum Metabolite Biomarkers in Diabetic Retinopathy Subjects. *Adv. Sci.* **2020**, *7*, 2001714. [CrossRef]
59. Yun, J.H.; Kim, J.M.; Jeon, H.J.; Oh, T.; Choi, H.J.; Kim, B.J. Metabolomics profiles associated with diabetic retinopathy in type 2 diabetes patients. *PLoS ONE* **2020**, *15*, e0241365. [CrossRef]
60. Quek, D.Q.Y.; He, F.; Sultana, R.; Banu, R.; Chee, M.L.; Nusinovici, S.; Thakur, S.; Qian, C.; Cheng, C.Y.; Wong, T.Y.; et al. Novel Serum and Urinary Metabolites Associated with Diabetic Retinopathy in Three Asian Cohorts. *Metabolites* **2021**, *11*, 614. [CrossRef]
61. Zuo, J.; Lan, Y.; Hu, H.; Hou, X.; Li, J.; Wang, T.; Zhang, H.; Zhang, N.; Guo, C.; Peng, F.; et al. Metabolomics-based multidimensional network biomarkers for diabetic retinopathy identification in patients with type 2 diabetes mellitus. *BMJ Open Diabetes Res. Care* **2021**, *9*, e001443. [CrossRef] [PubMed]
62. Guo, C.; Jiang, D.; Xu, Y.; Peng, F.; Zhao, S.; Li, H.; Jin, D.; Xu, X.; Xia, Z.; Che, M.; et al. High-Coverage Serum Metabolomics Reveals Metabolic Pathway Dysregulation in Diabetic Retinopathy: A Propensity Score-Matched Study. *Front. Mol. Biosci.* **2022**, *9*, 822647. [CrossRef] [PubMed]
63. Li, J.; Guo, C.; Wang, T.; Xu, Y.; Peng, F.; Zhao, S.; Li, H.; Jin, D.; Xia, Z.; Che, M.; et al. Interpretable machine learning-derived nomogram model for early detection of diabetic retinopathy in type 2 diabetes mellitus: A widely targeted metabolomics study. *Nutr. Diabetes* **2022**, *12*, 36. [CrossRef] [PubMed]
64. Wang, Z.; Tang, J.; Jin, E.; Zhong, Y.; Zhang, L.; Han, X.; Liu, J.; Cheng, Y.; Hou, J.; Shi, X.; et al. Serum Untargeted Metabolomics Reveal Potential Biomarkers of Progression of Diabetic Retinopathy in Asians. *Front. Mol. Biosci.* **2022**, *9*, 871291. [CrossRef]
65. Yang, J.; Liu, D.; Liu, Z. Integration of Metabolomics and Proteomics in Exploring the Endothelial Dysfunction Mechanism Induced by Serum Exosomes From Diabetic Retinopathy and Diabetic Nephropathy Patients. *Front. Endocrinol.* **2022**, *13*, 830466. [CrossRef]
66. Shen, Y.; Wang, H.; Fang, J.; Liu, K.; Xu, X. Novel insights into the mechanisms of hard exudate in diabetic retinopathy: Findings of serum lipidomic and metabolomics profiling. *Heliyon* **2023**, *9*, e15123. [CrossRef]
67. Lin, H.T.; Cheng, M.L.; Lo, C.J.; Lin, G.; Lin, S.F.; Yeh, J.T.; Ho, H.Y.; Lin, J.R.; Liu, F.C. (1)H Nuclear Magnetic Resonance (NMR)-Based Cerebrospinal Fluid and Plasma Metabolomic Analysis in Type 2 Diabetic Patients and Risk Prediction for Diabetic Microangiopathy. *J. Clin. Med.* **2019**, *8*, 874. [CrossRef]
68. Kar, S.S.; Maity, S.P. Automatic Detection of Retinal Lesions for Screening of Diabetic Retinopathy. *IEEE. Trans. Biomed. Eng.* **2018**, *65*, 608–618. [CrossRef]
69. Funatsu, H.; Yamashita, T.; Yamashita, H. Vitreous fluid biomarkers. *Adv. Clin. Chem.* **2006**, *42*, 111–166.
70. Haines, N.R.; Manoharan, N.; Olson, J.L.; D'Alessandro, A.; Reisz, J.A. Metabolomics Analysis of Human Vitreous in Diabetic Retinopathy and Rhegmatogenous Retinal Detachment. *J. Proteome Res.* **2018**, *17*, 2421–2427. [CrossRef]

71. Wang, H.; Fang, J.; Chen, F.; Sun, Q.; Xu, X.; Lin, S.H.; Liu, K. Metabolomic profile of diabetic retinopathy: A GC-TOFMS-based approach using vitreous and aqueous humor. *Acta diabetologica* **2020**, *57*, 41–51. [CrossRef] [PubMed]
72. Tomita, Y.; Cagnone, G.; Fu, Z.; Cakir, B.; Kotoda, Y.; Asakage, M.; Wakabayashi, Y.; Hellström, A.; Joyal, J.S.; Talukdar, S.; et al. Vitreous metabolomics profiling of proliferative diabetic retinopathy. *Diabetologia* **2021**, *64*, 70–82. [CrossRef] [PubMed]
73. Pietrowska, K.; Dmuchowska, D.A.; Krasnicki, P.; Bujalska, A.; Samczuk, P.; Parfieniuk, E.; Kowalczyk, T.; Wojnar, M.; Mariak, Z.; Kretowski, A.; et al. An exploratory LC-MS-based metabolomics study reveals differences in aqueous humor composition between diabetic and non-diabetic patients with cataract. *Electrophoresis* **2018**, *39*, 1233–1240. [CrossRef] [PubMed]
74. Kunikata, H.; Ida, T.; Sato, K.; Aizawa, N.; Sawa, T.; Tawarayama, H.; Murayama, N.; Fujii, S.; Akaike, T.; Nakazawa, T. Metabolomic profiling of reactive persulfides and polysulfides in the aqueous and vitreous humors. *Sci. Rep.* **2017**, *7*, 41984. [CrossRef]
75. Jin, H.; Zhu, B.; Liu, X.; Jin, J.; Zou, H. Metabolic characterization of diabetic retinopathy: An (1)H-NMR-based metabolomic approach using human aqueous humor. *J. Pharm. Biomed. Anal.* **2019**, *174*, 414–421. [CrossRef]
76. Khamis, M.M.; Adamko, D.J.; El-Aneed, A. Mass spectrometric based approaches in urine metabolomics and biomarker discovery. *Mass Spectrom. Rev.* **2017**, *36*, 115–134. [CrossRef]
77. Wang, X.; Li, Y.; Xie, M.; Deng, L.; Zhang, M.; Yin, Y. Urine metabolomics study of Bushen Huoxue Prescription on diabetic retinopathy rats by UPLC-Q-exactive Orbitrap-MS. *Biomed. Chromatogr.* **2020**, *34*, e4792. [CrossRef]
78. Quan, W.; Jiao, Y.; Xue, C.; Li, Y.; Liu, G.; He, Z.; Qin, F.; Zeng, M.; Chen, J. The Effect of Exogenous Free N^{ε}-(Carboxymethyl)Lysine on Diabetic-Model Goto-Kakizaki Rats: Metabolomics Analysis in Serum and Urine. *J. Agric. Food Chem.* **2021**, *69*, 783–793. [CrossRef]
79. Iatcu, C.O.; Steen, A.; Covasa, M. Gut Microbiota and Complications of Type-2 Diabetes. *Nutrients* **2021**, *14*, 166.
80. Liu, K.; Zou, J.; Fan, H.; Hu, H.; You, Z. Causal effects of gut microbiota on diabetic retinopathy: A Mendelian randomization study. *Front. Immunol.* **2022**, *13*, 930318. [CrossRef]
81. Zierer, J.; Jackson, M.A.; Kastenmüller, G.; Mangino, M.; Long, T.; Telenti, A.; Mohney, R.P.; Small, K.S.; Bell, J.T.; Steves, C.J.; et al. The fecal metabolome as a functional readout of the gut microbiome. *Nat. Genet.* **2018**, *50*, 790–795. [CrossRef] [PubMed]
82. Li, L.; Yang, K.; Li, C.; Zhang, H.; Yu, H.; Chen, K.; Yang, X.; Liu, L. Metagenomic shotgun sequencing and metabolomic profiling identify specific human gut microbiota associated with diabetic retinopathy in patients with type 2 diabetes. *Front. Immunol.* **2022**, *13*, 943325. [CrossRef] [PubMed]
83. Ye, P.; Zhang, X.; Xu, Y.; Xu, J.; Song, X.; Yao, K. Alterations of the Gut Microbiome and Metabolome in Patients With Proliferative Diabetic Retinopathy. *Front. Immunol.* **2021**, *12*, 667632. [CrossRef] [PubMed]
84. Zhou, Z.; Zheng, Z.; Xiong, X.; Chen, X.; Peng, J.; Yao, H.; Pu, J.; Chen, Q.; Zheng, M. Gut Microbiota Composition and Fecal Metabolic Profiling in Patients With Diabetic Retinopathy. *Front. Cell Dev. Biol.* **2021**, *9*, 732204. [CrossRef]
85. Wang, R.; Jian, Q.; Hu, G.; Du, R.; Xu, X.; Zhang, F. Integrated Metabolomics and Transcriptomics Reveal Metabolic Patterns in Retina of STZ-Induced Diabetic Retinopathy Mouse Model. *Metabolites* **2022**, *12*, 1245.
86. Olivares, A.M.; Althoff, K.; Chen, G.F.; Wu, S.; Morrisson, M.A.; DeAngelis, M.M.; Haider, N. Animal Models of Diabetic Retinopathy. *Curr. Diab. Rep.* **2017**, *17*, 93. [CrossRef]
87. Preguiça, I.; Alves, A.; Nunes, S.; Gomes, P.; Fernandes, R.; Viana, S.D.; Reis, F. Diet-Induced Rodent Models of Diabetic Peripheral Neuropathy, Retinopathy and Nephropathy. *Nutrients* **2020**, *12*, 250. [CrossRef]
88. Baig, M.A.; Panchal, S.S. Streptozotocin-Induced Diabetes Mellitus in Neonatal Rats: An Insight into its Applications to Induce Diabetic Complications. *Curr. Diab. Rev.* **2019**, *16*, 26–39. [CrossRef]
89. Lv, K.; Ying, H.; Hu, G.; Hu, J.; Jian, Q.; Zhang, F. Integrated multi-omics reveals the activated retinal microglia with intracellular metabolic reprogramming contributes to inflammation in STZ-induced early diabetic retinopathy. *Front. Immunol.* **2022**, *13*, 942768. [CrossRef]
90. Ighodaro, O.M.; Adeosun, A.M.; Akinloye, O.A. Alloxan-induced diabetes, a common model for evaluating the glycemic-control potential of therapeutic compounds and plants extracts in experimental studies. *Medicina* **2017**, *53*, 365–374. [CrossRef]
91. Sheskey, S.R.; Antonetti, D.A.; Rentería, R.C.; Lin, C.-M. Correlation of Retinal Structure and Visual Function Assessments in Mouse Diabetes Models. *Investig. Ophthalmol. Vis. Sci.* **2021**, *62*, 20. [CrossRef] [PubMed]
92. Aubin, A.-M.; Lombard-Vadnais, F.; Collin, R.; Aliesky, H.A.; McLachlan, S.M.; Lesage, S. The NOD Mouse Beyond Autoimmune Diabetes. *Front. Immunol.* **2022**, *13*, 874769. [PubMed]
93. Nadif, R.; Dilworth, M.R.; Sibley, C.P.; Baker, P.N.; Davidge, S.T.; Gibson, J.M.; Aplin, J.D.; Westwood, M. The Maternal Environment Programs Postnatal Weight Gain and Glucose Tolerance of Male Offspring, but Placental and Fetal Growth Are Determined by Fetal Genotype in theLeprdb/+ Model of Gestational Diabetes. *Endocrinology* **2015**, *156*, 360–366. [CrossRef] [PubMed]
94. Ali Rahman, I.S.; Li, C.-R.; Lai, C.-M.; Rakoczy, E.P. In VivoMonitoring of VEGF-Induced Retinal Damage in the Kimba Mouse Model of Retinal Neovascularization. *Curr. Eye Res.* **2011**, *36*, 654–662. [CrossRef]
95. Van Hove, I.; De Groef, L.; Boeckx, B.; Modave, E.; Hu, T.-T.; Beets, K.; Etienne, I.; Van Bergen, T.; Lambrechts, D.; Moons, L.; et al. Single-cell transcriptome analysis of the Akimba mouse retina reveals cell-type-specific insights into the pathobiology of diabetic retinopathy. *Diabetologia* **2020**, *63*, 2235–2248. [CrossRef]
96. Katsuda, Y.; Ohta, T.; Miyajima, K.; Kemmochi, Y.; Sasase, T.; Tong, B.; Shinohara, M.; Yamada, T. Diabetic Complications in Obese Type 2 Diabetic Rat Models. *Exp. Anim.* **2014**, *63*, 121–132. [CrossRef]

97. Lu, Z.Y.; Bhutto, I.A.; Amemiya, T. Retinal changes in Otsuka long-evans Tokushima Fatty rats (spontaneously diabetic rat)–possibility of a new experimental model for diabetic retinopathy. *Jpn. J. Ophthalmol.* **2003**, *47*, 28–35. [CrossRef]
98. Wallis, R.H.; Wang, K.; Marandi, L.; Hsieh, E.; Ning, T.; Chao, G.Y.C.; Sarmiento, J.; Paterson, A.D.; Poussier, P. Type 1 Diabetes in the BB Rat: A Polygenic Disease. *Diabetes* **2009**, *58*, 1007–1017. [CrossRef]
99. Tsuji, N.; Matsuura, T.; Ozaki, K.; Sano, T.; Narama, I. Diabetic retinopathy and choroidal angiopathy in diabetic rats (WBN/Kob). *Exp. Anim.* **2009**, *58*, 481–487. [CrossRef]
100. Berdugo, M.; Delaunay, K.; Lebon, C.; Naud, M.C.; Radet, L.; Zennaro, L.; Picard, E.; Daruich, A.; Beltrand, J.; Kermorvant-Duchemin, E.; et al. Long-Term Oral Treatment with Non-Hypoglycemic Dose of Glibenclamide Reduces Diabetic Retinopathy Damage in the Goto-KakizakiRat Model. *Pharmaceutics* **2021**, *13*, 1095. [CrossRef]
101. Rojo Arias, J.E.; Englmaier, V.E.; Jászai, J. VEGF-Trap Modulates Retinal Inflammation in the Murine Oxygen-Induced Retinopathy (OIR) Model. *Biomedicines* **2022**, *10*, 201. [CrossRef] [PubMed]
102. Delioglu, E.N.E.; Ugurlu, N.; Erdal, E.; Malekghasemi, S.; Cagil, N. Evaluation of sphingolipid metabolism on diabetic retinopathy. *Indian J. Ophthalmol.* **2021**, *69*, 3376–3380. [CrossRef] [PubMed]
103. Sun, Y.; Kong, L.; Zhang, A.H.; Han, Y.; Sun, H.; Yan, G.L.; Wang, X.J. A Hypothesis From Metabolomics Analysis of Diabetic Retinopathy: Arginine-Creatine Metabolic Pathway May Be a New Treatment Strategy for Diabetic Retinopathy. *Front. Endocrinol.* **2022**, *13*, 858012.

Disclaimer/Publisher's Note: The statements, opinions and data contained in all publications are solely those of the individual author(s) and contributor(s) and not of MDPI and/or the editor(s). MDPI and/or the editor(s) disclaim responsibility for any injury to people or property resulting from any ideas, methods, instructions or products referred to in the content.

MDPI
St. Alban-Anlage 66
4052 Basel
Switzerland
www.mdpi.com

Metabolites Editorial Office
E-mail: metabolites@mdpi.com
www.mdpi.com/journal/metabolites

Disclaimer/Publisher's Note: The statements, opinions and data contained in all publications are solely those of the individual author(s) and contributor(s) and not of MDPI and/or the editor(s). MDPI and/or the editor(s) disclaim responsibility for any injury to people or property resulting from any ideas, methods, instructions or products referred to in the content.

www.ingramcontent.com/pod-product-compliance
Lightning Source LLC
LaVergne TN
LVHW070441100526
838202LV00014B/1644